# PROJECTIVE GEOMETRY

BY

OSWALD VEBLEN

PROFESSOR OF MATHEMATICS, PRINCETON UNIVERSITY

AND

JOHN WESLEY YOUNG

PROFESSOR OF MATHEMATICS, DARTMOUTH COLLEGE

VOLUME I

GINN AND COMPANY

BOSTON · NEW YORK · CHICAGO · LONDON

ATLANTA · DALLAS · COLUMBUS · SAN FRANCISCO

The Athenæum Press
GINN AND COMPANY · PROPRIETORS · BOSTON · U.S.A.

# PREFACE

Geometry, which had been for centuries the most perfect example of a deductive science, during the creative period of the nineteenth century outgrew its old logical forms. The most recent period has however brought a clearer understanding of the logical foundations of mathematics and thus has made it possible for the exposition of geometry to resume the purely deductive form. But the treatment in the books which have hitherto appeared makes the work of laying the foundations seem so formidable as either to require for itself a separate treatise, or to be passed over without attention to more than the outlines. This is partly due to the fact that in giving the complete foundation for ordinary real or complex geometry, it is necessary to make a study of linear order and continuity, — a study which is not only extremely delicate, but whose methods are those of the theory of functions of a real variable rather than of elementary geometry.

The present work, which is to consist of two volumes and is intended to be available as a text in courses offered in American universities to upper-class and graduate students, seeks to avoid this difficulty by deferring the study of order and continuity to the second volume. The more elementary part of the subject rests on a very simple set of assumptions which characterize what may be called "general projective geometry." It will be found that the theorems selected on this basis of logical simplicity are also elementary in the sense of being easily comprehended and often used.

Even the limited space devoted in this volume to the foundations may seem a drawback from the pedagogical point of view of some mathematicians. To this we can only reply that, in our opinion, an adequate knowledge of geometry cannot be obtained without attention to the foundations. We believe, moreover, that the abstract treatment is peculiarly desirable in projective geometry, because it is through the latter that the other geometric disciplines are most readily coördinated. Since it is more natural to derive the geometrical disciplines associated with the names of Euclid, Descartes, Lobatchewsky, etc., from projective geometry than it is to derive projective geometry from one of them, it is natural to take the foundations of projective geometry as the foundations of all geometry.

The deferring of linear order and continuity to the second volume has necessitated the deferring of the discussion of the metric geometries characterized by certain subgroups of the general projective group. Such elementary applications as the metric properties of conics will therefore be found in the second volume. This will be a disadvantage if the present volume is to be used for a short course in which it is desired to include metric applications. But the arrangement of the material will make it possible, when the second volume is ready, to pass directly from Chapter VIII of the first volume to the study of order relations (which may themselves be passed over without detailed discussion, if this is thought desirable), and thence to the development of Euclidean metric geometry. We think that much is to be gained pedagogically as well as scientifically by maintaining the sharp distinction between the projective and the metric.

The introduction of analytic methods on a purely synthetic basis in Chapter VI brings clearly to light the generality of the set of assumptions used in this volume. What we call "general projective geometry" is, analytically, the geometry associated with a general number field. All the theorems of this volume are valid, not alone in the ordinary real and the ordinary complex projective spaces, but also in the ordinary rational space and in the finite spaces. The bearing of this general theory once fully comprehended by the student, it is hoped that he will gain a vivid conception of the organic unity of mathematics, which recent developments of postulational methods have so greatly emphasized.

The form of exposition throughout the book has been conditioned by the purpose of keeping to the fore such general ideas as group, configuration, linear dependence, the correspondence between and the logical interchangeability of analytic and synthetic methods, etc. Between two methods of treatment we have chosen the more conventional in all cases where a new method did not seem to have unquestionable advantages. We have tried also to avoid in general the introduction of new terminology. The use of the word *on* in connection with duality was suggested by Professor Frank Morley.

We have included among the exercises many theorems which in a larger treatise would naturally have formed part of the text. The more important and difficult of these have been accompanied by references to other textbooks and to journals, which it is hoped will introduce the student to the literature in a natural way. There has been no systematic effort, however, to trace theorems to their original sources, so that the book may be justly criticized for not always giving due credit to geometers whose results have been used.

Our cordial thanks are due to several of our colleagues and students who have given us help and suggestions. Dr. H. H. Mitchell has made all the drawings. The proof sheets have been read in whole or in part by Professors Birkhoff, Eisenhart, and Wedderburn, of Princeton University, and by Dr. R. L. Börger of the University of Illinois. Finally, we desire to express to Ginn and Company our sincere appreciation of the courtesies extended to us.

O. VEBLEN
J. W. YOUNG

August, 1910

In the second impression we have corrected a number of typographical and other errors. We have also added (p. 343) two pages of "Notes and Corrections" dealing with inaccuracies or obscurities which could not be readily dealt with in the text. We wish to express our cordial thanks to those readers who have kindly called our attention to errors and ambiguities.

O.V.
J.W.Y.

August, 1916

# CONTENTS

## INTRODUCTION



## CHAPTER I

### THEOREMS OF ALIGNMENT AND THE PRINCIPLE OF DUALITY



## CHAPTER II

### PROJECTION, SECTION, PERSPECTIVITY. ELEMENTARY CONFIGURATIONS



## CHAPTER III

### PROJECTIVITIES OF THE PRIMITIVE GEOMETRIC FORMS OF ONE, TWO, AND THREE DIMENSIONS

## CHAPTER IV

### HARMONIC CONSTRUCTIONS AND THE FUNDAMENTAL THEOREM OF PROJECTIVE GEOMETRY



## CHAPTER V

### CONIC SECTIONS



## CHAPTER VI

### ALGEBRA OF POINTS AND ONE-DIMENSIONAL COÖRDINATE SYSTEMS

## CHAPTER VII

### COÖRDINATE SYSTEMS IN TWO- AND THREE-DIMENSIONAL FORMS



## CHAPTER VIII

### PROJECTIVITIES IN ONE-DIMENSIONAL FORMS



## CHAPTER IX

### GEOMETRIC CONSTRUCTIONS. INVARIANTS

## CHAPTER X

### PROJECTIVE TRANSFORMATIONS OF TWO-DIMENSIONAL FORMS



## CHAPTER XI

### FAMILIES OF LINES

# PROJECTIVE GEOMETRY

## INTRODUCTION

**1. Undefined elements and unproved propositions.** Geometry deals with the properties of figures in space. Every such figure is made up of various elements (points, lines, curves, planes, surfaces, etc.), and these elements bear certain relations to each other (a point lies on a line, a line passes through a point, two planes intersect, etc.). The propositions stating these properties are logically interdependent, and it is the object of geometry to discover such propositions and to exhibit their logical interdependence.

Some of the elements and relations, by virtue of their greater simplicity, are chosen as fundamental, and all other elements and relations are defined in terms of them. Since any defined element or relation must be defined in terms of other elements and relations, it is necessary that one or more of the elements and one or more of the relations between them remain entirely *undefined;* otherwise a vicious circle is unavoidable. Likewise certain of the propositions are regarded as fundamental, in the sense that all other propositions are derivable, as logical consequences, from these fundamental ones. But here again it is a logical necessity that one or more of the propositions remain entirely *unproved;* otherwise a vicious circle is again inevitable.

*The starting point of any strictly logical treatment of geometry (and indeed of any branch of mathematics) must then be a set of undefined elements and relations, and a set of unproved propositions involving them; and from these all other propositions (theorems) are to be derived by the methods of formal logic.* Moreover, since we assumed the point of view of formal (i.e. symbolic) logic, the undefined elements are to be regarded as mere symbols devoid of content, except as implied by the fundamental propositions. Since it is manifestly absurd to speak of a proposition involving these symbols as self-evident, the unproved propositions referred to above must be regarded as mere *assumptions.* It is customary to refer to these fundamental propositions as axioms or postulates, but we prefer to retain the term *assumption* as more expressive of their real logical character.

We understand the term *a mathematical science* to mean *any set of propositions arranged according to a sequence of logical deduction.* From the point of view developed above such a science is purely *abstract.* If any concrete system of things may be regarded as satisfying the fundamental assumptions, this system is a *concrete application* or *representation* of the abstract science. The practical importance or triviality of such a science depends simply on the importance or triviality of its possible applications. These ideas will be illustrated and further discussed in the next section, where it will appear that an abstract treatment has many advantages quite apart from that of logical rigor.

**2. Consistency, categoricalness, independence. Example of a mathematical science.** The notion of a *class* * of objects is fundamental in logic and therefore in any mathematical science. The objects which make up the class are called the *elements* of the class. The notion of a class, moreover, and the relation of *belonging to a class* (being included in a class, being an element of a class, etc.) are primitive notions of logic, the meaning of which is not here called in question.†

The developments of the preceding section may now be illustrated and other important conceptions introduced by considering a simple example of a mathematical science. To this end let S be a class, the elements of which we will denote by $A, B, C, \ldots$ Further, let there be certain undefined subclasses ‡ of S, any one of which we will call an *m-class.* Concerning the elements of S and the *m*-classes we now make the following

ASSUMPTIONS:

I. *If $A$ and $B$ are distinct elements of* S, *there is at least one m-class containing both $A$ and $B$.*

* Synonyms for *class* are *set, aggregate, assemblage, totality;* in German, *Menge;* in French, *ensemble.*

† Cf. B. Russell, The Principles of Mathematics, Cambridge, 1903; and L. Couturat, Les principes des mathématiques, Paris, 1905.

‡ A class S′ is said to be a *subclass* of another class S, if every element of S′ is an element of S.

II. *If A and B are distinct elements of* S, *there is not more than one m-class containing both A and B.*

III. *Any two m-classes have at least one element of* S *in common.*

IV. *There exists at least one m-class.*

V. *Every m-class contains at least three elements of* S.

VI. *All the elements of* S *do not belong to the same m-class.*

VII. *No m-class contains more than three elements of* S.

The reader will observe that in this set of assumptions we have just two undefined terms, viz., *element of* S and *m-class,* and one undefined relation, *belonging to a class.* The undefined terms, moreover, are entirely devoid of content except such as is implied in the assumptions.

Now the first question to ask regarding a set of assumptions is: *Are they logically consistent?* In the example above, of a set of assumptions, the reader will find that the assumptions are all true statements, if the class S is interpreted to mean the digits 0, 1, 2, 3, 4, 5, 6 and the *m*-classes to mean the columns in the following table:

| | | | | | | | |
|---|---|---|---|---|---|---|---|
| | 0 | 1 | 2 | 3 | 4 | 5 | 6 |
| (1) | 1 | 2 | 3 | 4 | 5 | 6 | 0 |
| | 3 | 4 | 5 | 6 | 0 | 1 | 2 |

This interpretation is a concrete representation of our assumptions. Every proposition derived from the assumptions must be true of this system of triples. Hence none of the assumptions can be logically inconsistent with the rest; otherwise contradictory statements would be true of this system of triples.

Thus, in general, *a set of assumptions is said to be consistent if a single concrete representation of the assumptions can be given.**

Knowing our assumptions to be consistent, we may proceed to derive some of the *theorems* of the mathematical science of which they are the basis:

*Any two distinct elements of* S *determine one and only one m-class containing both these elements* (Assumptions I, II).

* It will be noted that this test for the consistency of a set of assumptions merely shifts the difficulty from one domain to another. It is, however, at present the only test known. On the question as to the possibility of an absolute test of consistency, cf. Hilbert, Grundlagen der Geometrie, 2d ed., Leipzig (1903), p. 18, and Verhandlungen d. III. intern. math. Kongresses zu Heidelberg, Leipzig (1904), p. 174; Padoa, L'Enseignement mathématique, Vol. V (1903), p. 85.

The $m$-class containing the elements $A$ and $B$ may conveniently be denoted by the symbol $AB$.

*Any two m-classes have one and only one element of* S *in common* (Assumptions II, III).

*There exist three elements of* S *which are not all in the same m-class* (Assumptions IV, V, VI).

In accordance with the last theorem, let $A, B, C$ be three elements of S not in the same $m$-class. By Assumption V there must be a third element in each of the $m$-classes $AB$, $BC$, $CA$, and by Assumption II these elements must be distinct from each other and from $A$, $B$, and $C$. Let the new elements be $D$, $E$, $G$, so that each of the triples $ABD$, $BCE$, $CAG$ belongs to the same $m$-class. By Assumption III the $m$-classes $AE$ and $BG$, which are distinct from all the $m$-classes thus far obtained, have an element of S in common, which, by Assumption II, is distinct from those hitherto mentioned; let it be denoted by $F$, so that each of the triples $AEF$ and $BFG$ belong to the same $m$-class. No use has as yet been made of Assumption VII. We have, then, the theorem:

*Any class* S *subject to Assumptions I–VI contains at least seven elements.*

Now, making use of Assumption VII, we find that the $m$-classes thus far obtained contain only the elements mentioned. The $m$-classes $CD$ and $AEF$ have an element in common (by Assumption III) which cannot be $A$ or $E$, and must therefore (by Assumption VII) be $F$. Similarly, $ACG$ and the $m$-class $DE$ have the element $G$ in common. The seven elements $A, B, C, D, E, F, G$ have now been arranged into $m$-classes according to the table

$$(1') \qquad \begin{array}{ccccccc} A & B & C & D & E & F & G \\ B & C & D & E & F & G & A \\ D & E & F & G & A & B & C \end{array}$$

in which the columns denote $m$-classes. The reader may note at once that this table is, except for the substitution of letters for digits, entirely equivalent to Table (1); indeed (1′) is obtained from (1) by replacing 0 by $A$, 1 by $B$, 2 by $C$, etc. We can show, furthermore, that S can contain no other elements than $A, B, C, D, E, F, G$. For suppose there were another element, $T$. Then, by Assumption III, the $m$-classes $TA$ and $BFG$ would have an element in common. This element cannot be $B$, for then $ABTD$ would belong to the same $m$-class; it cannot be $F$, for then $AFTE$ would all belong to the same $m$-class; and it cannot be $G$, for then $AGTC$ would all belong to the same $m$-class. These three possibilities all contradict Assumption VII. Hence the existence of $T$ would imply the existence of four elements in the $m$-class $BFG$, which is likewise contrary to Assumption VII.

The properties of the class S and its $m$-classes may also be represented vividly by the accompanying figure (fig. 1). Here we have represented the elements of S by points (or spots) in a plane, and have joined by a line every triple of these points which form an $m$-class. It is seen that the points may be so chosen that all but one of these lines is a straight line. This suggests at once a similarity to ordinary plane geometry. Suppose we interpret the elements of S to be the points of a plane, and interpret the $m$-classes to be the straight lines of the plane, and let us reread our assumptions with this interpretation. Assumption VII is false, but all the others are true with the exception of Assumption III, which is also true except when the lines are parallel. How this exception can be removed we will discuss in the next section, so that we may also regard the ordinary plane geometry as a representation of Assumptions I–VI.

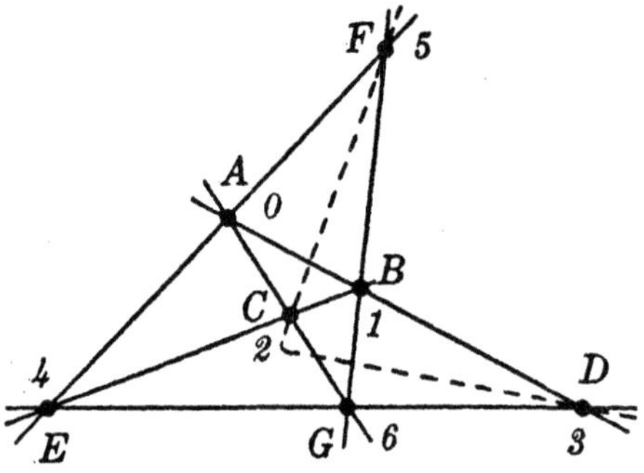

FIG. 1

Returning to our miniature mathematical science of triples, we are now in a position to answer another important question: *To what extent do Assumptions I–VII characterize the class* S *and the m-classes?* We have just seen that any class S satisfying these assumptions may be represented by Table (1′) merely by properly labeling the elements of S. In other words, if $S_1$ and $S_2$ are two classes S subject to these assumptions, every element of $S_1$ may be made to correspond * to a unique element of $S_2$, in such a way that every element of $S_2$ is the correspondent of a unique element of $S_1$, and that to every $m$-class of $S_1$ there corresponds an $m$-class of $S_2$. The two classes are

* The notion of *correspondence* is another primitive notion which we take over without discussion from the general logic of classes.

then said to be in *one-to-one reciprocal correspondence*, or to be *simply isomorphic*.* Two classes S are then abstractly equivalent; i.e. there exists essentially only one class S satisfying Assumptions I–VII. This leads to the following fundamental notion:

*A set of assumptions is said to be categorical, if there is essentially only one system for which the assumptions are valid; i.e. if any two such systems may be made simply isomorphic.*

We have just seen that the set of Assumptions I–VII is categorical. If, however, Assumption VII be omitted, the remaining set of six assumptions is not categorical. We have already observed the possibility of satisfying Assumptions I–VI by ordinary plane geomtry. Since Assumption III, however, occupies as yet a doubtful position in this interpretation, we give another, which, by virtue of its simplicity, is peculiarly adapted to make clear the distinction between categorical and noncategorical. The reader will find, namely, that each of the first six assumptions is satisfied by interpreting the class S to consist of the digits 0, 1, 2, · · ·, 12, arranged according to the following table of *m*-classes, every column constituting one *m*-class:

| | | | | | | | | | | | | | |
|---|---|---|---|---|---|---|---|---|---|---|---|---|---|
| | 0 | 1 | 2 | 3 | 4 | 5 | 6 | 7 | 8 | 9 | 10 | 11 | 12 |
| (2) | 1 | 2 | 3 | 4 | 5 | 6 | 7 | 8 | 9 | 10 | 11 | 12 | 0 |
| | 3 | 4 | 5 | 6 | 7 | 8 | 9 | 10 | 11 | 12 | 0 | 1 | 2 |
| | 9 | 10 | 11 | 12 | 0 | 1 | 2 | 3 | 4 | 5 | 6 | 7 | 8 |

Hence Assumptions I–VI are not sufficient to characterize completely the class S, for it is evident that Systems (1) and (2) cannot be made isomorphic. On the other hand, it should be noted that all theorems derivable from Assumptions I–VI are valid for both (1) and (2). These two systems are two essentially different concrete representations of the same mathematical science.

This brings us to a third question regarding our assumptions: *Are they independent?* That is, can any one of them be derived as a logical consequence of the others? Table (2) is an example which shows that Assumption VII is independent of the others, because it shows that they can all be true of a system in which Assumption VII is false. Again, if the class S is taken to mean the three letters *A*, *B*, *C*,

* The isomorphism of Systems (1) and (1′) is clearly exhibited in fig. 1, where each point is labeled both with a digit and with a letter. This isomorphism may, moreover, be established in 7·6·4 different ways.

and the *m*-classes to consist of the pairs *AB*, *BC*, *CA*, then it is clear that Assumptions I, II, III, IV, VI, VII are true of this class S, and therefore that any logical consequence of them is true with this interpretation. Assumption V, however, is false for this class, and cannot, therefore, be a logical consequence of the other assumptions. In like manner, other examples can be constructed to show that each of the Assumptions I–VII is independent of the remaining ones.

**3. Ideal elements in geometry.** The miniature mathematical science which we have just been studying suggests what we must do on a larger scale in a geometry which describes our ordinary space. We must first choose a set of undefined elements and a set of fundamental assumptions. This choice is in no way prescribed *a priori*, but, on the contrary, is very arbitrary. It is necessary only that the undefined symbols be such that all other elements and relations that occur are definable in terms of them; and the fundamental assumptions must satisfy the prime requirement of logical consistency, and be such that all other propositions are derivable from them by formal logic. It is desirable, further, that the assumptions be independent* and that certain sets of assumptions be categorical. There is, further, the desideratum of utmost symmetry and generality in the whole body of theorems. The latter means that the applicability of a theorem shall be as wide as possible. This has relation to the arrangement of the assumptions, and can be attained by using in the proof of each theorem a minimum of assumptions.†

Symmetry can frequently be obtained by a judicious choice of terminology. This is well illustrated by the concept of "points at infinity" which is fundamental in any treatment of projective geometry. Let us note first the reciprocal character of the relation expressed by the two statements:

A point lies on a line. A line passes through a point.

To exhibit clearly this reciprocal character, we agree to use the phrases

A point is *on* a line; A line is *on* a point

* This is obviously necessary for the precise distinction between an assumption and a theorem.

† If the set of assumptions used in the proof of a theorem is not categorical, the applicability of the theorem is evidently wider than in the contrary case. Cf. example of preceding section.

to express this relation. Let us now consider the following two propositions:

1. *Any two distinct points of a plane are on one and only one line.**

1′. *Any two distinct lines of a plane are on one and only one point.*

Either of these propositions is obtained from the other by simply interchanging the words *point* and *line.* The first of these propositions we recognize as true without exception in the ordinary Euclidean geometry. The second, however, has an exception when the two lines are parallel. In view of the symmetry of these two propositions it would clearly add much to the symmetry and generality of all propositions derivable from these two, if we could regard them both as true without exception. This can be accomplished by *attributing to two parallel lines a point of intersection.* Such a point is not, of course, a point in the ordinary sense; it is to be regarded as an *ideal point,* which we suppose two parallel lines to have in common. Its introduction amounts merely to a change in the ordinary terminology. Such an ideal point we call a *point at infinity;* and we suppose one such point to exist on every line.†

The use of this new term leads to a change in the statement, though not in the meaning, of many familiar propositions, and makes us modify the way in which we think of points, lines, etc. Two non-parallel lines cannot have in common a point at infinity without doing violence to propositions 1 and 1′; and since each of them has a point at infinity, there must be at least two such points. Proposition 1, then, requires that we attach a meaning to the notion of a line on two points at infinity. Such a line we call a *line at infinity,* and think of it as consisting of all the points at infinity in a plane. In like manner, if we do not confine ourselves to the points of a single plane, it is found desirable to introduce the notion of a plane through three points at infinity which are not all on the same line at infinity. Such a plane we call a *plane at infinity,* and we think

* By *line* throughout we mean *straight line.*

† It should be noted that (since we are taking the point of view of Euclid) we do not think of a line as containing more than one point at infinity; for the supposition that a line contains two such points would imply either that two parallels can be drawn through a given point to a given line, or that two distinct lines can have more than one point in common.

of it as consisting of all the points at infinity in space. Every ordinary plane is supposed to contain just one line at infinity; every system of parallel planes in space is supposed to have a line at infinity in common with the plane at infinity, etc.

The fact that we have difficulty in presenting to our imagination the notions of a point at infinity on a line, the line at infinity in a plane, and the plane at infinity in space, need not disturb us in this connection, provided we can satisfy ourselves that the new terminology is self-consistent and cannot lead to contradictions. The latter condition amounts, in the treatment that follows, simply to the condition that the assumptions on which we build the subsequent theory be consistent. That they are consistent will be shown at the time they are introduced. The use of the new terminology may, however, be justified on the basis of ordinary analytic geometry. This we do in the next section, the developments of which will, moreover, be used frequently in the sequel for proving the consistency of the assumptions there made.

**4. Consistency of the notion of points, lines, and plane at infinity.** We will now reduce the question of the consistency of our new terminology to that of the consistency of an algebraic system. For this purpose we presuppose a knowledge of the elements of analytic geometry of three dimensions.* In this geometry a point is equivalent to a set of three numbers $(x, y, z)$. The totality of all such sets of numbers constitute the analytic space of three dimensions. If the numbers are all real numbers, we are dealing with the ordinary "real" space; if they are any complex numbers, we are dealing with the ordinary "complex" space of three dimensions. The following discussion applies primarily to the real case.

A *plane* is the set of all points (number triads) which satisfy a single linear equation

$$ax + by + cz + d = 0.$$

A *line* is the set of all points which satisfy two linear equations,

$$a_1x + b_1y + c_1z + d_1 = 0,$$
$$a_2x + b_2y + c_2z + d_2 = 0,$$

* Such knowledge is not presupposed elsewhere in this book, except in the case of consistency proofs. The elements of analytic geometry are indeed developed from the beginning (cf. Chaps. VI, VII).

provided the relations

$$\frac{a_1}{a_2} = \frac{b_1}{b_2} = \frac{c_1}{c_2}$$

do not hold.*

Now the points $(x, y, z)$, with the exception of $(0, 0, 0)$, may also be denoted by the direction cosines of the line joining the point to the origin of coördinates and the distance of the point from the origin; say by

$$\left(l, m, n, \frac{1}{d}\right),$$

where $d = \sqrt{x^2 + y^2 + z^2}$, and $l = \frac{x}{d}$, $m = \frac{y}{d}$, $n = \frac{z}{d}$. The origin itself may be denoted by $(0, 0, 0, k)$, where $k$ is arbitrary. Moreover, any four numbers $(x_1, x_2, x_3, x_4)$ $(x_4 \neq 0)$, proportional respectively to $\left(l, m, n, \frac{1}{d}\right)$, will serve equally well to represent the point $(x, y, z)$, provided we agree that $(x_1, x_2, x_3, x_4)$ and $(cx_1, cx_2, cx_3, cx_4)$ represent the same point for all values of $c$ different from 0. For a point $(x, y, z)$ determines

$$x_1 = \frac{cx}{\sqrt{x^2 + y^2 + z^2}} = cl, \quad x_2 = \frac{cy}{\sqrt{x^2 + y^2 + z^2}} = cm,$$

$$x_3 = \frac{cz}{\sqrt{x^2 + y^2 + z^2}} = cn, \quad x_4 = \frac{c}{\sqrt{x^2 + y^2 + z^2}} = \frac{c}{d},$$

where $c$ is arbitrary $(c \neq 0)$, and $(x_1, x_2, x_3, x_4)$ determines

$$(1) \qquad x = \frac{x_1}{x_4}, \quad y = \frac{x_2}{x_4}, \quad z = \frac{x_3}{x_4},$$

provided $x_4 \neq 0$.

We have not assigned a meaning to $(x_1, x_2, x_3, x_4)$ when $x_4 = 0$, but it is evident that if the point $\left(cl, cm, cn, \frac{c}{d}\right)$ moves away from the origin an unlimited distance on the line whose direction cosines are $l, m, n$, its coördinates approach $(cl, cm, cn, 0)$. A little consideration will show that as a point moves on any other line with direction

* It should be noted that we are not yet, in this section, supposing anything known regarding points, lines, etc., at infinity, but are placing ourselves on the basis of elementary geometry.

cosines $l$, $m$, $n$, so that its distance from the origin increases indefinitely, its coördinates also approach $(cl, cm, cn, 0)$. Furthermore, these values are approached, no matter in which of the two opposite directions the point moves away from the origin. We now *define* $(x_1, x_2, x_3, 0)$ as a *point at infinity* or an *ideal point*. We have thus associated with every set of four numbers $(x_1, x_2, x_3, x_4)$ a point, ordinary or ideal, with the exception of the set $(0, 0, 0, 0)$, which we exclude entirely from the discussion. The ordinary points are those for which $x_4$ is not zero; their ordinary Cartesian coördinates are given by the equations (1). The ideal points are those for which $x_4 = 0$. The numbers $(x_1, x_2, x_3, x_4)$ we call the *homogeneous* coördinates of the point.

We now define a *plane* to be the set of all points $(x_1, x_2, x_3, x_4)$ which satisfy a linear homogeneous equation:

$$ax_1 + bx_2 + cx_3 + dx_4 = 0.$$

It is at once clear from the preceding discussion that as far as all ordinary points are concerned, this definition is equivalent to the one given at the beginning of this section. However, according to this definition all the ideal points constitute a plane $x_4 = 0$. This plane we call the *plane at infinity*. In like manner, we define a line to consist of all points $(x_1, x_2, x_3, x_4)$ which satisfy two distinct linear homogeneous equations:

$$a_1x_1 + b_1x_2 + c_1x_3 + d_1x_4 = 0,$$
$$a_2x_1 + b_2x_2 + c_2x_3 + d_2x_4 = 0.$$

Since these expressions are to be distinct, the corresponding coefficients throughout must not be proportional. According to this definition the points common to any plane (not the plane at infinity) and the plane $x_4 = 0$ constitute a line. Such a line we call a *line at infinity*, and there is one such in every ordinary plane. Finally, the line defined above by two equations contains one and only one point with coördinates $(x_1, x_2, x_3, 0)$; that is, an ordinary line contains one and only one point at infinity. It is readily seen, moreover, that with the above definitions two parallel lines have their points at infinity in common.

Our discussion has now led us to an analytic definition of what may be called, for the present, an analytic *projective space* of three dimensions. It may be defined, in a way which allows it to be either real or complex, as consisting of:

*Points:* All sets of four numbers $(x_1, x_2, x_3, x_4)$, except the set $(0, 0, 0, 0)$, where $(cx_1, cx_2, cx_3, cx_4)$ is regarded as identical with $(x_1, x_2, x_3, x_4)$, provided $c$ is not zero.

*Planes:* All sets of points satisfying one linear homogeneous equation.

*Lines:* All sets of points satisfying two distinct linear homogeneous equations.

Such a projective space cannot involve contradictions unless our ordinary system of real or complex algebra is inconsistent. The definitions here made of points, lines, and the plane at infinity are, however, precisely equivalent to the corresponding notions of the preceding section. We may therefore use these notions precisely in the same way that we consider ordinary points, lines, and planes. Indeed, the fact that no exceptional properties attach to our ideal elements follows at once from the symmetry of the analytic formulation; the coördinate $x_4$, whose vanishing gives rise to the ideal points, occupies no exceptional position in the algebra of the homogeneous equations. The ideal points, then, are not to be regarded as different from the ordinary points.

All the assumptions we shall make in our treatment of projective geometry will be found to be satisfied by the above analytic creation, which therefore constitutes a proof of the consistency of the assumptions in question. This the reader will verify later.

**5. Projective and metric geometry.** In projective geometry no distinction is made between ordinary points and points at infinity, and it is evident by a reference forward that our assumptions provide for no such distinction. We proceed to explain this a little more fully, and will at the same time indicate in a general way the difference between *projective* and the ordinary Euclidean *metric* geometry.

Confining ourselves first to the plane, let $m$ and $m'$ be two distinct lines, and $P$ a point not on either of the two lines. Then the points of $m$ may be made to correspond to the points of $m'$ as follows: To every point $A$ on $m$ let correspond that point $A'$ on $m'$ in which $m'$ meets the line joining $A$ to $P$ (fig. 2). In this way every point on either line is assigned a unique corresponding point on the other line. This type of correspondence is called *perspective,* and the points on one line are said to be transformed into the points of the other by a *perspective transformation with center P.* If the points of a line $m$ be transformed into the points of a line $m'$ by a perspective transformation with center $P$, and then the points of $m'$ be transformed into the points of a third line $m''$ by a perspective transformation with a new center $Q$; and if this be continued any finite number of times, ultimately the points of the line $m$ will have been brought into correspondence with the points of a line $m^{(n)}$, say, in such a way that every point of $m$ corresponds to a unique point of $m^{(n)}$. A correspondence obtained in this way is called *projective,* and the points of $m$ are said

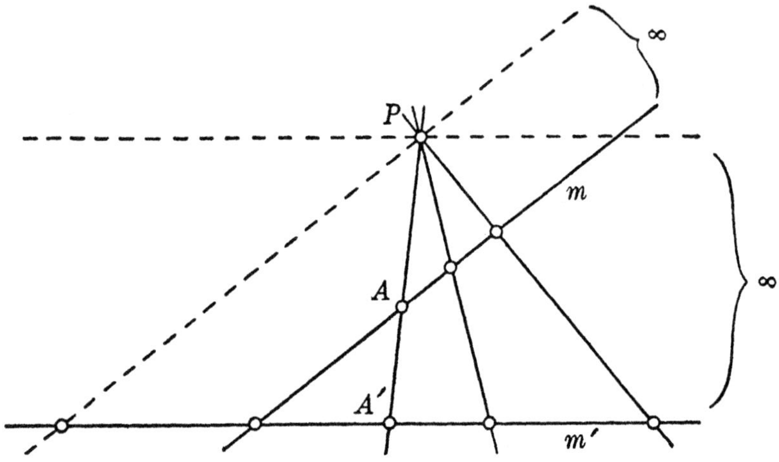


FIG. 2

to have been transformed into the points of $m^{(n)}$ by a *projective transformation.*

Similarly, in three-dimensional space, if lines are drawn joining every point of a plane figure to a fixed point $P$ not in the plane $\pi$ of the figure, then the points in which this totality of lines meets another plane $\pi'$ will form a new figure, such that to every point of $\pi$ will correspond a unique point of $\pi'$, and to every line of $\pi$ will correspond a unique line of $\pi'$. We say that the figure in $\pi$ has been transformed into the figure in $\pi'$ by a *perspective transformation with center P.* If a plane figure be subjected to a succession of such perspective transformations with different centers, the final figure will still be such that its points and lines correspond uniquely to the points and lines of the original figure. Such a transformation is again called a *projective transformation.* In projective geometry two figures that may be made to correspond to each other by means of a projective transformation are not regarded as different. In other words, *projective geometry is concerned with those properties of figures that are left unchanged when the figures are subjected to a projective transformation.*

It is evident that no properties that involve essentially the notion of measurement can have any place in projective geometry as such;* hence the term *projective*, to distinguish it from the ordinary geometry, which is almost exclusively concerned with properties involving the idea of measurement. In case of a plane figure, a perspective transformation is clearly equivalent to the change brought about in the aspect of a figure by looking at it from a different angle, the observer's eye being the center of the perspective transformation. The properties of the aspect of a figure that remain unaltered when the observer changes his position will then be properties with which projective geometry concerns itself. For this reason von Staudt called this science *Geometrie der Lage*.

In regard to the points and lines at infinity, we can now see why they cannot be treated as in any way different from the ordinary points and lines of a figure. For, in the example given of a perspective transformation between lines, it is clear that to the point at infinity on $m$ corresponds in general an ordinary point on $m'$, and conversely. And in the example given of a perspective transformation between planes we see that to the line at infinity in one plane corresponds in general an ordinary line in the other. In projective geometry, then, there can be no distinction between the ordinary and the ideal elements of space.

* The theorems of metric geometry may however be regarded as special cases of projective theorems.

# CHAPTER I

## THEOREMS OF ALIGNMENT AND THE PRINCIPLE OF DUALITY

**6. The assumptions of alignment.** In the following treatment of projective geometry we have chosen the point and the line as undefined elements. We consider a *class* (cf. § 2, p. 2) the elements of which we call *points*, and certain *undefined classes of points* which we call *lines*. Here the words *point* and *line* are to be regarded as mere symbols devoid of all content except as implied in the assumptions (presently to be made) concerning them, and which *may represent any elements for which the latter may be valid propositions.* In other words, these elements are not to be considered as having properties in common with the points and lines of ordinary Euclidean geometry, except in so far as such properties are formal logical consequences of explicitly stated assumptions.

We shall in the future generally use the capital letters of the alphabet, as $A$, $B$, $C$, $P$, etc., as names for points, and the small letters, as $a$, $b$, $c$, $l$, etc., as names for lines. If $A$ and $B$ denote the same point, this will be expressed by the relation $A = B$; if they represent distinct points, by the relation $A \neq B$. If $A = B$, it is sometimes said that *$A$ coincides with $B$*, or that *$A$ is coincident with $B$.* The same remarks apply to two lines, or indeed to any two elements of the same kind.

All the relations used are defined in general logical terms, mainly by means of the relation of *belonging to a class* and the notion of *one-to-one correspondence.* In case a point is an element of one of the classes of points which we call lines, we shall express this relation by any one of the phrases: the point *is on* or *lies on* or *is a point of* the line, or *is united with* the line; the line *passes through* or *contains* or *is united with* the point. We shall often find it convenient to use also the phrase *the line is on the point* to express this relation. Indeed, all the assumptions and theorems in this chapter will be stated consistently in this way. The reader will quickly become accustomed to this "on" language, which is introduced with the purpose of exhibiting in its most elegant form one of the most far-reaching theorems of projective geometry (Theorem 11). Two lines which have a point in common are said to *intersect in* or to *meet in* that point, or to *be on a common point.* Also, if two distinct points lie on the same line, the line is said to *join* the points. Points which are on the same line are said to be *collinear;* points which are not on the same line are said to be *noncollinear.* Lines which are on the same point (i.e. contain the same point) are said to be *copunctal,* or *concurrent.** 

Concerning points and lines we now make the following assumptions:

THE ASSUMPTIONS OF ALIGNMENT, A:

A 1. *If A and B are distinct points, there is at least one line on both A and B.*

A 2. *If A and B are distinct points, there is not more than one line on both A and B.*

A 3. *If A, B, C are points not all on the same line, and D and E* ($D \neq E$) *are points such that B, C, D are on a line and C, A, E are on a line, there is a point F such that A, B, F are on a line and also D, E, F are on a line* (fig. 3).†

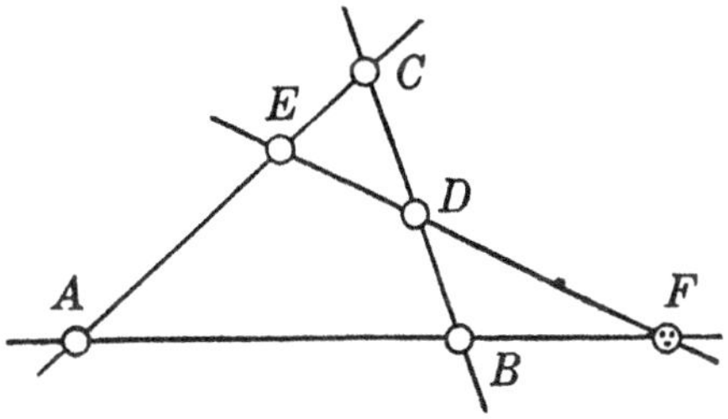

FIG. 3

It should be noted that this set of assumptions is satisfied by the triple system (1), p. 3, and also by the system of quadruples (2), p. 6, as well as by the points and lines of ordinary Euclidean geometry with the notion of "points at infinity" (cf. § 3, p. 8), and by

* The object of this paragraph is simply to define the terms in common use in terms of the general logical notion of belonging to a class. In later portions of this book we may omit the explicit definition of such common terms when such definition is obvious.

† The figures are to be regarded as a concrete representation of our science, in which the undefined "points" and "lines" of the science are represented by points and lines of ordinary Euclidean geometry (this requires the notion of ideal points; cf. § 3, p. 8). Their function is not merely to exhibit one of the many possible concrete representations, but also to help keep in mind the various relations in question. In using them, however, great care must be exercised not to use any properties of such figures that are not formal logical consequences of the assumptions; in other words, care must be taken that all deductions are made formally from the assumptions and theorems previously derived from the assumptions.

the "analytic projective space" described in § 4. Any one of these representations shows that *our set of Assumptions* A *is consistent.**

The following three theorems are immediate consequences of the first two assumptions.

THEOREM 1. *Two distinct points are on one and only one line.* (A 1, A 2) †

The line determined by the points $A$, $B$ $(A \neq B)$ will often be denoted by the symbol or name $AB$.

THEOREM 2. *If $C$ and $D$ $(C \neq D)$ are points on the line $AB$, $A$ and $B$ are points on the line $CD$.* (A 1, A 2)

THEOREM 3. *Two distinct lines cannot be on more than one common point.* (A 1, A 2)

Assumption A 3 will be used in the derivation of the next theorem. It may be noted that under Assumptions A 1, A 2 it may be stated more conveniently as follows: If $A$, $B$, $C$ are points not all on the same line, the line joining any point $D$ on the line $BC$ to any point $E$ $(D \neq E)$ on the line $CA$ meets the line $AB$ in a point $F$. This is the form in which this assumption is generally used in the sequel.

**7. The plane.** DEFINITION. If $P$, $Q$, $R$ are three points not on the same line, and $l$ is a line joining $Q$ and $R$, the class $S_2$ of all points on the lines joining $P$ to the points of $l$ is called the *plane* determined by $P$ and $l$.

We shall use the small letters of the Greek alphabet, $\alpha$, $\beta$, $\gamma$, $\pi$, etc., as names for planes. It follows at once from the definition that $P$ and every point of $l$ are points of the plane determined by $P$ and $l$.

THEOREM 4. *If $A$ and $B$ are points on a plane $\pi$, then every point on the line $AB$ is on $\pi$.* (A)

*Proof.* Let the plane $\pi$ under consideration be determined by the point $P$ and the line $l$.

* In the multiplicity of the possible concrete representations is seen one of the great advantages of the formal treatment quite aside from that of logical rigor. It is clear that there is a great gain in generality as long as the fundamental assumptions are not categorical (cf. p. 6). In the present treatment our assumptions are not made categorical until very late.

† The symbols placed in parentheses after a theorem indicate the assumptions needed in its proof. The symbol A will be used to denote the whole set of Assumptions A 1, A 2, A 3.

1. If both $A$ and $B$ are on $l$, or if the line $AB$ contains $P$, the theorem is immediate.

2. Suppose $A$ is on $l$, $B$ not on $l$, and $AB$ does not contain $P$ (fig. 4). Since $B$ is a point of $\pi$, there is a point $B'$ on $l$ collinear with $B$ and $P$. If $C$ be any point on $AB$, the line joining $C$ on $AB$ to $P$ on $BB'$ will have a point $T$ in common with $AB' = l$ (A 3). Hence $C$ is a point of $\pi$.

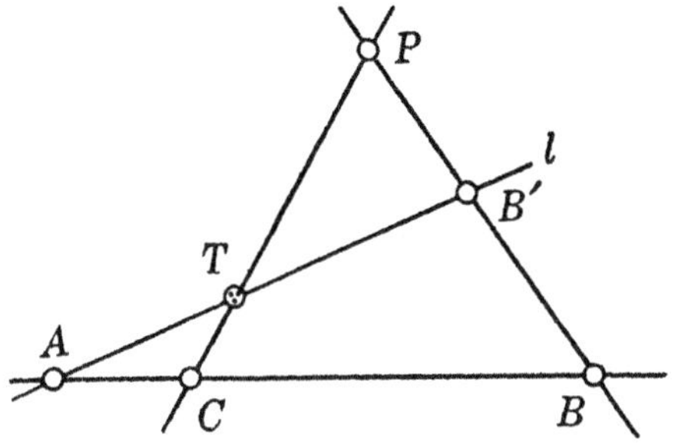

FIG. 4

3. Suppose neither $A$ nor $B$ is on $l$ and that $AB$ does not contain $P$ (fig. 5). Since $A$ and $B$ are points of $\pi$, there exist two points $A'$ and $B'$ on $l$ collinear with $A, P$ and $B, P$ respectively. The line joining $A$ on $A'P$ to $B$ on $PB'$ has a point $Q$ in common with $B'A'$ (A 3). Hence every point of the line $AB = AQ$ is a point of $\pi$, by the preceding case.

This completes the proof.

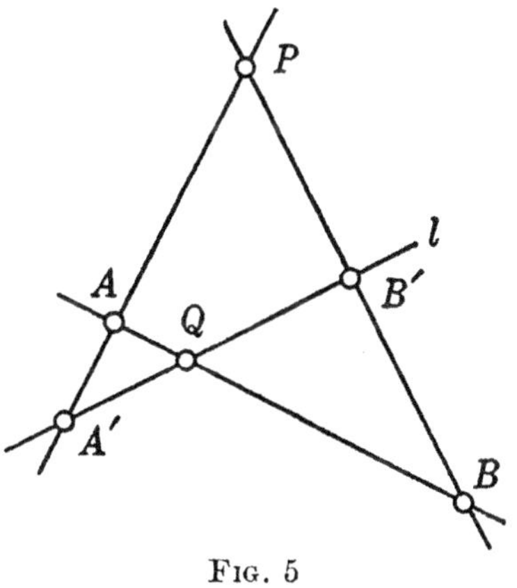

FIG. 5

If all the points of a line are points of a plane, the line is said to be a *line of the plane*, or to *lie in* or to *be in* or to *be on* the plane; the plane is said to *pass through*, or to *contain* the line, or we may also say *the plane is on the line*. Further, a point of a plane is said to *be in* or to *lie in* the plane, and *the plane is on the point*.

**8. The first assumption of extension.** The theorems of the preceding section were stated and proved on the assumption (explicitly stated in each case) that the necessary points and lines exist. The *assumptions of extension, E*, insuring the existence of all the points which we consider, will be given presently. The first of these, however, it is desirable to introduce at this point.

AN ASSUMPTION OF EXTENSION:

E 0. *There are at least three points on every line.*

This assumption is needed in the proof of the following

THEOREM 5. *Any two lines on the same plane $\pi$ are on a common point.* (A, E 0)

*Proof.* Let the plane $\pi$ be determined by the point $P$ and the line $l$, and let $a$ and $b$ be two distinct lines of $\pi$.

1. Suppose $a$ coincides with $l$ (fig. 6). If $b$ contains $P$, any point $B$ of $b$ (E 0) is collinear with $P$ and some point of $l = a$, which proves the theorem when $b$ contains $P$. If $b$ does not contain $P$, there exist on $b$ two points $A$ and $B$ not on $l$ (E 0), and since they are points of $\pi$, they are collinear with $P$ and two points $A'$ and $B'$ of $l$ respectively. The line joining $A$ on $A'P$ to $B$ on $PB'$ has a point $R$ in common with $A'B'$ (A 3), i.e. $l = a$ and $b$ have a point in common. Hence *every line in the plane* $\pi$ *has a point in common with* $l$.

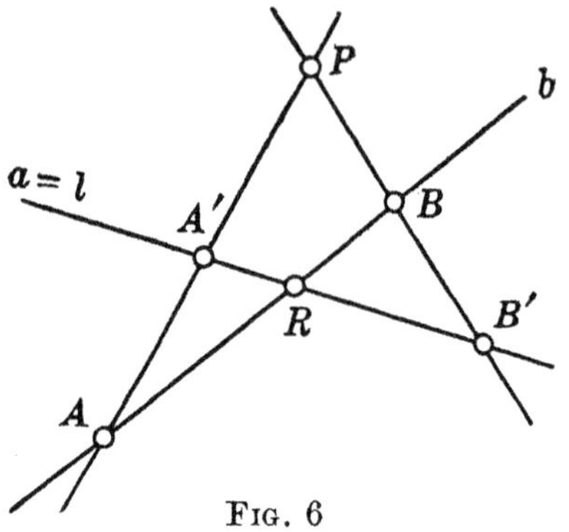


FIG. 6

2. Let $a$ and $b$ both be distinct from $l$. (i) Let $a$ contain $P$ (fig. 7). The line joining $P$ to any point $B$ of $b$ (E 0) has a point $B'$ in common with $l$ (Case 1 of this proof). Also the lines $a$ and $b$ have points $A'$ and $R$ respectively in common with $l$ (Case 1). Now the line $A'P = a$ contains the points $A'$ of $RB'$ and $P$ of $B'B$, and hence has a point $A$ in common with $BR = b$. Hence *every line of* $\pi$ *has a point in common with any line of* $\pi$ *through* $P$. (ii) Let neither $a$ nor $b$ contain $P$ (fig. 8). As before, $a$ and $b$ meet $l$ in two points $Q$ and $R$ respectively. Let $B'$ be a point of $l$ distinct from $Q$ and $R$ (E 0). The line $PB'$ then meets $a$ and $b$ in two points $A$ and $B$ respectively (Case 2, (i)). If $A = B$, the theorem is proved. If $A \neq B$, the line $b$ has the point $R$ in common with $QB'$ and the point $B$ in common with $B'A$, and hence has a point in common with $AQ = a$ (A 3).

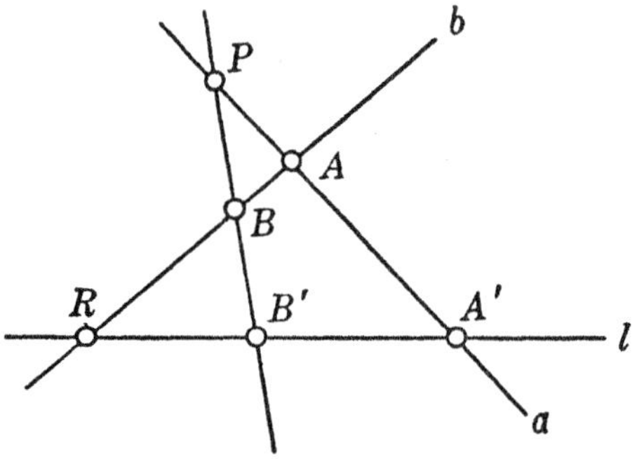


FIG. 7

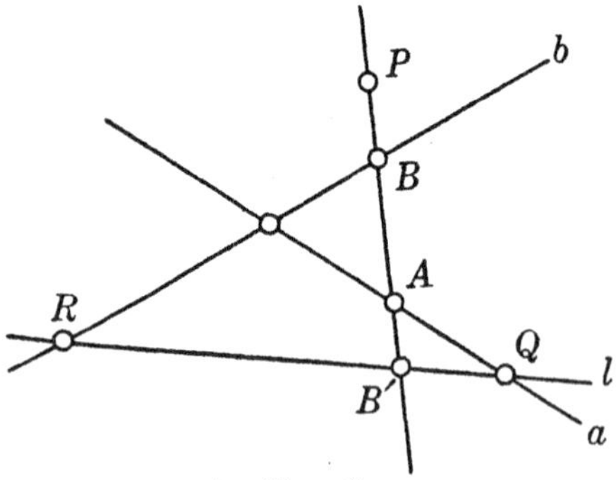


FIG. 8

THEOREM 6. *The plane $\alpha$ determined by a line $l$ and a point $P$ is identical with the plane $\beta$ determined by a line $m$ and a point $Q$, provided $m$ and $Q$ are on $\alpha$.* (A, E 0)

*Proof.* Any point $B$ of $\beta$ is collinear with $Q$ and a point $A$ of $m$ (fig. 9). $A$ and $Q$ are both points of $\alpha$, and hence every point of the line $AQ$ is a point of $\alpha$ (Theorem 4). Hence every point of $\beta$ is a point of $\alpha$. Conversely, let $B$ be any point of $\alpha$. The line $BQ$ meets $m$ in a point (Theorem 5). Hence every point of $\alpha$ is also a point of $\beta$.

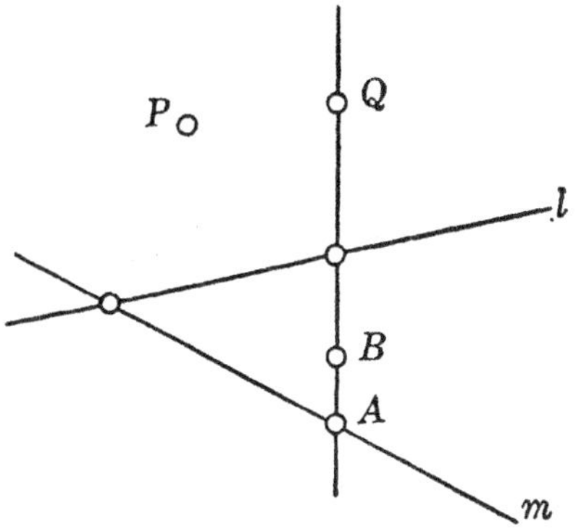

FIG. 9

COROLLARY. *There is one and only one plane determined by three non-collinear points, or by a line and a point not on the line, or by two intersecting lines.* (A, E 0)

The data of the corollary are all equivalent by virtue of E 0. We will denote by $ABC$ the plane determined by the points $A$, $B$, $C$; by $aA$ the plane determined by the line $a$ and the point $A$, etc.

THEOREM 7. *Two distinct planes which are on two common points $A$, $B$ $(A \neq B)$ are on all the points of the line $AB$, and on no other common points.* (A, E 0)

*Proof.* By Theorem 4 the line $AB$ lies in each of the two planes, which proves the first part of the proposition. Suppose $C$, not on $AB$, were a point common to the two planes. Then the plane determined by $A$, $B$, $C$ would be identical with each of the given planes (Theorem 6), which contradicts the hypothesis that the planes are distinct.

COROLLARY. *Two distinct planes cannot be on more than one common line.* (A, E 0)

**9. The three-space.** DEFINITION. If $P$, $Q$, $R$, $T$ are four points not in the same plane, and if $\pi$ is a plane containing $Q$, $R$, and $T$, the class $S_3$ of all points on the lines joining $P$ to the points of $\pi$ is called the *space of three dimensions*, or the *three-space* determined by $P$ and $\pi$.

If a point belongs to a three-space or is a point of a three-space, it is said to *be in* or to *lie in* or to *be on* the three-space. If all the points of a line or plane are points of a three-space $S_3$, the line or plane is said to *lie in* or to *be in* or to *be on* the $S_3$. Also the three-space is said to *be on* the point, line, or plane. It is clear from the definition that $P$ and every point of $\pi$ are points of the three-space determined by $P$ and $\pi$.

THEOREM 8. *If $A$ and $B$ are distinct points on a three-space $S_3$, every point on the line $AB$ is on $S_3$.* (A)

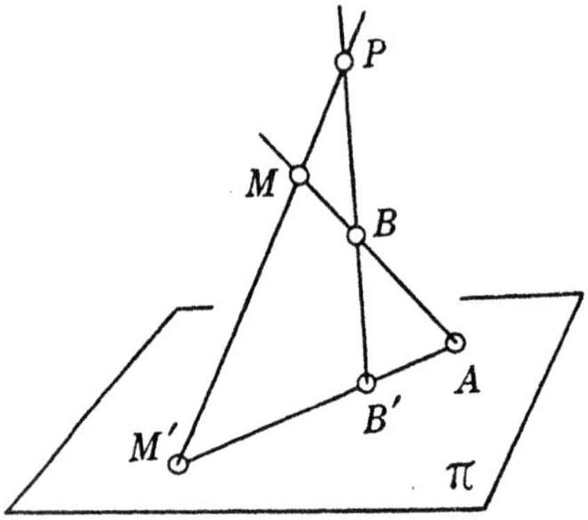


FIG. 10

*Proof.* Let $S_3$ be determined by a plane $\pi$ and a point $P$.

1. If $A$ and $B$ are both in $\pi$, the theorem is an immediate consequence of Theorem 4.

2. If the line $AB$ contains $P$, the theorem is obvious.

3. Suppose $A$ is in $\pi$, $B$ not in $\pi$, and $AB$ does not contain $P$ (fig. 10). There then exists a point $B'$ ($\neq A$) of $\pi$ collinear with $B$ and $P$ (def.). The line joining any point $M$ on $AB$ to $P$ on $BB'$ has a point $M'$ in common with $B'A$ (A 3). But $M'$ is a point of $\pi$, since it is a point of $AB'$. Hence $M$ is a point of $S_3$ (def.).

4. Let neither $A$ nor $B$ lie in $\pi$, and let $AB$ not contain $P$ (fig. 11). The lines $PA$ and $PB$ meet $\pi$ in two points $A'$ and $B'$ respectively. But the line joining $A$ on $A'P$ to $B$ on $PB'$ has a point $C$ in common with $B'A'$. $C$ is a point of $\pi$, which reduces the proof to Case 3.

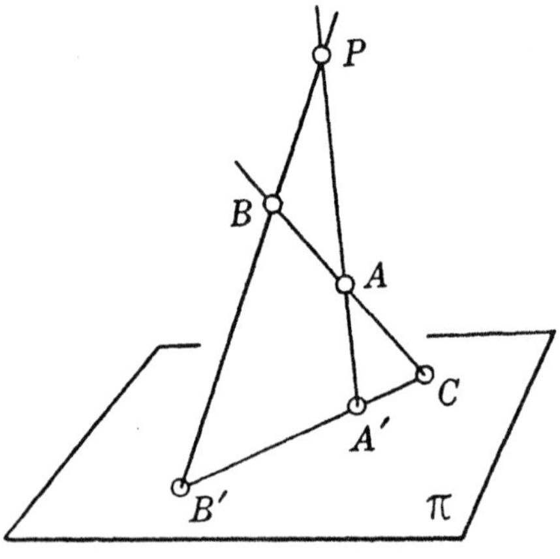


FIG. 11

It may be noted that in this proof no use has been made of E 0.

In discussing Case 4 we have proved incidentally, in connection with E 0 and Theorem 4, the following corollary:

COROLLARY 1. *If $S_3$ is a three-space determined by a point $P$ and a plane $\pi$, then $\pi$ and any line on $S_3$ but not on $\pi$ are on one and only one common point.* (A, E 0)

COROLLARY 2. *Every point on any plane determined by three non-collinear points on a three-space $S_3$ is on $S_3$.* (A)

*Proof.* As before, let the three-space be determined by $\pi$ and $P$, and let the three noncollinear points be $A, B, C$. Every point of the line $BC$ is a point of $S_3$ (Theorem 8), and every point of the plane $ABC$* is collinear with $A$ and some point of $BC$.

COROLLARY 3. *If a three-space $S_3$ is determined by a point $P$ and a plane $\pi$, then $\pi$ and any plane on $S_3$ distinct from $\pi$ are on one and only one common line.* (A, E 0)

*Proof.* Any plane contains at least three lines not passing through the same point (def., A 1). Two of these lines must meet $\pi$ in two distinct points, which are also points of the plane of the lines (Cor. 1). The result then follows from Theorem 7.

THEOREM 9. *If a plane $\alpha$ and a line $a$ not on $\alpha$ are on the same three-space $S_3$, then $\alpha$ and $a$ are on one and only one common point.* (A, E 0)

*Proof.* Let $S_3$ be determined by the plane $\pi$ and the point $P$.

1. If $\alpha$ coincides with $\pi$, the theorem reduces to Cor. 1 of Theorem 8.

2. If $\alpha$ is distinct from $\pi$, it has a line $l$ in common with $\pi$ (Theorem 8, Cor. 3). Let $A$ be any point on $\alpha$ not on $a$ (E 0) (fig. 12). The plane $aA$, determined by $A$ and $a$, meets $\pi$ in a line $m \neq l$ (Theorem 8, Cor. 3). The lines $l$, $m$ have a point $B$ in common (Theorem 5). The line $AB$ in $aA$ meets $a$ in a point $Q$ (Theorem 5), which is on $\alpha$, since $AB$ is on $\alpha$. That $\alpha$ and $a$ have no other point in common follows from Theorem 4.

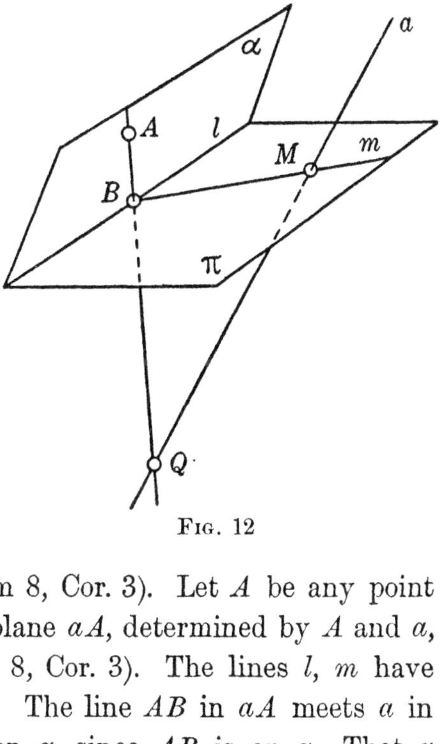

FIG. 12

COROLLARY 1. *Any two distinct planes on a three-space are on one and only one common line.* (A, E 0)

The proof is similar to that of Theorem 8, Cor. 3, and is left as an exercise.

COROLLARY 2. *Conversely, if two planes are on a common line, there exists a three-space on both.* (A, E 0)

* The proof can evidently be so worded as not to imply Theorem 6.

*Proof.* If the planes $\alpha$ and $\beta$ are distinct and have a line $l$ in common, any point $P$ of $\beta$ not on $l$ will determine with $\alpha$ a three-space containing $l$ and $P$ and hence containing $\beta$ (Theorem 8, Cor. 2).

COROLLARY 3. *Three planes on a three-space which are not on a common line are on one and only one common point.* (A, E 0)

*Proof.* This follows without difficulty from the theorem and Cor. 1.

Two planes are said to *determine* the line which they have in common, and to *intersect* or *meet* in that line. Likewise if three planes have a point in common, they are said to *intersect* or *meet* in the point.

COROLLARY 4. *If $\alpha$, $\beta$, $\gamma$ are three distinct planes on the same $S_3$ but not on the same line, and if a line $l$ is on each of two planes $\mu, \nu$ which are on the lines $\beta\gamma$ and $\gamma\alpha$ respectively, then it is on a plane $\lambda$ which is on the line $\alpha\beta$.* (A, E 0)

*Proof.* By Cor. 3 the planes $\alpha$, $\beta$, $\gamma$ have a point $P$ in common, so that the lines $\beta\gamma$, $\gamma\alpha$, $\alpha\beta$ all contain $P$. The line $l$, being common to planes through $\beta\gamma$ and $\gamma\alpha$, must pass through $P$, and the lines $l$ and $\alpha\beta$ therefore intersect in $P$ and hence determine a plane $\lambda$ (Theorem 6, Cor.).

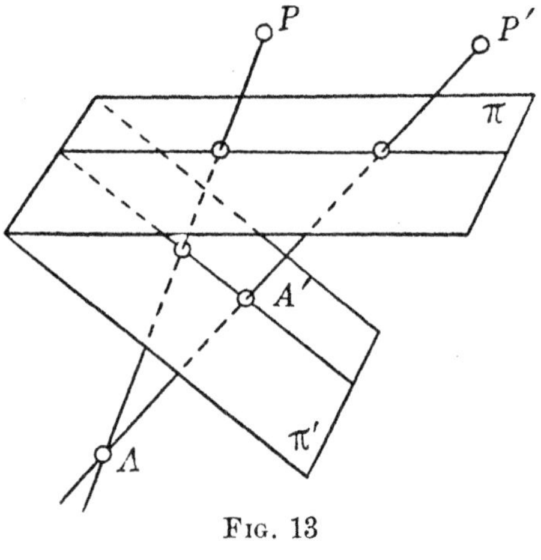


FIG. 13

THEOREM 10. *The three-space $S_3$ determined by a plane $\pi$ and a point $P$ is identical with the three-space $S_3'$ determined by a plane $\pi'$ and a point $P'$, provided $\pi'$ and $P'$ are on $S_3$.* (A, E 0)

*Proof.* Any point $A$ of $S_3'$ (fig. 13) is collinear with $P'$ and some point $A'$ of $\pi'$; but $P'$ and $A'$ are both points of $S_3$ and hence $A$ is a point of $S_3$ (Theorem 8). Hence every point of $S_3'$ is a point of $S_3$. Conversely, if $A$ is any point of $S_3$, the line $AP'$ meets $\pi'$ in a point (Theorem 9). Hence every point of $S_3$ is also a point of $S_3'$.

COROLLARY. *There is one and only one three-space on four given points not on the same plane, or a plane and a point not on the plane, or two nonintersecting lines.* (A, E 0)

The last part of the corollary follows from the fact that two nonintersecting lines are equivalent to four points not in the same plane (E 0).

It is convenient to use the term *coplanar* to describe points in the same plane. And we shall use the term *skew* lines for lines that have no point in common. Four noncoplanar points or two skew lines are said to *determine* the three-space in which they lie.

**10. The remaining assumptions of extension for a space of three dimensions.** In § 8 we gave a first assumption of extension. We will now add the assumptions which insure the existence of a space of three dimensions, and will exclude from our consideration spaces of higher dimensionality.

ASSUMPTIONS OF EXTENSION, E:

E 1. *There exists at least one line.*

E 2. *All points are not on the same line.*

E 3. *All points are not on the same plane.*

E 3′. *If* $S_3$ *is a three-space, every point is on* $S_3$.

The last may be called an assumption of *closure*.*

The last assumption might be replaced by any one of several equivalent propositions, such as for example:

*Every set of five points lie on the same three-space;* or

*Any two distinct planes have a line in common.* (Cf. Cor. 2, Theorem 9)

There is no logical difficulty, moreover, in replacing the assumption (E 3′) of closure given above by an assumption that all the points are not on the same three-space, and then to define a "four-space" in a manner entirely analogous to the definitions of the plane and to the three-space already given. And indeed a meaning can be given to the words *point* and *line* such that this last assumption is satisfied as well as those that precede it (excepting E 3′ of course). We could thus proceed step by step to define the notion of a linear space of any number of dimensions and derive the fundamental properties of alignment for such a space. But that is aside from our present purpose. The derivation of these properties for a four-space will furnish an excellent exercise, however, in the formal reasoning here emphasized (cf. Ex. 4, p. 25). The treatment for the $n$-dimensional case will be found in § 12, p. 29.

* The terms *extension* and *closure* in this connection were suggested by N. J. Lennes. It will be observed that the notation has been so chosen that E$i$ insures the existence of a space of $i$ dimensions, the line and the plane being regarded as spaces of one and two dimensions respectively.

The following *corollaries of extension* are readily derived from the assumptions just made. The proofs are left as exercises.

COROLLARY 1. *At least three coplanar lines are on every point.*

COROLLARY 2. *At least three distinct planes are on every line.*

COROLLARY 3. *All planes are not on the same line.*

COROLLARY 4. *All planes are not on the same point.*

COROLLARY 5. *If $S_3$ is a three-space, every plane is on $S_3$.*

## EXERCISES

**1.** Prove that through a given point $P$ not on either of two skew lines $l$ and $l'$ there is one and only one line meeting both the lines $l$, $l'$.

**2.** Prove that any two lines, each of which meets three given skew lines, are skew to each other.

**3.** Our assumptions do not as yet determine whether the number of points on a line is finite or infinite. Assuming that the number of points on one line is finite and equal to $n + 1$, prove that

i. the number of ~~points~~ on every ~~line~~ is $n + 1$;

ii. the number of points on every plane is $n^2 + n + 1$;

iii. the number of points on every three-space is $n^3 + n^2 + n + 1$;

iv. the number of lines on a three-space is $(n^2 + 1)(n^2 + n + 1)$;

v. the number of lines meeting any two skew lines on a three-space is $(n + 1)^2$;

vi. the number of lines on a point or on a plane is $n^2 + n + 1$.

**4.** Using the definition below, prove the following *theorems of alignment for a four-space* on the basis of Assumptions A and E 0:

DEFINITION. If $P$, $Q$, $R$, $S$, $T$ are five points not on the same three-space, and $S_3$ is a three-space on $Q$, $R$, $S$, $T$, the class $S_4$ of all points on the lines joining $P$ to the points of $S_3$ is called the *four-space* determined by $P$ and $S_3$.

i. If $A$ and $B$ are distinct points on a four-space, every point on the line $AB$ is on the four-space.

ii. Every line on a four-space $PQRST$ which is not on the three-space $QRST$ has one and only one point in common with the three-space.

iii. Every point on any plane determined by three noncollinear points on a four-space is on the four-space.

iv. Every point on a three-space determined by four noncoplanar points of a four-space is on the four-space.

v. Every plane of a four-space determined by a point $P$ and a three-space $S_3$ has one and only one line in common with $S_3$, provided the plane is not on $S_3$.

vi. Every three-space on a four-space determined by a point $P$ and a three-space $S_3$ has one and only one plane in common with $S_3$, provided it does not coincide with $S_3$.

vii. If a three-space $S_3$ and a plane $\alpha$ not on $S_3$ are on the same four-space, $S_3$ and $\alpha$ have one and only one line in common.

viii. If a three-space $S_3$ and a line $l$ not on $S_3$ are on the same four-space, $S_3$ and $l$ have one and only one point in common.

ix. Two planes on the same four-space but not on the same three-space have one and only one point in common.

x. Any two distinct three-spaces on the same four-space have one and only one plane in common.

xi. If two three-spaces have a plane in common, they lie in the same four-space.

xii. The four-space $S_4$ determined by a three-space $S_3$ and a point $P$ is identical with the four-space determined by a three-space $S_3'$ and a point $P'$, provided $S_3'$ and $P'$ are on $S_4$.

**5.** On the assumption that a line contains $n+1$ points, extend the results of Ex. 3 to a four-space.

**11. The principle of duality.** It is in order to exhibit the theorem of duality as clearly as possible that we have introduced the symmetrical, if not always elegant, terminology:

| | |
|---|---|
| A point is on a line. | A line is on a point. |
| A point is on a plane. | A plane is on a point. |
| A line is on a plane. | A plane is on a line. |
| A point is on a three-space. | A three-space is on a point. |
| A line is on a three-space. | A three-space is on a line. |
| A plane is on a three-space. | A three-space is on a plane. |

The theorem in question rests on the following observation: If any one of the preceding assumptions, theorems, or corollaries is expressed by means of this "on" terminology and then a new proposition is formed by simply interchanging the words *point* and *plane*, then this new proposition will be valid, i.e. will be a logical consequence of the Assumptions A and E. We give below, on the left, a complete list of the assumptions thus far made, expressed in the "on" terminology, and have placed on the right, opposite each, the corresponding proposition obtained by interchanging the words *point* and *plane* together with the reference to the place where the latter proposition occurs in the preceding sections:

| | |
|---|---|
| ASSUMPTIONS A 1, A 2. If $A$ and $B$ are distinct points, there is one and only one line on $A$ and $B$. | THEOREM 9, COR. 1. If $\alpha$ and $\beta$ are distinct planes, there is one and only one line on $\alpha$ and $\beta$.* |

* By virtue of Assumption E 3′ it is not necessary to impose the condition that the elements to be considered are in the same three-space. This observation should emphasize, however, that the assumption of closure is essential in the theorem to be proved.

| | |
|---|---|
| ASSUMPTION A 3. If $A, B, C$ are points not all on the same line, and $D$ and $E$ $(D \neq E)$ are points such that $B, C, D$ are on a line and $C, A, E$ are on a line, then there is a point $F$ such that $A, B, F$ are on a line and also $D, E, F$ are on a line. | THEOREM 9, COR. 4. If $\alpha, \beta, \gamma$ are planes not all on the same line, and $\mu$ and $\nu$ $(\mu \neq \nu)$ are planes such that $\beta, \gamma, \mu$ are on a line and $\gamma, \alpha, \nu$ are on a line, then there is a plane $\lambda$ such that $\alpha, \beta, \lambda$ are on a line and also $\mu, \nu, \lambda$ are on a line. |
| ASSUMPTION E 0. There are at least three points on every line. | COR. 2, p. 25. There are at least three planes on every line. |
| ASSUMPTION E 1. There exists at least one line. | ASSUMPTION E 1. There exists at least one line. |
| ASSUMPTION E 2. All points are not on the same line. | COR. 3, p. 25. All planes are not on the same line. |
| ASSUMPTION E 3. All points are not on the same plane. | COR. 4, p. 25. All planes are not on the same point. |
| ASSUMPTION E 3′. If $S_3$ is a three-space, every point is on $S_3$. | COR. 5, p. 25. If $S_3$ is a three-space, every plane is on $S_3$. |

In all these propositions it is to be noted that a line is a class of points whose properties are determined by the assumptions, while a plane is a class of points specified by a definition. This definition in the "on" language is given below on the left, together with a definition obtained from it by the interchange of *point* and *plane.* Two statements in this relation to one another are referred to as (space) *duals* of one another.

| | |
|---|---|
| If $P, Q, R$ are points not on the same line, and $l$ is a line on $Q$ and $R$, the class $S_2$ of all points such that every point of $S_2$ is on a line with $P$ and some point on $l$ is called the *plane* determined by $P$ and $l$. | If $\lambda, \mu, \nu$ are planes not on the same line, and $l$ is a line on $\mu$ and $\nu$, the class $B_2$ of all planes such that every plane of $B_2$ is on a line with $\lambda$ and some plane on $l$ is called the *bundle* determined by $\lambda$ and $l$. |

Now it is evident that, since $\lambda, \mu, \nu$ and $l$ all pass through a point $O$, the bundle determined by $\lambda$ and $l$ is simply *the class of all planes on the point O.* In like manner, it is evident that the dual of the definition of a three-space is simply a definition *of the class of all planes on a three-space.* Moreover, dual to the class of all planes on a line we have the class of all points on a line, i.e. the line itself, and conversely.

With the aid of these observations we are now ready to establish the so-called principle of duality:

THEOREM 11. THE THEOREM OF DUALITY FOR A SPACE OF THREE DIMENSIONS. *Any proposition deducible from Assumptions* A *and* E *concerning points, lines, and planes of a three-space remains valid, if stated in the "on" terminology, when the words "point" and "plane" are interchanged.* (A, E)

*Proof.* Any proposition deducible from Assumptions A and E is obtained from the assumptions given above on the left by a certain sequence of formal logical inferences. Clearly the same sequence of logical inferences may be applied to the corresponding propositions given above on the right. They will, of course, refer to the class of all planes on a line when the original argument refers to the class of all points on a line, i.e. to a line, and to a bundle of planes when the original argument refers to a plane. The steps of the original argument lead to a conclusion necessarily stated in terms of some or all of the twelve types of "on" statements enumerated at the beginning of this section. The derived argument leads in the same way to a conclusion which, whenever the original states that a point $P$ is on a line $l$, says that a plane $\pi'$ is one of the class of planes on a line $l'$, i.e. that $\pi'$ is on $l'$; or which, whenever the original argument states that a plane $\pi$ is on a point $P$, says that a bundle of planes on a point $P'$ contains a plane $\pi'$, i.e. that $P'$ is on $\pi'$. Applying similar considerations to each of the twelve types of "on" statements in succession, we see that to each statement in the conclusion arrived at by the original argument corresponds a statement arrived at by the derived argument in which the words *point* and *plane* in the original statement have been simply interchanged.

Any proposition obtained in accordance with the principle of duality just proved is called the *space dual* of the original proposition. The point and plane are said to be *dual* elements; the line is *self-dual.* We may derive from the above similar theorems on duality in a plane and at a point. For, consider a plane $\pi$ and a point $P$ not on $\pi$, together with all the lines joining $P$ with every point of $\pi$. Then to every point of $\pi$ will correspond a line through $P$, and to every line of $\pi$ will correspond a plane through $P$. Hence every proposition concerning the points and lines of $\pi$ is also valid for the corresponding lines and planes through $P$. The space dual of the latter proposition is a new proposition concerning lines and points on a plane, which could have been obtained directly by interchanging the words *point* and *line* in the original proposition, supposing the latter to be expressed in the "on" language. This gives

THEOREM 12. THE THEOREM OF DUALITY IN A PLANE. *Any proposition deducible from Assumptions* A *and* E *concerning the points and lines of a plane remains valid, if stated in the "on" terminology, when the words "point" and "line" are interchanged.* (A, E)

The space dual of this theorem then gives

THEOREM 13. THE THEOREM OF DUALITY AT A POINT. *Any proposition deducible from Assumptions* A *and* E *concerning the planes and lines through a point remains valid, if stated in the "on" terminology, when the words "plane" and "line" are interchanged.* (A, E)

The principle of duality was first stated explicitly by Gergonne (1826), but was led up to by the writings of Poncelet and others during the first quarter of the nineteenth century. It should be noted that this principle was for several years after its publication the subject of much discussion and often acrimonious dispute, and the treatment of this principle in many standard texts is far from convincing. The method of formal inference from explicitly stated assumptions makes the theorems appear almost self-evident. This may well be regarded as one of the important advantages of this method.

It is highly desirable that the reader gain proficiency in forming the duals of given propositions. It is therefore suggested as an exercise that he state the duals of each of the theorems and corollaries in this chapter. He should in this case state both the original and the dual proposition in the ordinary terminology in order to gain facility in dualizing propositions without first stating them in the often cumbersome "on" language. It is also desirable that he dualize several of the proofs by writing out in order the duals of each proposition used in the proofs in question.

### EXERCISE

Prove the theorem of duality for a space of four dimensions: Any proposition derivable from the assumptions of alignment and extension and closure for a space of four dimensions concerning points, lines, planes, and three-spaces remains valid when stated in the "on" terminology, if the words *point* and *three-space* and the words *line* and *plane* be interchanged.

*** 12. The theorems of alignment for a space of *n* dimensions.** We have already called attention to the fact that Assumption E 3′, whereby we limited ourselves to the consideration of a space of only

* This section may be omitted on a first reading.

three dimensions, is entirely arbitrary. This section is devoted to the discussion of the theorems of alignment, i.e. theorems derivable from Assumptions A and E 0, for a space of any number of dimensions. In this section, then, we make use of Assumptions A and E 0 only.

DEFINITION. If $P_0, P_1, P_2, \cdots, P_n$ are $n+1$ points not on the same $(n-1)$-space, and $S_{n-1}$ is an $(n-1)$-space on $P_1, P_2, \cdots, P_n$, the class $S_n$ of all points on the lines joining $P_0$ to the points of $S_{n-1}$ is called the *n-space* determined by $P_0$ and $S_{n-1}$.

As a three-space has already been defined, this definition clearly determines the meaning of "$n$-space" for every positive integral value of $n$. We shall use $S_n$ as a symbol for an $n$-space, calling a plane a 2-space, a line a 1-space, and a point a 0-space, when this is convenient. $S_0$ is then a symbol for a point.

DEFINITION. An $S_r$ is *on* an $S_t$ and an $S_t$ is *on* an $S_r$ $(r < t)$, provided that every point of $S_r$ is a point of $S_t$.

DEFINITION. $k$ points are said to be *independent*, if there is no $S_{k-2}$ which contains them all.

Corresponding to the theorems of §§ 6–9 we shall now establish the propositions contained in the following Theorems $S_n 1$, $S_n 2$, $S_n 3$. As these propositions have all been proved for the case $n = 3$, it is sufficient to prove them on the hypothesis that they have already been proved for the cases $n = 3, 4, \cdots, n-1$; i.e. we assume that the propositions contained in Theorem $S_{n-1} 1$, $a, b, c, d, e, f$ have been proved, and derive Theorem $S_n 1$, $a, \cdots, f$ from them. By the principle of mathematical induction this establishes the theorem for any $n$.

THEOREM $S_n 1$. *Let the n-space $S_n$ be defined by the point $R_0$ and the $(n-1)$-space $R_{n-1}$.*

*a. There is an n-space on any $n+1$ independent points.*

*b. Any line on two points of $S_n$ has one point in common with $R_{n-1}$, and is on $S_n$.*

*c. Any $S_r (r < n)$ on $r+1$ independent points of $S_n$ is on $S_n$.*

*d. Any $S_r (r < n)$ on $r+1$ independent points of $S_n$ has an $S_{r-1}$ in common with $R_{n-1}$, provided the $r+1$ points are not all on $R_{n-1}$.*

*e. Any line $l$ on two points of $S_n$ has one point in common with any $S_{n-1}$ on $S_n$.*

*f. If $T_0$ and $T_{n-1}$ ($T_0$ not on $T_{n-1}$) are any point and any $(n-1)$-space respectively of the n-space determined by $R_0$ and $R_{n-1}$, the latter n-space is the same as that determined by $T_0$ and $T_{n-1}$.*

*Proof. a.* Let the $n+1$ independent points be $P_0, P_1, \cdots, P_n$. Then the points $P_1, P_2, \cdots, P_n$ are independent; for, otherwise, there would exist an $S_{n-2}$ containing them all (definition), and this $S_{n-2}$ with $P_0$ would determine an $S_{n-1}$ containing all the points $P_0, P_1, \cdots, P_n$, contrary to the hypothesis that they are independent. Hence, by Theorem $S_{n-1}1\,a$, there is an $S_{n-1}$ on the points $P_1, P_2, \cdots, P_n$; and this $S_{n-1}$ with $P_0$ determines an $n$-space which is on the points $P_0, P_1, P_2, \cdots, P_n$.

*b.* If the line $l$ is on $R_0$ or $R_{n-1}$, the proposition is evident from the definition of $S_n$. If $l$ is not on $R_0$ or $R_{n-1}$, let $A$ and $B$ be the given points of $l$ which are on $S_n$. The lines $R_0A$ and $R_0B$ then meet $R_{n-1}$ in two points $A'$ and $B'$ respectively. The line $l$ then meets the two lines $B'R_0$, $R_0A'$; and hence, by Assumption A 3, it must meet the line $A'B'$ in a point $P$ which is on $R_{n-1}$ by Theorem $S_{n-1}1\,b$. To show that every point of $l$ is on $S_n$, consider the points $A, A', P$. Any line joining an arbitrary point $Q$ of $l$ to $R_0$, meets the two lines $PA$ and $AA'$, and hence, by Assumption A 3, meets the third line $A'P$. But every point of $A'P$ is on $R_{n-1}$ (Theorem $S_{n-1}1\,b$), and hence $Q$ is, by definition, a point of $S_n$.

*c.* This may be proved by induction with respect to $r$. For $r=1$ it reduces to Theorem $S_n1\,b$. If the proposition is true for $r=k-1$, all the points of an $S_k$ on $k+1$ independent points of $S_n$ are, by definition and Theorem $S_k1\,f$, on lines joining one of these points to the points of the $S_{k-1}$ determined by the remaining $k$ points. But under the hypothesis of the induction this $S_{k-1}$ is on $S_n$, and hence, by Theorem $S_n1\,b$, all points of $S_k$ are on $S_n$.

*d.* Let $r+1$ independent points of $S_n$ be $P_0, P_1, \cdots, P_r$ and let $P_0$ be not on $R_{n-1}$. Each of the lines $P_0P_k$ $(k=1, \cdots, r)$ has a point $Q_k$ in common with $R_{n-1}$ (by $S_n1\,b$). The points $Q_1, Q_2, \cdots, Q_r$ are independent; for if not, they would all be on the same $S_{r-2}$, which, together with $P_0$, would determine an $S_{r-1}$ containing all the points $P_k$ (by $S_{r-1}1\,b$). Hence, by $S_{r-1}1\,a$, there is an $S_{r-1}$ on $Q_1, Q_2, \cdots, Q_r$ which, by $c$, is on both $S_r$ and $S_n$.

*e.* We will suppose, first, that one of the given points is $R_0$. Let the other be $A$. By definition $l$ then meets $R_{n-1}$ in a point $A'$, and, by $S_{n-1}1\,b$, in only one such point. If $R_0$ is on $S_{n-1}$, no proof is required for this case. Suppose, then, that $R_0$ is not on $S_{n-1}$, and let $C$ be any point of $S_{n-1}$. The line $R_0C$ meets $R_{n-1}$ in a point $C'$ (by definition). By $d$, $S_{n-1}$ has in common with $R_{n-1}$ an $(n-2)$-space, $S_{n-2}$, and, by

Theorem $S_{n-1}1\,e$, this has in common with the line $A'C'$ at least one point $D'$. All points of the line $D'C$ are then on $S_{n-1}$, by $S_{n-1}1\,b$. Now the line $l$ meets the two lines $C'D'$ and $CC'$; hence it meets the line $CD'$ (Assumption A 3), and has at least one point on $S_{n-1}$.

We will now suppose, secondly, that both of the given points are distinct from $R_0$. Let them be denoted by $A$ and $B$, and suppose that $R_0$ is not on $S_{n-1}$. By the case just considered, the lines $R_0A$ and $R_0B$ meet $S_{n-1}$ in two points $A'$ and $B'$ respectively. The line $l$, which meets $R_0A'$ and $R_0B'$ must then meet $A'B'$ in a point which, by Theorem $S_{n-1}1\,b$, is on $S_{n-1}$.

Suppose, finally, that $R_0$ is on $S_{n-1}$, still under the hypothesis that $l$ is not on $R_0$. By $d$, $S_{n-1}$ meets $R_{n-1}$ in an $(n-2)$-space $Q_{n-2}$, and the plane $R_0l$ meets $R_{n-1}$ in a line $l'$. By Theorem $S_{n-1}1\,e$, $l'$ and $Q_{n-2}$ have in common at least one point $P$. Now the lines $l$ and $R_0P$ are on the plane $R_0l$, and hence have in common a point $Q$ (by Theorem $S_21\,e$ = Theorem 5). By $S_{n-1}1\,b$ the point $Q$ is common to $S_{n-1}$ and $l$.

*f*. Let the $n$-space determined by $T_0$ and $T_{n-1}$ be denoted by $T_n$. Any point of $T_n$ is on a line joining $T_0$ with some point of $T_{n-1}$. Hence, by $b$, every point of $T_n$ is on $S_n$. Let $P$ be any point of $S_n$ distinct from $T_0$. The line $T_0P$ meets $T_{n-1}$ in a point, by $e$. Hence every point of $S_n$ is a point of $T_n$.

COROLLARY. *On $n+1$ independent points there is one and but one $S_n$.*

This is a consequence of Theorem $S_n1\,a$ and $S_n1f$. The formal proof is left as an exercise.

THEOREM $S_n2$. *An $S_r$ and an $S_k$ having in common an $S_p$, but not an $S_{p+1}$, are on a common $S_{r+k-p}$ and are not both on the same $S_n$, if $n < r+k-p$.*

*Proof.* If $k=p$, $S_k$ is on $S_r$. If $k>p$, let $P_1$ be a point on $S_k$ not on $S_p$. Then $P_1$ and $S_r$ determine an $S_{r+1}$, and $P_1$ and $S_p$ an $S_{p+1}$, such that $S_{p+1}$ is contained in $S_{r+1}$ and $S_k$. If $k>p+1$, let $P_2$ be a point of $S_k$ not on $S_{p+1}$. Then $P_2$ and $S_{r+1}$ determine an $S_{r+2}$, while $P_2$ and $S_{p+1}$ determine an $S_{p+2}$, which is on $S_{r+2}$ and $S_k$. This process can be continued until there results an $S_{p+i}$ containing all the points of $S_k$. By Theorem $S_n1$, Cor., we have $i=k-p$. At this stage in the process we obtain an $S_{r+k-p}$ which contains both $S_r$ and $S_k$.

The argument just made shows that $P_1, P_2, \cdots, P_{k-p}$, together with any set $Q_1, Q_2, \cdots, Q_{r+1}$, of $r+1$ independent points of $S_r$, constitute a set of $r+k-p+1$ independent points, each of which is either in $S_r$ or $S_k$. If $S_r$ and $S_k$ were both on an $S_n$, where $n < r+k-p$, these could not be independent.

THEOREM $S_n 3$. *An* $S_r$ *and an* $S_k$ *contained in an* $S_n$ *are both on the same* $S_{r+k-n}$.

*Proof.* If there were less than $r+k-n+1$ independent points common to $S_r$ and $S_k$, say $r+k-n$ points, they would, by Theorem $S_n 2$, determine an $S_q$, where $q = r+k-(r+k-n-1) = n+1$.

Theorems $S_n 2$ and $S_n 3$ can be remembered and applied very easily by means of a diagram in which $S_n$ is represented by $n+1$ points. Thus, if $n = 3$, we have a set of four points. That any two $S_2$'s have an $S_1$ in common corresponds to the fact that any two sets of three must have at least two points in common. In the general case a set of $r+1$ points and a set of $k+1$ selected from the same set of $n+1$ have in common at least $r+k-n+1$ points, and this corresponds to the last theorem. This diagram is what our assumptions would describe directly, if Assumption E 0 were replaced by the assumption:

*Every line contains two and only two points.*

If one wishes to confine one's attention to the geometry in a space of a given number of dimensions, Assumptions E 2, E 3, and E 3′ may be replaced by the following:

En. *Not all points are on the same* $S_k$, *if* $k < n$.

En′. *If* $S$ *is an* $S_n$, *all points are on* $S$.

For every $S_n$ there is a principle of duality analogous to that which we have discussed for $n = 3$. In $S_n$ the duality is between $S_k$ and $S_{n-k-1}$ (counting a point as an $S_0$), for all $k$'s from 0 to $n-1$. If $n$ is odd, there is a self-dual space in $S_n$; if $n$ is even, $S_n$ contains no self-dual space.

## EXERCISES

**1.** State and prove the theorems of duality in $S_5$; in $S_n$.

**2.** If $m+1$ is the number of points on a line, how many $S_k$'s are there in an $S_n$?

* **3.** State the assumptions of extension by which to replace Assumption En and En′ for spaces of an infinite number of dimensions. Make use of the transfinite numbers.

* Exercises marked * are of a more advanced or difficult character.

# CHAPTER II

## PROJECTION, SECTION, PERSPECTIVITY. ELEMENTARY CONFIGURATIONS

**13. Projection, section, perspectivity.** The point, line, and plane are the *simple elements* of space *; we have seen in the preceding chapter that the relation expressed by the word *on* is a reciprocal relation that may exist between any two of these simple elements. In the sequel we shall have little occasion to return to the notion of a line as being a class of points, or to the definition of a plane; but shall regard these elements simply as entities for which the relation "on" has been defined. The theorems of the preceding chapter are to be regarded as expressing the fundamental properties of this relation.† We proceed now to the study of certain sets of these elements, and begin with a series of definitions.

DEFINITION. A *figure* is any set of points, lines, and planes in space. A *plane figure* is any set of points and lines on the same plane. A *point figure* is any set of planes and lines on the same point.

It should be observed that the notion of a point figure is the space dual of the notion of a plane figure. In the future we shall frequently place dual definitions and theorems side by side. By virtue of the principle of duality it will be necessary to give the proof of only one of two dual theorems.

DEFINITION. Given a figure F and a point $P$; every point of F distinct from $P$ determines with $P$ a line, and every line of F not on $P$ determines with $P$ a plane; the set of these lines and planes through $P$ is called the *projection*

DEFINITION. Given a figure F and a plane $\pi$; every plane of F distinct from $\pi$ determines with $\pi$ a line, and every line of F not on $\pi$ determines with $\pi$ a point; the set of these lines and points on $\pi$ is called the *section* ‡ of F

* The word *space* is used in place of the three-space in which are all the elements considered.

† We shall not in future, however, confine ourselves to the "on" terminology, but shall also use the more common expressions.

‡ A section by a plane is often called a *plane section*.

of F from $P$. The individual lines and planes of the projection are also called the *projectors* of the respective points and lines of F.

by $\pi$. The individual lines and points of the section are also called the *traces* of the respective planes and lines of F.

If F is a plane figure and the point $P$ is in the plane of the figure, the definition of the projection of F from $P$ has the following *plane dual*:

DEFINITION. Given a plane figure F and a line $l$ in the plane of F; the set of points in which the lines of F distinct from $l$ meet $l$ is called the *section* of F by $l$. The line $l$ is called a *transversal*, and the points are called the *traces* of the respective lines of F.

As examples of these definitions we mention the following: The projection of three mutually intersecting nonconcurrent lines from a point $P$ not in the plane of the lines consists of three planes through $P$; the lines of intersection of these planes are part of the projection only if the points of intersection of the lines are thought of as part of the projected figure. The section of a set of planes all on the same line by a plane not on this line consists of a set of concurrent lines, the traces of the planes. The section of this set of concurrent lines in a plane by a line in the plane not on their common point consists of a set of points on the transversal, the points being the traces of the respective lines.

DEFINITION. Two figures $F_1$, $F_2$ are said to be *in* (1, 1) *correspondence* or to *correspond in a one-to-one reciprocal way*, if every element of $F_1$ corresponds (cf. footnote, p. 5) to a unique element of $F_2$ in such a way that every element of $F_2$ is the correspondent of a unique element of $F_1$. A figure is *in* (1, 1) *correspondence with itself*, if every element of the figure corresponds to a unique element of the same figure in such a way that every element of the figure is the correspondent of a unique element. Two elements that are associated in this way are said to be *corresponding* or *homologous* elements.

A correspondence of fundamental importance is described in the following definitions:

DEFINITION. If any two homologous elements of two corresponding figures have the same projector from a fixed point $O$, such that all the projectors are

DEFINITION. If any two homologous elements of two corresponding figures have the same trace in a fixed plane $\omega$, such that all the traces of either distinct, the figures are said to be *perspective* from $O$. The point $O$ is called the *center of perspectivity.*

figure are distinct, the figures are said to be *perspective from* $\omega$. The plane $\omega$ is called the *plane of perspectivity.*

DEFINITION. If any two homologous lines in two corresponding figures in the same plane have the same trace on a line $l$, such that all the traces of either figure are distinct, the figures are said to be *perspective from* $l$. The line $l$ is called the *axis of perspectivity.*

Additional definitions of perspective figures will be given in the next chapter (p. 56). These are sufficient for our present purpose.

DEFINITION. *To project a figure in a plane* $\alpha$ *from a point* $O$ *onto a plane* $\alpha'$, distinct from $\alpha$, is to form the section by $\alpha'$ of the projection of the given figure from $O$. *To project a set of points of a line* $l$ *from a point* $O$ *onto a line* $l'$, distinct from $l$ but in the same plane with $l$ and $O$, is to form the section by $l'$ of the projection of the set of points from $O$.

Clearly in either case the two figures are perspective from $O$, provided $O$ is not on either of the planes $\alpha$, $\alpha'$ or the lines $l$, $l'$.

### EXERCISE

What is the dual of the process described in the last definition?

The notions of projection and section and perspectivity are fundamental in all that follows.* They will be made use of almost immediately in deriving one of the most important theorems of projective geometry. We proceed first, however, to define an important class of figures.

**14. The complete n-point, etc.** DEFINITION. A *complete n-point in space* or a *complete space n-point* is the figure formed by $n$ points, no four of which lie in the same plane, together with the $n(n-1)/2$ lines joining every pair of the points and the $n(n-1)(n-2)/6$ planes joining every set of three of the points. The points, lines, and planes of this figure are called the *vertices, edges,* and *faces* respectively of the complete $n$-point.

* The use of these notions in deriving geometrical theorems goes back to early times. Thus, e.g., B. Pascal (1623–1662) made use of them in deriving the theorem on a hexagon inscribed in a conic which bears his name. The systematic treatment of these notions is due to Poncelet; cf. his Traité des propriétés projectives des figures, Paris, 1822.

The simplest complete $n$-point in space is the complete space four-point. It consists of four vertices, six edges, and four faces, and is called a *tetrahedron.* It is a self-dual figure.

### EXERCISE

Define the *complete n-plane in space* by dualizing the last definition. The planes, lines, and points of the complete $n$-plane are also called the *faces*, *edges*, and *vertices* of the $n$-plane.

DEFINITION. A *complete n-point in a plane* or a *complete plane n-point* is the figure formed by $n$ points of a plane, no three of which are collinear, together with the $n(n-1)/2$ lines joining every pair of the points. The points are called the *vertices* and the lines are called the *sides* of the $n$-point. The plane dual of a complete plane $n$-point is called a *complete plane n-line.* It has $n$ *sides* and $n(n-1)/2$ *vertices.* The simplest complete plane $n$-point consists of three vertices and three sides and is called a *triangle.*

DEFINITION. A *simple space n-point* is a set of $n$ points $P_1, P_2, P_3, \cdots, P_n$ *taken in a certain order,* in which no four consecutive points are coplanar, together with the $n$ lines $P_1P_2$, $P_2P_3$, $\cdots$, $P_nP_1$ joining successive points and the $n$ planes $P_1P_2P_3$, $\cdots$, $P_nP_1P_2$ determined by successive lines. The points, lines, and planes are called the *vertices, edges,* and *faces* respectively of the figure. The space dual of a simple space $n$-point is a *simple space n-plane.*

DEFINITION. A *simple plane n-point* is a set of $n$ points $P_1, P_2, P_3, \cdots P_n$ of a plane *taken in a certain order* in which no three consecutive points are collinear, together with the $n$ lines $P_1P_2$, $P_2P_3$, $\cdots$, $P_nP_1$ joining successive points. The points and lines are called the *vertices* and *sides* respectively of the figure. The plane dual of a simple plane $n$-point is called a *simple plane n-line.*

Evidently the simple space $n$-point and the simple space $n$-plane are identical figures, as likewise the simple plane $n$-point and the simple plane $n$-line. Two sides of a simple $n$-line which meet in one of its vertices are *adjacent.* Two vertices are adjacent if in the dual relation. Two vertices of a simple $n$-point $P_1P_2 \cdots P_n$ ($n$ even) are *opposite* if, in the order $P_1P_2 \cdots P_n$, as many vertices follow one and precede the other as precede the one and follow the other. If $n$ is odd, a vertex and a side are opposite if, in the order $P_1P_2 \cdots P_n$, as many vertices follow the side and precede the vertex as follow the vertex and precede the side.

The space duals of the complete plane $n$-point and the complete plane $n$-line are the *complete $n$-plane on a point* and the *complete $n$-line on a point* respectively. They are the projections from a point, of the plane $n$-line and the plane $n$-point respectively.

**15. Configurations.** The figures defined in the preceding section are examples of a more general class of figures of which we will now give a general definition.

DEFINITION. A figure is called a *configuration*, if it consists of a finite number of points, lines, and planes, with the property that each point is on the same number $a_{12}$ of lines and also on the same number $a_{13}$ of planes; each line is on the same number $a_{21}$ of points and the same number $a_{23}$ of planes; and each plane is on the same number $a_{31}$ of points and the same number $a_{32}$ of lines.

A configuration may conveniently be described by a square matrix:

| | 1<br>point | 2<br>line | 3<br>plane |
|---|---|---|---|
| 1 point | $a_{11}$ | $a_{12}$ | $a_{13}$ |
| 2 line | $a_{21}$ | $a_{22}$ | $a_{23}$ |
| 3 plane | $a_{31}$ | $a_{32}$ | $a_{33}$ |

In this notation, if we call a point an element of the first kind, a line an element of the second kind, and a plane one of the third kind, the number $a_{ij}$ $(i \neq j)$ gives the number of elements of the $j$th kind on every element of the $i$th kind. The numbers $a_{11}$, $a_{22}$, $a_{33}$ give the total number of points, lines, and planes respectively. Such a square matrix is called the *symbol* of the configuration.

A tetrahedron, for example, is a figure consisting of four points, six lines, and four planes; on every line of the figure are two points of the figure, on every plane are three points, through every point pass three lines and also three planes, every plane contains three lines, and through every line pass two planes. A tetrahedron is therefore a configuration of the symbol

| | | |
|---|---|---|
| 4 | 3 | 3 |
| 2 | 6 | 2 |
| 3 | 3 | 4 |

The symmetry shown in this symbol is due to the fact that the figure in question is self-dual. A triangle evidently has the symbol

| | |
|---|---|
| 3 | 2 |
| 2 | 3 |

Since all the numbers referring to planes are of no importance in case of a plane figure, they are omitted from the symbol for a plane configuration.

In general, a complete plane $n$-point is of the symbol

| | |
|---|---|
| $n$ | $n-1$ |
| 2 | $\frac{1}{2}n(n-1)$ |

and a complete space $n$-point of the symbol

| | | |
|---|---|---|
| $n$ | $n-1$ | $\frac{1}{2}(n-1)(n-2)$ |
| 2 | $\frac{1}{2}n(n-1)$ | $n-2$ |
| 3 | 3 | $\frac{1}{6}n(n-1)(n-2)$ |

Further examples of configurations are figs. 14 and 15, regarded as plane figures.

### EXERCISE

Prove that the numbers in a configuration symbol must satisfy the condition

$$a_{ij}a_{ii} = a_{ji}a_{jj} \qquad (i, j = 1, 2, 3)$$

**16. The Desargues configuration.** A very important configuration is obtained by taking the plane section of a complete space five-point. The five-point is clearly a configuration with the symbol

| | | |
|---|---|---|
| 5 | 4 | 6 |
| 2 | 10 | 3 |
| 3 | 3 | 10 |

and it is clear that the section by a plane not on any of the vertices is a configuration whose symbol may be obtained from the one just given by removing the first column and the first row. This is due to the fact that every line of the space figure gives rise to a point in the plane, and every plane gives rise to a line. The configuration in the plane has then the symbol

| | |
|---|---|
| 10 | 3 |
| 3 | 10 |

We proceed to study in detail the properties of the configuration just obtained. It is known as the *configuration of Desargues.*

We may consider the vertices of the complete space five-point as consisting of the vertices of a triangle $A$, $B$, $C$ and of two points $O_1$, $O_2$

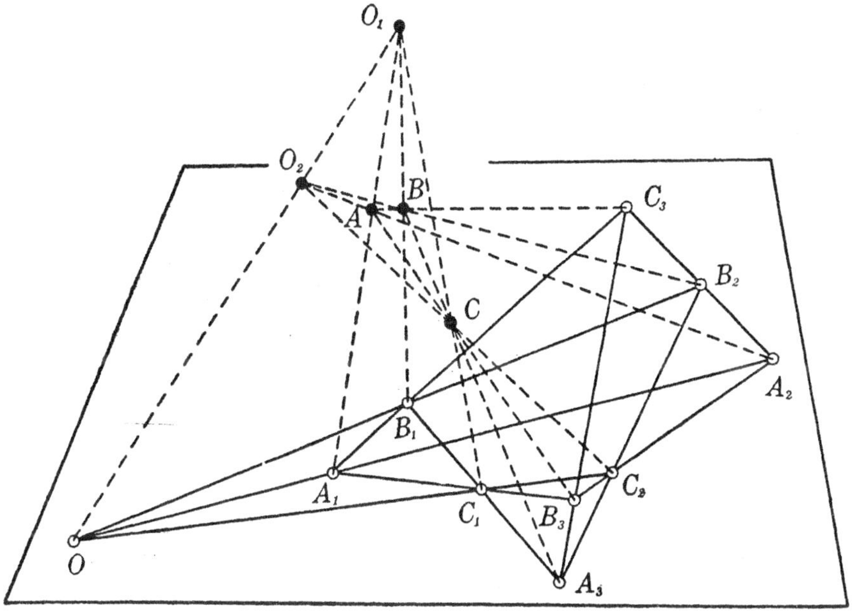


FIG. 14

not coplanar with any two vertices of the triangle (fig. 14). The section by a plane $\alpha$ not passing through any of the vertices will then consist of the following:

A triangle $A_1B_1C_1$, the projection of the triangle $ABC$ from $O_1$ on $\alpha$.

A triangle $A_2B_2C_2$, the projection of the triangle $ABC$ from $O_2$ on $\alpha$.

The trace $O$ of the line $O_1O_2$.

The traces $A_3$, $B_3$, $C_3$ of the lines $BC$, $CA$, $AB$ respectively.

The trace of the plane $ABC$, which contains the points $A_3$, $B_3$, $C_3$.

The traces of the three planes $AO_1O_2$, $BO_1O_2$, $CO_1O_2$, which contain respectively the triples of points $OA_1A_2$, $OB_1B_2$, $OC_1C_2$.

The configuration may then be considered (in ten ways) as consisting of two triangles $A_1B_1C_1$ and $A_2B_2C_2$, perspective from a point $O$ and having homologous sides meeting in three collinear points $A_3$, $B_3$, $C_3$. These considerations lead to the following fundamental theorem:

THEOREM 1. THE THEOREM OF DESARGUES.* *If two triangles in the same plane are perspective from a point, the three pairs of homologous sides meet in collinear points; i.e. the triangles are perspective from a line.* (A, E)

*Proof.* Let the two triangles be $A_1B_1C_1$ and $A_2B_2C_2$ (fig. 14), the lines $A_1A_2$, $B_1B_2$, $C_1C_2$ meeting in the point $O$. Let $B_1A_1$, $B_2A_2$ intersect in the point $C_3$; $A_1C_1$, $A_2C_2$ in $B_3$; $B_1C_1$, $B_2C_2$ in $A_3$. It is required to prove that $A_3$, $B_3$, $C_3$ are collinear. Consider any line through $O$ which is not in the plane of the triangles, and denote by $O_1$, $O_2$ any two distinct points on this line other than $O$. Since the lines $A_2O_2$ and $A_1O_1$ lie in the plane $(A_1A_2, O_1O_2)$, they intersect in a point $A$. Similarly, $B_1O_1$ and $B_2O_2$ intersect in a point $B$, and likewise $C_1O_1$ and $C_2O_2$ in a point $C$. Thus $ABCO_1O_2$, together with the lines and planes determined by them, form a complete five-point in space of which the perspective triangles form a part of a plane section. The theorem is proved by completing the plane section. Since $AB$ lies in a plane with $A_1B_1$, and also in a plane with $A_2B_2$, the lines $A_1B_1$, $A_2B_2$, and $AB$ meet in $C_3$. So also $A_1C_1$, $A_2C_2$, and $AC$ meet in $B_3$; and $B_1C_1$, $B_2C_2$, and $BC$ meet in $A_3$. Since $A_3$, $B_3$, $C_3$ lie in the plane $ABC$ and also in the plane of the triangles $A_1B_1C_1$ and $A_2B_2C_2$, they are collinear.

THEOREM 1'. *If two triangles in the same plane are perspective from a line, the lines joining pairs of homologous vertices are concurrent; i.e. the triangles are perspective from a point.* (A, E)

This, the converse of Theorem 1, is also its plane dual, and hence requires no further proof.

COROLLARY. *If two triangles not in the same plane are perspective from a point, the pairs of homologous sides intersect in collinear points; and conversely.* (A, E)

A more symmetrical and for many purposes more convenient notation for the Desargues configuration may be obtained as follows: Let the vertices of the space five-point be denoted by $P_1$, $P_2$, $P_3$, $P_4$, $P_5$ (fig. 15). The trace of the line $P_1P_2$ in the plane section is then naturally denoted by $P_{12}$, — in general, the trace of the line $P_iP_j$ by $P_{ij}$ $(i, j = 1, 2, 3, 4, 5, i \neq j)$. Likewise the trace of the plane $P_iP_jP_k$ may

* Girard Desargues, 1593–1662.

be denoted by $l_{ijk}$ ($i, j, k = 1, 2, 3, 4, 5$). This notation makes it possible to tell at a glance which lines and points are united. Clearly a point is on a line of the configuration if and only if the suffixes of the point are both among the suffixes of the line. Also the third point on the line joining $P_{ij}$ and $P_{jk}$ is the point $P_{ki}$; two points are on the same line if and only if they have a suffix in common, etc.

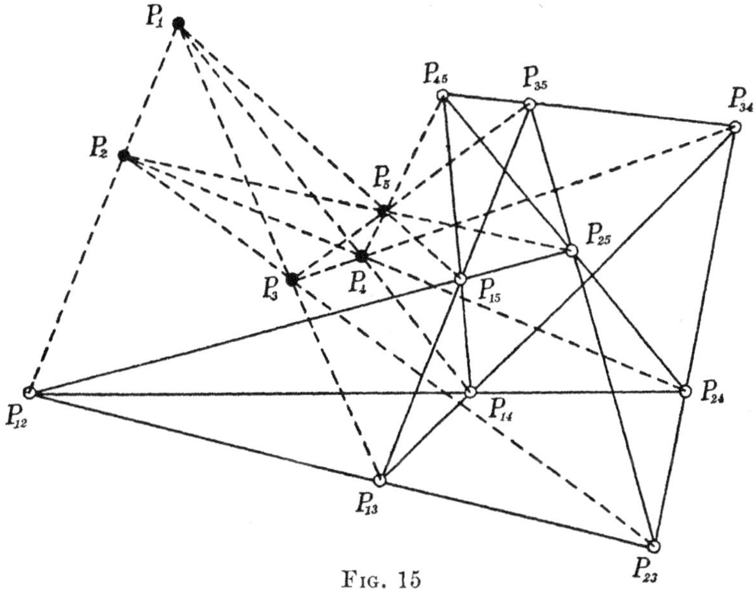


FIG. 15

## EXERCISES

**1.** Prove Theorem 1′ without making use of the principle of duality.

**2.** If two complete $n$-points in different planes are perspective from a point, the pairs of homologous sides intersect in collinear points. What is the dual theorem? What is the corresponding theorem concerning any two plane figures in different planes?

**3.** State and prove the converse of the theorems in Ex. 2.

**4.** If two complete $n$-points in the same plane correspond in such a way that homologous sides intersect in points of a straight line, the lines joining homologous vertices are concurrent; i.e. the two $n$-points are perspective from a point. Dualize.

**5.** What is the figure formed by two complete $n$-points in the same plane when they are perspective from a point? Consider particularly the cases $n = 4$ and $n = 5$. Show that the figure corresponding to the general case is a plane section of a complete space $(n + 2)$-point. Give the configuration symbol and dualize.

**6.** If three triangles are perspective from the same point, the three axes of perspectivity of the three pairs of triangles are concurrent; and conversely. Dualize, and compare the configuration of the dual theorem with the case $n = 4$ of Ex. 5 (cf. fig. 15, regarded as a plane figure).

**17. Perspective tetrahedra.** As an application of the corollary of the last theorem we may now derive a theorem in space analogous to the theorem of Desargues in the plane.

THEOREM 2. *If two tetrahedra are perspective from a point, the six pairs of homologous edges intersect in coplanar points, and the four pairs of homologous faces intersect in coplanar lines; i.e. the tetrahedra are perspective from a plane.* (A, E)

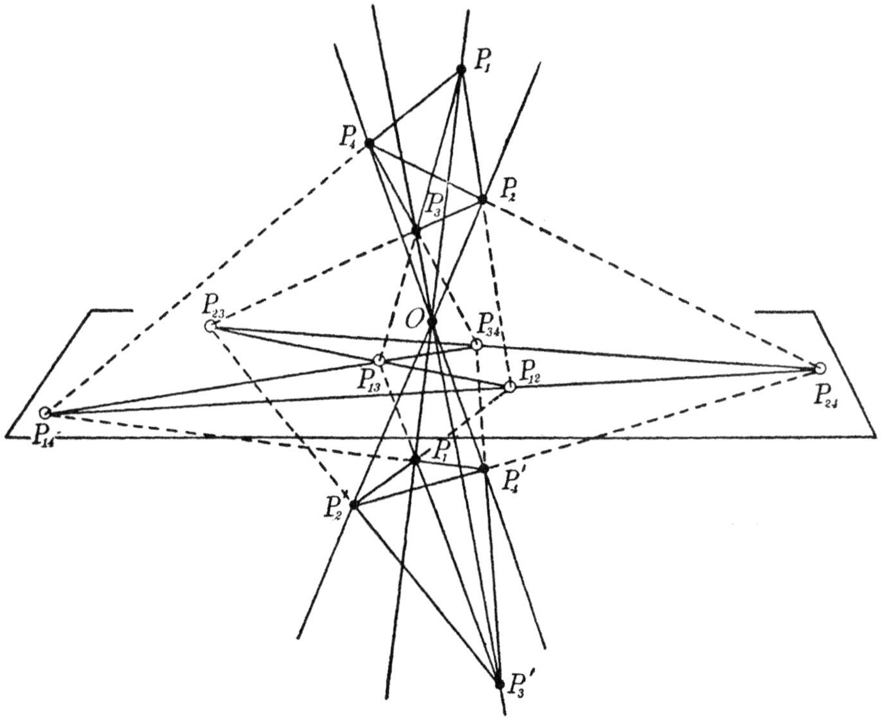


FIG. 16

*Proof.* Let the two tetrahedra be $P_1P_2P_3P_4$ and $P_1'P_2'P_3'P_4'$, and let the lines $P_1P_1'$, $P_2P_2'$, $P_3P_3'$, $P_4P_4'$ meet in the center of perspectivity $O$. Two homologous edges $P_iP_j$ and $P_i'P_j'$ then clearly intersect; call the point of intersection $P_{ij}$. The points $P_{12}$, $P_{13}$, $P_{23}$ lie on the same line, since the triangles $P_1P_2P_3$ and $P_1'P_2'P_3'$ are perspective from $O$ (Theorem 1, Cor.). By similar reasoning applied to the other pairs of perspective triangles we find that the following triples of points are collinear:

$$P_{12}, P_{13}, P_{23}; \quad P_{12}, P_{14}, P_{24}; \quad P_{13}, P_{14}, P_{34}; \quad P_{23}, P_{24}, P_{34}.$$

The first two triples have the point $P_{12}$ in common, and hence determine a plane; each of the other two triples has a point in common with each of the first two. Hence all the points $P_{ij}$ lie in the same plane. The lines of the four triples just given are the lines of intersection of the pairs of homologous faces of the tetrahedra. The theorem is therefore proved.

THEOREM 2'. *If two tetrahedra are perspective from a plane, the lines joining pairs of homologous vertices are concurrent, as likewise the planes determined by pairs of homologous edges; i.e. the tetrahedra are perspective from a point.* (A, E)

This is the space dual and the converse of Theorem 2.

EXERCISE

Write the symbols for the configurations of the last two theorems.

**18. The quadrangle-quadrilateral configuration.**

DEFINITION. A complete plane four-point is called a *complete quadrangle.* It consists of four vertices and six sides. Two sides not on the same vertex are called *opposite.* The intersection of two opposite sides is called a *diagonal point.* If the three diagonal points are not collinear, the triangle formed by them is called the *diagonal triangle* of the quadrangle.*

DEFINITION. A complete plane four-line is called a *complete quadrilateral.* It consists of four sides and six vertices. Two vertices not on the same side are called *opposite.* The line joining two opposite vertices is called a *diagonal line.* If the three diagonal lines are not concurrent, the triangle formed by them is called the *diagonal triangle* of the quadrilateral.*

The assumptions A and E on which all our reasoning is based do not suffice to prove that there are more than three points on any line. In fact, they are all satisfied by the triple system (1), p. 3 (cf. fig. 17). In a case like this the diagonal points of a complete quadrangle are collinear and the diagonal lines of a complete quadrilateral concurrent, as may readily be verified. Two perspective triangles cannot exist in such a plane, and hence the Desargues theorem becomes

* In general, the intersection of two sides of a complete plane $n$-point which do not have a vertex in common is called a *diagonal point* of the $n$-point, and the line joining two vertices of a complete plane $n$-line which do not lie on the same side is called a *diagonal line* of the $n$-line. A complete plane $n$-point ($n$-line) then has $n(n-1)(n-2)(n-3)/8$ diagonal points (lines). Diagonal points and lines are sometimes called *false vertices* and *false sides* respectively.

trivial. Later on we shall add an assumption* which excludes all such cases as this, and, in fact, provides for the existence of an infinite number of points on a line. A part of what is contained in this assumption is the following:

ASSUMPTION $H_0$. *The diagonal points of a complete quadrangle are noncollinear.*

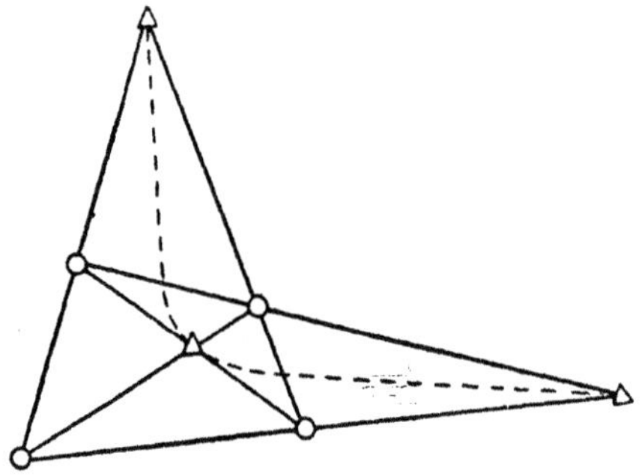

FIG. 17

Many of the important theorems of geometry, however, require the existence of no more than a finite number of points. We shall therefore proceed without the use of further assumptions than A and E, understanding that in order to give our theorems meaning *there must be postulated the existence of the points specified in their hypotheses.* In most cases the existence of a sufficient number of points is insured by Assumption $H_0$, and the reader who is taking up the subject for the first time may well take it as having been added to A and E. It is to be used in the solution of problems.

We return now to a further study of the Desargues configuration. A complete space five-point may evidently be regarded (in five ways) as a tetrahedron and a complete four-line at a point. A plane section of a four-line is a quadrangle and the plane section of a tetrahedron is a quadrilateral. It follows that (in five ways) the Desargues configuration may be regarded as a quadrangle and a quadrilateral. Moreover, it is clear that the six sides of the quadrangle pass through the six vertices of the quadrilateral. In the notation described on page 41 one such quadrangle is $P_{12}$, $P_{13}$, $P_{14}$, $P_{15}$ and the corresponding quadrilateral is $l_{234}$, $l_{235}$, $l_{245}$, $l_{345}$.

The question now naturally arises as to placing the figures thus obtained in special relations. As an application of the theorem of Desargues we will show how to construct † a quadrilateral which has the same diagonal triangle as a given quadrangle. We will assume in our discussion that the diagonal points of any quadrangle form a triangle.

* Merely saying that there are more than three points on a line does not insure that the diagonal points of a quadrangle are noncollinear. Cases where the diagonal points are collinear occur whenever the number of points on a line is $2^n + 1$.

† To *construct* a figure is to determine its elements in terms of certain given elements.

Let $P_1$, $P_2$, $P_3$, $P_4$ be the vertices of the given complete quadrangle, and let $D_{12}$, $D_{13}$, $D_{14}$ be the vertices of the diagonal triangle, $D_{12}$ being on the side $P_1P_2$, $D_{13}$ on the side $P_1P_3$, and $D_{14}$ on the side $P_1P_4$ (fig. 18). We observe first that *the diagonal triangle is perspective with each of the four triangles formed by a set of three of the vertices of the quadrangle, the center of perspectivity being in each case the fourth vertex.* This gives rise to four axes of perspectivity (Theorem 1), one corresponding to each vertex of the quadrangle.* These four lines clearly form the sides of a complete quadrilateral whose diagonal triangle is $D_{12}$, $D_{13}$, $D_{14}$.

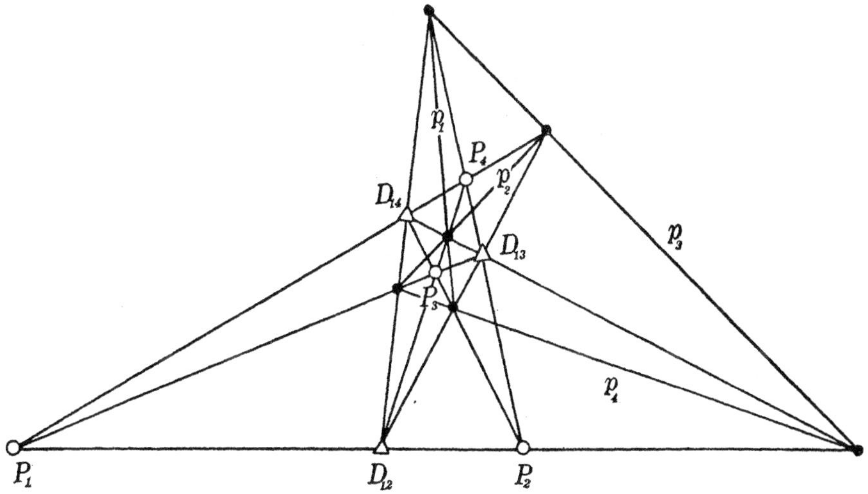


FIG. 18

It may readily be verified, by selecting two perspective triangles, that the figure just formed is, indeed, a Desargues configuration. This special case of the Desargues configuration is called the *quadrangle-quadrilateral configuration.*†

## EXERCISES

**1.** If $p$ is the polar of $P$ with regard to the triangle $ABC$, then $P$ is the pole of $p$ with regard to the same triangle; that is, $P$ is obtained from $p$ by a construction dual to that used in deriving $p$ from $P$. From this theorem it follows that the relation between the quadrangle and quadrilateral in this

* The line thus uniquely associated with a vertex is called the *polar* of the point with respect to the triangle formed by the remaining three vertices. The plane dual process leads to a point associated with any line. This point is called the *pole* of the line with respect to the triangle.

† A further discussion of this configuration and its generalizations will be found in the thesis of H. F. McNeish. Some of the results in this paper are indicated in the exercises.

configuration is mutual; that is, if either is given, the other is determined. For a reason which will be evident later, either is called a covariant of the other.

**2.** Show that the configuration consisting of two perspective tetrahedra, their center and plane of perspectivity, and the projectors and traces may be regarded in six ways as consisting of a complete 5-point $P_{12}$, $P_{13}$, $P_{14}$, $P_{15}$, $P_{16}$ and a complete 5-plane $\pi_{3456}$, $\pi_{2456}$, $\pi_{2356}$, $\pi_{2346}$, $\pi_{2345}$, the notation being analogous to that used on page 41 for the Desargues configuration. Show that the vertices of the 5-plane are on the faces of the 5-point.

**3.** If $P_1$, $P_2$, $P_3$, $P_4$, $P_5$, are vertices of a complete space 5-point, the ten points $D_{ij}$, in which an edge $p_{ij}$ meets a face $P_kP_lP_m$ ($i$, $j$, $k$, $l$, $m$ all distinct), are called *diagonal points*. The tetrahedra $P_2P_3P_4P_5$ and $D_{12}D_{13}D_{14}D_{15}$ are perspective with $P_1$ as center. Their plane of perspectivity, $\pi_1$, is called the *polar* of $P_1$ with regard to the four vertices. In like manner, the points $P_2$, $P_3$, $P_4$, $P_5$ determine their polar planes $\pi_2$, $\pi_3$, $\pi_4$, $\pi_5$. Prove that the 5-point and the polar 5-plane form the configuration of two perspective tetrahedra; that the plane section of the 5-point by any of the five planes is a quadrangle-quadrilateral configuration; and that the dual of the above construction applied to the 5-plane determines the original 5-point.

**4.** If $P$ is the pole of $\pi$ with regard to the tetrahedron $A_1A_2A_3A_4$, then is $\pi$ the polar of $P$ with regard to the same tetrahedron?

## 19. The fundamental theorem on quadrangular sets.

THEOREM 3. *If two complete quadrangles $P_1P_2P_3P_4$ and $P_1'P_2'P_3'P_4'$ correspond — $P_1$ to $P_1'$, $P_2$ to $P_2'$, etc. — in such a way that five of the pairs of homologous sides intersect in points of a line $l$, then the sixth pair of homologous sides will intersect in a point of $l$.* (A, E)

This theorem holds whether the quadrangles are in the same or in different planes.

*Proof.* Suppose, first, that none of the vertices or sides of one of the quadrangles coincide with any vertex or side of the other. Let $P_1P_2$, $P_1P_3$, $P_1P_4$, $P_2P_3$, $P_2P_4$ be the five sides which, by hypothesis, meet their homologous sides $P_1'P_2'$, $P_1'P_3'$, $P_1'P_4'$, $P_2'P_3'$, $P_2'P_4'$ in points of $l$ (fig. 19). We must show that $P_3P_4$ and $P_3'P_4'$ meet in a point of $l$. The triangles $P_1P_2P_3$ and $P_1'P_2'P_3'$ are, by hypothesis, perspective from $l$; as also the triangles $P_1P_2P_4$ and $P_1'P_2'P_4'$. Each pair is therefore (Theorem 1') perspective from a point, and this point is in each case the intersection $O$ of the lines $P_1P_1'$ and $P_2P_2'$. Hence the triangles $P_2P_3P_4$ and $P_2'P_3'P_4'$ are perspective from $O$ and their pairs of homologous sides intersect in the points of a line, which is evidently $l$, since it contains two points of $l$. But $P_3P_4$ and $P_3'P_4'$ are two homologous sides of these last two triangles. Hence they intersect in a point of the line $l$.

If a vertex or side of one quadrangle coincides with a vertex or side of the other, the proof is made by considering a third quadrangle* whose vertices and sides are distinct from those of both of the others, and which has five of its sides passing through the five given points

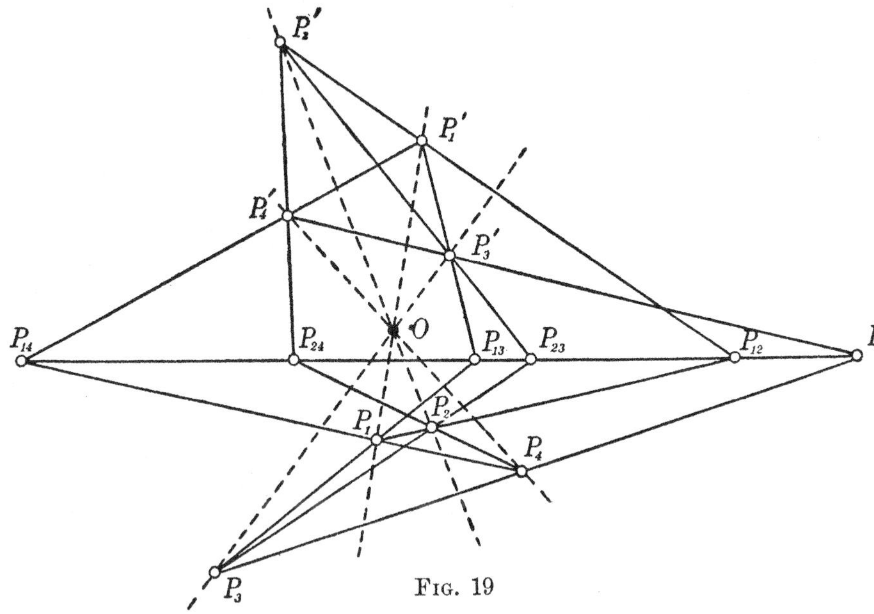


FIG. 19

of intersection of homologous sides of the two given quadrangles. By the argument above, its sixth side will meet the sixth side respectively of each of the two given quadrangles in the same point of $l$. This completes the proof of the theorem.

NOTE 1. It should be noted that the theorem is still valid if the line $l$ contains one or more of the diagonal points of the quadrangles. The case in which $l$ contains two diagonal points is of particular importance and will be discussed in Chap. IV, § 31.

NOTE 2. It is of importance to note in how far the quadrangle $P_1'P_2'P_3'P_4'$ is determined when the quadrangle $P_1P_2P_3P_4$ and the line $l$ are given. It may be readily verified that in such a case it is possible to choose any point $P_1'$ to correspond to any one of the vertices $P_1$, $P_2$, $P_3$, $P_4$, say $P_1$; and that if $m$ is any line of the plane $lP_1'$ (not passing through $P_1'$) which meets one of the sides, say $a$, of $P_1P_2P_3P_4$ (not passing through $P_1$) in a point of $l$, then $m$ may be chosen as the side homologous to $a$. But then the remainder of the figure is uniquely determined.

* This evidently exists whenever the theorem is not trivially obvious.

THEOREM 3′. *If two complete quadrilaterals* $a_1a_2a_3a_4$ *and* $a'_1a'_2a'_3a'_4$ *correspond* — $a_1$ to $a'_1$, $a_2$ *to* $a'_2$, *etc.* — *in such a way that five of the lines joining homologous vertices pass through a point* $P$, *the line joining the sixth pair of homologous vertices will also pass through* $P$. (A, E)

This is the plane dual of Theorem 3 regarded as a plane theorem.

DEFINITION. A set of points in which the sides of a complete quadrangle meet a line $l$ is called a *quadrangular set* of points.

Any three sides of a quadrangle either form a triangle or meet in a vertex; in the former case they are said to form a *triangle triple*, in the latter a *point triple* of lines. In a quadrangular set of points on a line $l$ any three points in which the lines of a triangle triple meet $l$ is called a *triangle triple of points* in the set; three points in which the lines of a point triple meet $l$ are called a *point triple of points.* A quadrangular set of points will be denoted by

$$Q(ABC, DEF),$$

where $ABC$ is a point triple and $DEF$ is a triangle triple, and where $A$ and $D$, $B$ and $E$, and $C$ and $F$ are respectively the intersections with the line of the set of the pairs of opposite sides of the quadrangle.

The notion of a quadrangular set is of great importance in much that follows. It should be noted again in this connection that one or two * of the pairs $A, D$ or $B, E$ or $C, F$ may consist of coincident points; this occurs when the line of the set passes through one or two of the diagonal points.†

We have just seen (Theorem 3) that if we have a quadrangular set of points obtained from a given quadrangle, there exist other quadrangles that give rise to the same quadrangular set. In the quadrangles mentioned in Theorem 3 there corresponded to every triangle triple of one a triangle triple of the other.

DEFINITION. When two quadrangles giving rise to the same quadrangular set are so related with reference to the set that to a triangle triple of one corresponds a triangle triple of the other, the

* All three may consist of coincident points in a space in which the diagonal points of a complete quadrangle are collinear.

† It should be kept in mind that similar remarks and a similar definition may be made to the effect that the lines joining the vertices of a quadrilateral to a point $P$ form a quadrangular set of lines, etc. (cf. § 30, Chap. IV).

quadrangles are said to be *similarly placed* (fig. 20); if a point triple of one corresponds to a triangle triple of the other, they are said to be *oppositely placed* (fig. 21).

It will be shown later (Chap. IV) that quadrangles oppositely placed with respect to a quadrangular set are indeed possible.

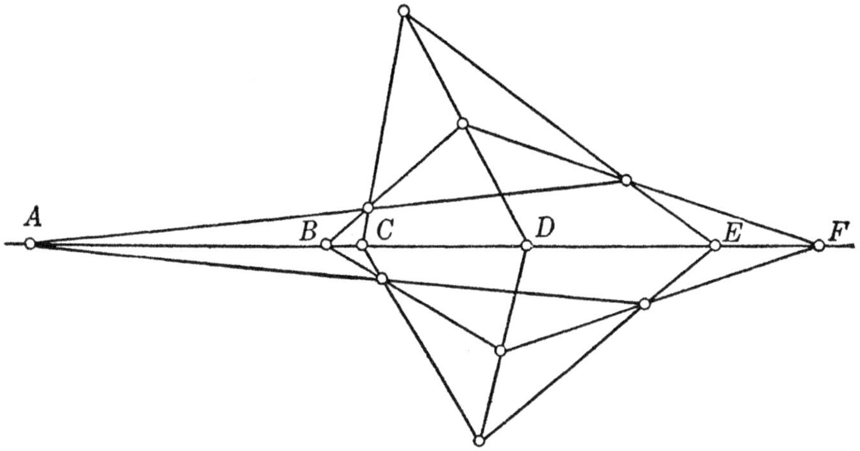


FIG. 20

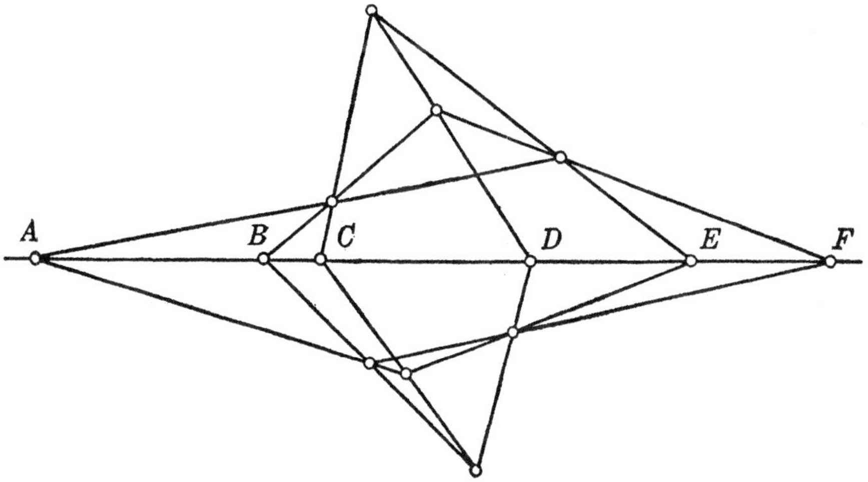


FIG. 21

With the notation for quadrangular sets defined above, the last theorem leads to the following

COROLLARY. *If all but one of the points of a quadrangular set* Q (*ABC, DEF*) *are given, the remaining one is uniquely determined.* (A, E)

For two quadrangles giving rise to the same quadrangular set *with the same notation* must be similarly placed, and must hence be in correspondence as described in the theorem.

The quadrangular set which is the section by a 1-space of a complete 4-point in a 2-space, the Desargues configuration which is the section by a 2-space of a complete 5-point in a 3-space, the configuration of two perspective tetrahedra which may be considered as the section by a 3-space of a complete 6-point in a 4-space are all special cases of the section by an $n$-space of a complete $(n+3)$-point in an $(n+1)$-space. The theorems which we have developed for the three cases here considered are not wholly parallel. The reader will find it an entertaining and far from trivial exercise to develop the analogy in full.

### EXERCISES

**1.** A necessary and sufficient condition that three lines containing the vertices of a triangle shall be concurrent is that their intersections $P$, $Q$, $R$ with a line $l$ form, with intersections $E$, $F$, $G$ of corresponding sides of the triangle with $l$, a quadrangular set $Q(PQR, EFG)$.

**2.** If on a given transversal line two quadrangles determine the same quadrangular set and are similarly placed, their diagonal triangles are perspective from the center of perspectivity of the two quadrangles.

**3.** The polars of a point $P$ on a line $l$ with regard to all triangles which meet $l$ in three fixed points pass through a common point $P'$ on $l$.

**4.** In a plane $\pi$ let there be given a quadrilateral $a_1$, $a_2$, $a_3$, $a_4$ and a point $O$ not on any of these lines. Let $A_1$, $A_2$, $A_3$, $A_4$ be any tetrahedron whose four faces pass through the lines $a_1$, $a_2$, $a_3$, $a_4$ respectively. The polar planes of $O$ with respect to all such tetrahedra pass through the same line of $\pi$.

**20. Additional remarks concerning the Desargues configuration.** The ten edges of a complete space five-point may be regarded (in six ways) as the edges of two simple space five-points. Two such five-points are, for example, $P_1P_2P_3P_4P_5$ and $P_1P_3P_5P_2P_4$. Corresponding thereto, the Desargues configuration may be regarded in six ways as a pair of simple plane pentagons (five-points). In our previous notation the two corresponding to the two simple space five-points just given are $P_{12}P_{23}P_{34}P_{45}P_{51}$ and $P_{13}P_{35}P_{52}P_{24}P_{41}$. Every vertex of each of these pentagons is on a side of the other.

Every point, $P_{12}$ for instance, has associated with it a unique line of the configuration, viz. $l_{345}$ in the example given, whose notation does not contain the suffixes occurring in the notation of the point. The line may be called the polar of the point in the configuration, and the point the pole of the line. It is then readily seen that the polar of any point is the axis of perspectivity of two triangles whose center of perspectivity is the point. In case we regard the configuration as consisting of a complete quadrangle and complete quadrilateral, it is found that a pole and polar are homologous vertex and side of the quadrilateral and quadrangle. If we consider the configuration as consisting of two simple pentagons, a pole and polar are a vertex and its opposite side, e.g. $P_{12}$ and $P_{34}P_{45}$.

The Desargues configuration is one of a class of configurations having similar properties. These configurations have been studied by a number of writers.* Some of the theorems contained in these memoirs appear in the exercises below.

## EXERCISES

*In discussing these exercises the existence should be assumed of a sufficient number of points on each line so that the figures in question do not degenerate. In some cases it may also be assumed that the diagonal points of a complete quadrangle are not collinear. Without these assumptions our theorems are true, indeed, but trivial.*

**1.** What is the peculiarity of the Desargues configuration obtained as the section of a complete space five-point by a plane which contains the point of intersection of an edge of the five-point with the face not containing this edge? also by a plane containing two or three such points?

**2.** Given a simple pentagon in a plane, construct another pentagon in the same plane, whose vertices lie on the sides of the first and whose sides contain the vertices of the first (cf. p. 51). Is the second uniquely determined when the first and one side of the second are given?

**3.** If two sets of three points $A$, $B$, $C$ and $A'$, $B'$, $C'$ on two coplanar lines $l$ and $l'$ respectively are so related that the lines $AA'$, $BB'$, $CC'$ are concurrent, then the points of intersection of the pairs of lines $AB'$ and $BA'$, $BC'$ and $CB'$, $CA'$ and $AC'$ are collinear with the point $ll'$. The line thus determined is called the *polar* of the point $(AA', BB')$ with respect to $l$ and $l'$. Dualize.

**4.** Using the theorem of Ex. 3, give a construction for a line joining any given point in the plane of two lines $l$, $l'$ to the point of intersection of $l$, $l'$ without making use of the latter point.

**5.** Using the definition in Ex. 3, show that if the point $P'$ is on the polar $p$ of a point $P$ with respect to two lines $l$, $l'$, then the point $P$ is on the polar $p'$ of $P'$ with respect to $l$, $l'$.

**6.** If the vertices $A_1$, $A_2$, $A_3$, $A_4$ of a simple plane quadrangle are respectively on the sides $a_1$, $a_2$, $a_3$, $a_4$ of a simple plane quadrilateral, and if the intersection of the pair of opposite sides $A_1A_2$, $A_3A_4$ is on the line joining the pair of opposite points $a_1a_4$, $a_2a_3$, the remaining pair of opposite sides of the quadrangle will meet on the line joining the remaining pair of opposite vertices of the quadrilateral. Dualize.

* A. Cayley, Collected Works, Vol. I (1846), p. 317. G. Veronese, Mathematische Annalen, Vol. XIX (1882). Further references will be found in a paper by W. B. Carver, Transactions of the American Mathematical Society, Vol. VI (1905), p. 534.

**7.** If two complete plane $n$-points $A_1, A_2, \cdots, A_n$ and $A_1', A_2', \cdots, A_n'$ are so related that the side $A_1A_2$ and the remaining $2(n-2)$ sides passing through $A_1$ and $A_2$ meet the corresponding sides of the other $n$-point in points of a line $l$, the remaining pairs of homologous sides of the two $n$-points meet on $l$ and the two $n$-points are perspective from a point. Dualize.

**8.** If five sides of a complete quadrangle $A_1A_2A_3A_4$ pass through five vertices of a complete quadrilateral $a_1a_2a_3a_4$ in such a way that $A_1A_2$ is on $a_3a_4$, $A_2A_3$ on $a_4a_1$, etc., then the sixth side of the quadrangle passes through the sixth vertex of the quadrilateral. Dualize.

**9.** If on each of three concurrent lines $a, b, c$ two points are given, — $A_1, A_2$ on $a$; $B_1, B_2$ on $b$; $C_1, C_2$ on $c$, — there can be formed four pairs of triangles $A_iB_jC_k$ $(i, j, k = 1, 2)$ and the pairs of corresponding sides meet in six points which are the vertices of a complete quadrilateral (Veronese, Atti dei Lincei, 1876–1877, p. 649).

**10.** With nine points situated in sets of three on three concurrent lines are formed 36 sets of three perspective triangles. For each set of three distinct triangles the axes of perspectivity meet in a point; and the 36 points thus obtained from the 36 sets of triangles lie in sets of four on 27 lines, giving a configuration $\boxed{\begin{array}{cc} 36 & 3 \\ 4 & 27 \end{array}}$ (Veronese, loc. cit.).

**11.** A plane section of a 6-point in space can be considered as 3 triangles perspective in pairs from 3 collinear points with corresponding sides meeting in 3 collinear points.

**12.** A plane section of a 6-point in space can be considered as 2 perspective complete quadrangles with corresponding sides meeting in the vertices of a complete quadrilateral.

**13.** A plane section of an $n$-point in space gives the configuration* $\boxed{\begin{array}{cc} {}_nC_2 & n-2 \\ 3 & {}_nC_3 \end{array}}$ which may be considered (in ${}_nC_{n-k}$ ways) as a set of $(n-k)$ $k$-points perspective in pairs from ${}_{n-k}C_2$ points, which form a configuration $\boxed{\begin{array}{cc} {}_{n-k}C_2 & n-k-2 \\ 3 & {}_{n-k}C_3 \end{array}}$ and the points of intersection of corresponding sides form a configuration $\boxed{\begin{array}{cc} {}_kC_2 & k-2 \\ 3 & {}_kC_3 \end{array}}$.

**14.** A plane section of a 7-point in space can be considered (in 120 ways) as composed of three simple heptagons (7-points) cyclically circumscribing each other.

**15.** A plane section of an 11-point in space can be considered (in $|\underline{9}$ ways) as composed of five 11-points cyclically circumscribing each other.

**16.** A plane section of an $n$-point in space for $n$ prime can be considered (in $|\underline{n-2}$ ways) as $\dfrac{n-1}{2}$ simple $n$-points cyclically circumscribing each other.

* The symbol ${}_nC_r$ is used to denote the number of combinations of $n$ things taken $r$ at a time.

**17.** A plane section of a 6-point in space gives (in six ways) a 5-point whose sides pass through the points of a configuration $\boxed{\begin{array}{cc} 10 & 3 \\ 3 & 10 \end{array}}$.

**18.** A plane section of an $n$-point in space gives a complete $(n-1)$-point whose sides pass through the points of a configuration $\boxed{\begin{array}{cc} {}_{n-1}C_2 & n-3 \\ 3 & {}_{n-1}C_3 \end{array}}$.

* **19.** The $n$-space section of an $m$-point $(m \geqq n+2)$ in an $(n+1)$-space can be considered in the $n$-space as $(m-k)$ $k$-points (in ${}_mC_{m-k}$ ways) perspective in pairs from the vertices of the $n$-space section of one $(m-k)$-point; the $r$-spaces of the $k$-point figures meet in $(r-1)$-spaces $(r = 1, 2, \cdots, n-1)$ which form the $n$-space section of a $k$-point.

* **20.** The figure of two perspective $(n+1)$-points in an $n$-space separates (in $n+3$ ways) into two dual figures, respectively an $(n+2)$-point circumscribing the figure of $(n+2)$ $(n-1)$-spaces.

* **21.** The section by a 3-space of an $n$-point in 4-space is a configuration

$$\boxed{\begin{array}{ccc} {}_nC_2 & n-2 & {}_{n-2}C_2 \\ 3 & {}_nC_3 & n-3 \\ 6 & 4 & {}_nC_4 \end{array}}.$$

The plane section of this configuration is

$$\boxed{\begin{array}{cc} {}_nC_3 & n-3 \\ 4 & {}_nC_4 \end{array}}.$$

**22.** Let there be three points on each of two concurrent lines $l_1$, $l_2$. The nine lines joining points of one set of three to points of the other determine six triangles whose vertices are not on $l_1$ or $l_2$. The point of intersection of $l_1$ and $l_2$ has the same polar with regard to all six of these triangles.

**23.** If two triangles are perspective, then are perspective also the two triangles whose vertices are points of intersection of each side of the given triangles with a line joining a fixed point of the axis of perspectivity to the opposite vertex.

* **24.** Show that the configuration of the two perspective tetrahedra of Theorem 2 can be obtained as the section by a 3-space of a complete 6-point in a 4-space.

* **25.** If two 5-points in a 4-space are perspective from a point, the corresponding edges meet in the vertices, the corresponding plane faces meet in the lines, and the corresponding 3-space faces in the planes of a complete 5-plane in a 3-space.

* **26.** If two $(n+1)$-points in an $n$-space are perspective from a point, their corresponding $r$-spaces meet in $(r-1)$-spaces which lie in the same $(n-1)$-space $(r = 1, 2 \cdots, n-1)$ and form a complete configuration of $(n+1)$ $(n-2)$-spaces in $(n-1)$-space.

# CHAPTER III

## PROJECTIVITIES OF THE PRIMITIVE GEOMETRIC FORMS OF ONE, TWO, AND THREE DIMENSIONS

### 21. The nine primitive geometric forms.

DEFINITION. A *pencil of points* or a *range* is the figure formed by the set of all points on the same line. The line is called the *axis* of the pencil.

DEFINITION. A *pencil of planes* or an *axial pencil** is the figure formed by the set of all planes on the same line. The line is called the *axis* of the pencil.

As indicated, the pencil of points is the space dual of the pencil of planes.

DEFINITION. A *pencil of lines* or a *flat pencil* is the figure formed by the set of all lines which are at once on the same point and the same plane; the point is called the *vertex* or *center* of the pencil.

The pencil of lines is clearly self-dual in space, while it is the plane dual of the pencil of points. The pencil of points, the pencil of lines, and the pencil of planes are called the *primitive geometric forms of the first grade* or *of one dimension.*

DEFINITION. The following are known as the *primitive geometric forms of the second grade* or *of two dimensions:*

The set of all points on a plane is called a *plane of points.* The set of all lines on a plane is called a *plane of lines.* The plane is called the *base* of the two forms. The figure composed of a plane of points and a plane of lines with the same base is called a *planar field.*

The set of all planes on a point is called a *bundle of planes.* The set of all lines on a point is called a *bundle of lines.* The point is called the *center* of the bundles. The figure composed of a bundle of lines and a bundle of planes with the same center is called simply a *bundle.*

DEFINITION. The set of all planes in space and the set of all points in space are called the *primitive geometric forms of the third grade* or *of three dimensions.*

* The pencil of planes is also called by some writers a *sheaf.*

There are then, all told, nine primitive geometric forms in a space of three dimensions.*

**22. Perspectivity and projectivity.** In Chap. II, § 13, we gave a definition of perspectivity. This definition we will now apply to the case of two primitive forms and will complete it where needed. We note first that, according to the definition referred to, two pencils of points in the same plane are perspective provided every two homologous points of the pencils are on a line of a flat pencil, for they then have the same projection from a point. Two planes of points (lines) are perspective, if every two homologous elements are on a line (plane) of a bundle of lines (planes). Two pencils of lines in the same plane are perspective, if every two homologous lines intersect in a point of the same pencil of points. Two pencils of planes are perspective, if every two homologous planes are on a point of a pencil of points (they then have the same section by a line). Two bundles of lines (planes) are perspective, if every two homologous lines (planes) are on a point (line) of a plane of points (lines) (they then have the same section by a plane), etc. Our previous definition does not, however, cover all possible cases. In the first place, it does not allow for the possibility of two forms of different kinds being perspective, such as a pencil of points and a pencil of lines, a plane of points and a bundle of lines, etc. This lack of completeness is removed for the case of one-dimensional forms by the following definition. It should be clearly noted that it is in complete agreement with the previous definition of perspectivity; as far as one-dimensional forms are concerned it is wider in its application.

DEFINITION. Two one-dimensional primitive forms of different kinds, not having a common axis, are *perspective,* if and only if they correspond in such a (1, 1) way that each element of one is on its homologous element in the other; two one-dimensional primitive forms of the same kind are *perspective,* if and only if every two homologous elements are on an element of a third one-dimensional form not having an axis in common with one of the given forms. If the third form is a pencil of lines with vertex $P$, the perspectivity is said to be

* Some writers enumerate only six, by defining the set of all points and lines on a plane as a single form, and by regarding the set of all planes and lines at a point and the set of all points and planes in space each as a single form. We have followed the usage of Enriques, Vorlesungen über Projektive Geometrie.

*central* with center $P$; if the third form is a pencil of points or a pencil of planes with axis $l$, the perspectivity is said to be *axial* with axis $l$.

As examples of this definition we mention the following: Two pencils of points on skew lines are perspective, if every two homologous elements are on a plane of a pencil of planes; two pencils of lines in different planes are perspective, if every two homologous lines are on a point of a pencil of points or a plane of a pencil of planes (either of the latter conditions is a consequence of the other); two pencils of planes are perspective, if every two homologous planes are on a point of a pencil of points or a line of a pencil of lines (in the latter case the axes of the pencils of planes are coplanar). A pencil of points and a pencil of lines are perspective, if every point is on its homologous line, etc.

It is of great importance to note that our definitions of perspective primitive forms are dual throughout; i.e. that if two forms are perspective, the dual figure will consist of perspective forms. Hence any theorem proved concerning perspectivities can at once be dualized; in particular, any theorem concerning the perspectivity of two forms of the same kind is true of any other two forms of the same kind.

We use the notation $[P]$ to denote a class of elements of any kind and denote individuals of the class by $P$ alone or with an index or subscript. Thus two ranges of points may be denoted by $[P]$ and $[Q]$. To indicate a perspective correspondence between them we write

$$[P] \underset{\wedge}{=} [Q].$$

The same symbol, $\underset{\wedge}{=}$, is also used to indicate a perspectivity between any two one-dimensional forms. If the two forms are of the same kind, it implies that there exists a third form such that every pair of homologous elements of the first two forms is on an element of the third form. The third form may also be exhibited in the notation by placing a symbol representing the third form immediately over the sign of perspectivity, $\underset{\wedge}{=}$.

Thus the symbols

$$[P] \overset{A}{\underset{\wedge}{=}} [Q] \underset{\wedge}{=} [r] \overset{a}{\underset{\wedge}{=}} [s]$$

denote that the range $[P]$ is perspective by means of the center $A$ with the range $[Q]$, that each $Q$ is on a line $r$ of the flat pencil $[r]$, and that the pencil $[r]$ is perspective by the axis $a$ with the flat pencil $[s]$.

A class of elements containing a finite number of elements can be indicated by the symbols for the several elements. When this notation is used, the symbol of perspectivity indicates that elements appearing in corresponding places in the two sequences of symbols are homologous. Thus

$$1\ 2\ 3\ 4 \doublebarwedge A\ B\ C\ D$$

implies that 1 and $A$, 2 and $B$, 3 and $C$, 4 and $D$ are homologous.

DEFINITION.* Two one-dimensional primitive forms $[\sigma]$ and $[\sigma']$ (of the same or different kinds) are said to be *projective*, provided there exists a sequence of forms $[\tau]$, $[\tau']$, $\cdots$, $[\tau^{(n)}]$ such that

$$[\sigma] \doublebarwedge [\tau] \doublebarwedge [\tau'] \doublebarwedge \cdots \doublebarwedge [\tau^{(n)}] \doublebarwedge [\sigma'].$$

The correspondence thus established between $[\sigma]$ and $[\sigma']$ is called a *projective correspondence* or *projectivity*, or also a *projective transformation*. Any element $\sigma$ is said to be *projected into* its homologous element $\sigma'$ by the sequence of perspectivities.

Thus a projectivity is the resultant of a sequence of perspectivities. It is evident that $[\sigma]$ and $[\sigma']$ may be the same form, in which case the projectivity effects a permutation of the elements of the form. For example, it is proved later in this chapter that any four points $A, B, C, D$ of a line can be projected into $B, A, D, C$ respectively.

A projectivity establishes a one-to-one correspondence between the elements of two one-dimensional forms, which correspondence we may consider abstractly without direct reference to the sequence of perspectivities by which it is defined. Such a correspondence we denote by

$$[\sigma] \barwedge [\sigma'].$$

Projectivities we will, in general, denote by letters of the Greek alphabet, such as $\pi$. If a projectivity $\pi$ makes an element $\sigma$ of a form homologous with an element $\sigma'$ of another or the same form, we will sometimes denote this by the relation $\pi(\sigma) = \sigma'$. In this case we may say the projectivity *transforms* $\sigma$ into $\sigma'$. Here the symbol $\pi(\ )$ is used as a functional symbol† acting on the *variable*‡ $\sigma$, which represents any one of the elements of a given form.

* This is Poncelet's definition of a projectivity.

† Just like $F(x)$, $\sin(x)$, $\log(x)$, etc.

‡ The definition of *variable* is "a symbol $x$ which represents any one of a class of elements $[x]$." It is in this sense that we speak of "a variable point."

**23. The projectivity of one-dimensional primitive forms.** The projectivity of one-dimensional primitive forms will be discussed with reference to the projectivity of pencils of points. The corresponding properties for the other one-dimensional primitive forms will then follow immediately by the theorems of duality (Theorems 11–13, Chap. I).

THEOREM 1. *If $A$, $B$, $C$ are three points of a line $l$ and $A'$, $B'$, $C'$ three points of another line $l'$, then $A$ can be projected into $A'$, $B$ into $B'$, and $C$ into $C'$ by means of two centers of perspectivity.* (The lines may be in the same or in different planes.) (A, E)

*Proof.* If the points in any one of the pairs $AA'$, $BB'$, or $CC'$ are coincident, one center is sufficient, viz., the intersection of the lines determined by the other two pairs. If each of these pairs consists of distinct points, let $S$ be any point of the line $AA'$, distinct from $A$ and $A'$ (fig. 22). From $S$ project $A$, $B$, $C$ on any line $l''$ distinct from $l$ and $l'$, but containing $A'$ and a point of $l$. If $B''$, $C''$ are the points of $l''$ corresponding to $B$, $C$ respectively, the point of intersection $S'$ of the lines $B'B''$ and $C'C''$ is the second center of perspectivity. This argument holds without modification, if one of the points $A$, $B$, $C$ coincides with one of the points $A'$, $B'$, $C'$ other than its corresponding point.

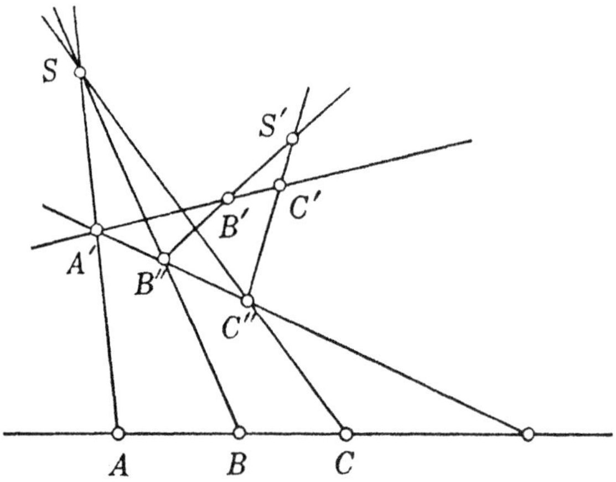

FIG. 22

COROLLARY 1. *If $A$, $B$, $C$ and $A'$, $B'$, $C'$ are on the same line, three centers of perspectivity are sufficient to project $A$, $B$, $C$ into $A'$, $B'$, $C'$ respectively.* (A, E)

COROLLARY 2. *Any three distinct elements of a one-dimensional primitive form are projective with any three distinct elements of another or the same one-dimensional primitive form.* (A, E)

For, when the two forms are of the same kind, the result is obtained from the theorem and the first corollary directly from the theorems of duality (Theorems 11–13, Chap. I). If they are of different kinds, a projection or section is sufficient to reduce them to the same kind.

THEOREM 2. *The projectivity* $ABCD \,\overline{\wedge}\, BADC$ *holds for any four distinct points* $A$, $B$, $C$, $D$ *of a line.* (A, E)

*Proof.* From a point $S$, not on the line $l = AB$, project $ABCD$ into $AB'C'D'$ on a line $l'$ through $A$ and distinct from $l$ (fig. 23). From $D$ project $AB'C'D'$ on the line $SB$. The last four points will then project into $BADC$ by means of the center $C'$. In fig. 23 we have

$$ABCD \overset{S}{\underset{\wedge}{=}} AB'C'D' \overset{D}{\underset{\wedge}{=}} BB'C''S \overset{C'}{\underset{\wedge}{=}} BADC.$$

It is to be noted that a *geometrical order* of the points $ABCD$ has no bearing on the theorem. In fact, the notion of such order has not yet been introduced into our geometry and, indeed, cannot be introduced on the basis of the present assumptions alone. The theorem merely states that *the correspondence obtained by interchanging any two of four collinear points and also interchanging the remaining two is projective.* The notion of *order* is, however, implied in our notation of projectivity and perspectivity. Thus, for example, we introduce the following definition:

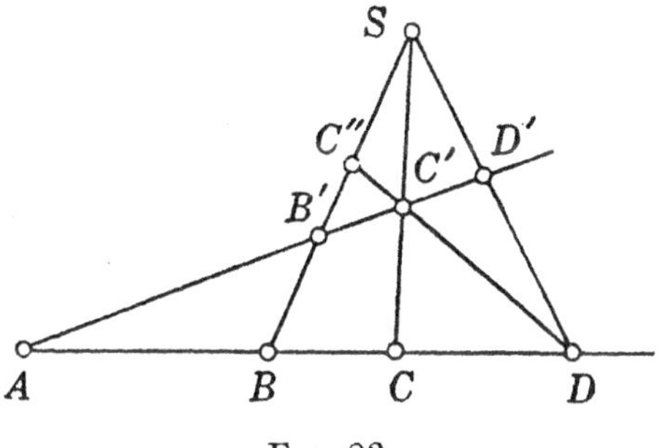


FIG. 23

DEFINITION. Two ordered pairs of elements of any one-dimensional form are called a *throw;* if the pairs are $AB$, $CD$, this is denoted by $\mathsf{T}(AB, CD)$. Two throws are said to be *equal*, provided they are projective; in symbols, $\mathsf{T}(AB, CD) = \mathsf{T}(A'B', C'D')$, provided we have $ABCD \,\overline{\wedge}\, A'B'C'D'$.

The last theorem then states the equality of throws:

$$\mathsf{T}(AB, CD) = \mathsf{T}(BA, DC) = \mathsf{T}(CD, AB) = \mathsf{T}(DC, BA).$$

The results of the last two theorems may be stated in the following form:

THEOREM 1'. *If 1, 2, 3 are elements of any one-dimensional primitive form, there exist projective transformations which will effect any one of the six permutations of these three elements.*

THEOREM 2′. *If 1, 2, 3, 4 are any four distinct elements of a one-dimensional primitive form, there exist projective transformations which will transform 1234 into any one of the following permutations of itself: 1234, 2143, 3412, 4321.*

A projective transformation has been defined as the resultant of any sequence of perspectivities. We proceed now to the proof of a chain of theorems, which lead to the fundamental result that any projective transformation between two distinct one-dimensional primitive forms of the same kind can be obtained as the resultant of *two* perspectivities.

THEOREM 3. *If* $[P]$, $[P']$, $[P'']$ *are pencils of points on three distinct concurrent lines* $l$, $l'$, $l''$ *respectively, such that* $[P] \overset{S}{\underset{\wedge}{=}} [P']$ *and* $[P'] \overset{S'}{\underset{\wedge}{=}} [P'']$, *then likewise* $[P] \overset{S''}{\underset{\wedge}{=}} [P'']$, *and the three centers of perspectivity* $S$, $S'$ $S''$ *are collinear.* (A, E)

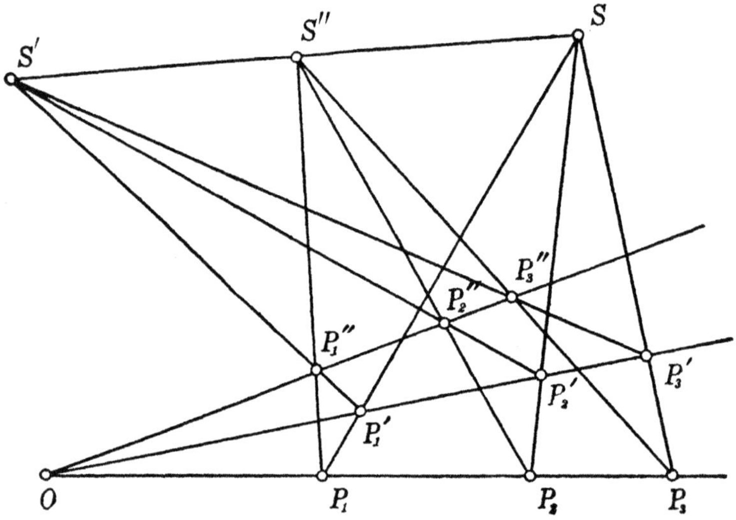


FIG. 24

*Proof.* Let $O$ be the common point of the lines $l, l', l''$. If $P_1, P_2, P_3$ are three points of $[P]$, and $P_1'P_2'P_3'$ and $P_1''P_2''P_3''$ the corresponding points of $[P']$, $[P'']$ (fig. 24), it is clear that the triangles $P_1P_1'P_1''$, $P_2P_2'P_2''$, $P_3P_3'P_3''$ are perspective from $O$.* By Desargues's theorem (Theorem 1, Chap. II) homologous sides of any pair of these three triangles meet in collinear points. The conclusion of the theorem then follows readily from the hypotheses.

* If the points in each of these sets of three are collinear, the theorem is obvious and the three centers of perspectivity coincide.

COROLLARY. *If n concurrent lines* $l_1, l_2, l_3, \cdots, l_n$ *are connected by perspectivities* $[P_1] \overset{S_{12}}{\underset{\wedge}{=}} [P_2] \overset{S_{23}}{\underset{\wedge}{=}} [P_3] \overset{S_{34}}{\underset{\wedge}{=}} \cdots \overset{S_{n-1,n}}{\underset{\wedge}{=}} [P_n]$, *and if* $l_1$ *and* $l_n$ *are distinct lines, then we have* $[P_1] \underset{\wedge}{=} [P_n]$. (A, E)

*Proof.* This follows almost immediately from the theorem, except when it happens that a set of four successive lines of the set $l_1 l_2 l_3 \cdots l_n$ are such that the first and third coincide and likewise the second and fourth. That this case forms no exception to the corollary may be shown as follows: Consider the perspectivities connecting the pencils of points on the lines $l_1, l_2, l_3, l_4$ on the hypothesis that $l_1 = l_3$, $l_2 = l_4$ (fig. 25.) Let $l_1$, $l_2$ meet in $O$, and let the line $S_{12}S_{23}$ meet $l_1$ in $A_1$,

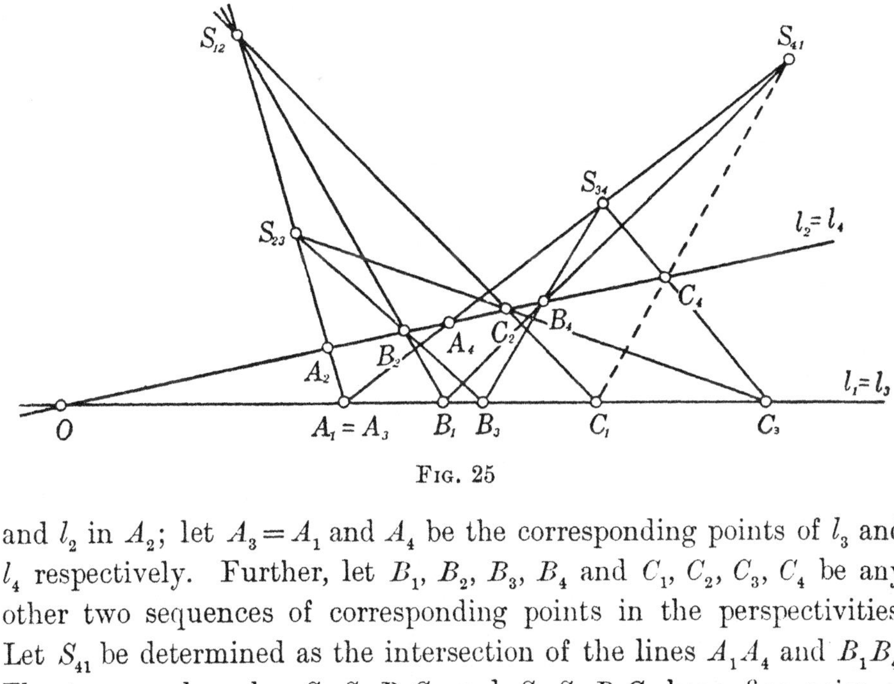


FIG. 25

and $l_2$ in $A_2$; let $A_3 = A_1$ and $A_4$ be the corresponding points of $l_3$ and $l_4$ respectively. Further, let $B_1, B_2, B_3, B_4$ and $C_1, C_2, C_3, C_4$ be any other two sequences of corresponding points in the perspectivities. Let $S_{41}$ be determined as the intersection of the lines $A_1A_4$ and $B_1B_4$. The two quadrangles $S_{12}S_{23}B_2C_2$ and $S_{41}S_{34}B_4C_4$ have five pairs of homologous sides meeting $l_1 = l_3$ in the points $OA_1B_1B_3C_3$. Hence the side $S_{41}C_4$ meets $l_1$ in $C_1$ (Theorem 3, Chap. II).

THEOREM 4. *If* $[P_1]$, $[P_2]$, $[P]$ *are pencils of points on distinct lines* $l_1, l_2, l$ *respectively, such that* $[P_1] \overset{S_1}{\underset{\wedge}{=}} [P] \overset{S_2}{\underset{\wedge}{=}} [P_2]$, *and if* $[P']$ *is the pencil of points on any line* $l'$ *containing the intersection of* $l_1, l$ *and also a point of* $l_2$, *but not containing* $S_2$, *then there exists a point* $S_1'$ *on* $S_1S_2$, *such that* $[P_1] \overset{S_1'}{\underset{\wedge}{=}} [P'] \overset{S_2}{\underset{\wedge}{=}} [P_2]$. (A, E)

*Proof.* Clearly we have

$$[P_1] \overset{S_1}{\underset{\wedge}{=}} [P] \overset{S_2}{\underset{\wedge}{=}} [P'] \overset{S_2}{\underset{\wedge}{=}} [P_2].$$

But by the preceding theorem and the conditions on $l'$ we have $[P_1] \overset{S_1'}{\underset{\wedge}{=}} [P']$, where $S_1'$ is a point of $S_1S_2$. Hence we have

$$[P_1] \overset{S_1'}{\underset{\wedge}{=}} [P'] \overset{S_2}{\underset{\wedge}{=}} [P_2].$$

This theorem leads readily to the next theorem, which is the result toward which we have been working. We prove first the following lemmas:

LEMMA 1. *Any axial perspectivity between the points of two skew lines is equivalent to (and may be replaced by) two central perspectivities.* (A, E)

For let $[P]$, $[P']$ be the pencils of points on the skew lines. Then if $S$ and $S'$ are any two points on the axis $s$ of the axial perspectivity, the pencils of lines $S[P]$, $S'[P']$* are so related that pairs of homologous lines intersect in points of the line common to the planes of the two pencils $S[P]$ and $S'[P']$, since each pair of homologous lines lie, by hypothesis, in a plane of the axial pencil $s[P] = s[P']$.

LEMMA 2. *Any projectivity between pencils of points may be defined by a sequence of central perspectivities.*

For any noncentral perspectivities occurring in the sequence defining a projectivity may, in consequence of Lemma 1, be replaced by sequences of central perspectivities.

THEOREM 5. *If two pencils of points $[P]$ and $[P']$ on distinct lines are projective, there exists a pencil of points $[Q]$ and two points $S$, $S'$ such that we have* $[P] \overset{S}{\underset{\wedge}{=}} [Q] \overset{S'}{\underset{\wedge}{=}} [P']$. (A, E)

*Proof.* By hypothesis and the two preceding lemmas we have a sequence of perspectivities

$$[P] \overset{S_1}{\underset{\wedge}{=}} [P_1] \overset{S_2}{\underset{\wedge}{=}} [P_2] \overset{S_3}{\underset{\wedge}{=}} [P_3] \overset{S_4}{\underset{\wedge}{=}} \cdots \overset{S_n}{\underset{\wedge}{=}} [P'].$$

* Given a class of elements $[P]$; the symbol $S[P]$ is used to denote the class of elements $SP$ determined by a given element $S$ and any element of $[P]$. Hence, if $[P]$ is a pencil of points and $S$ a point not in $[P]$, $S[P]$ is a pencil of lines with center $S$; if $s$ is a line not on any $P$, $s[P]$ is a pencil of planes with axis $s$.

We assume the number of these perspectivities to be greater than two, since otherwise the theorem is proved. By applying the corollary of Theorem 3, when necessary, this sequence of perspectivities may be so modified that no three successive axes are concurrent. We may also assume that no two of the axes $l, l_1, l_2, l_3, \cdots, l'$ of the pencils $[P], [P_1], [P_2], [P_3], \cdots [P']$ are coincident; for Theorem 4 may evidently be used to replace any $l_k (= l_i)$ by a line $l_k'' (\neq l_i)$. Now let $l_1'$ be the line joining the points $ll_1$ and $l_2l_3$, and let us suppose that it does not contain the center $S_2$ (fig. 26). If then $[P_1']$ is the pencil of points on $l_1'$, we may (by Theorem 4) replace the given sequence of perspectivities by $[P] \overset{S_1'}{\underset{\wedge}{=}} [P_1'] \overset{S_2}{\underset{\wedge}{=}} [P_2] \overset{S_3}{\underset{\wedge}{=}} [P_3] \overset{S_4}{\underset{\wedge}{=}} \cdots$ and this sequence may in turn be replaced by

$$[P] \overset{S_1'}{\underset{\wedge}{=}} [P_1'] \overset{S_3'}{\underset{\wedge}{=}} [P_3] \overset{S_4}{\underset{\wedge}{=}} \cdots$$

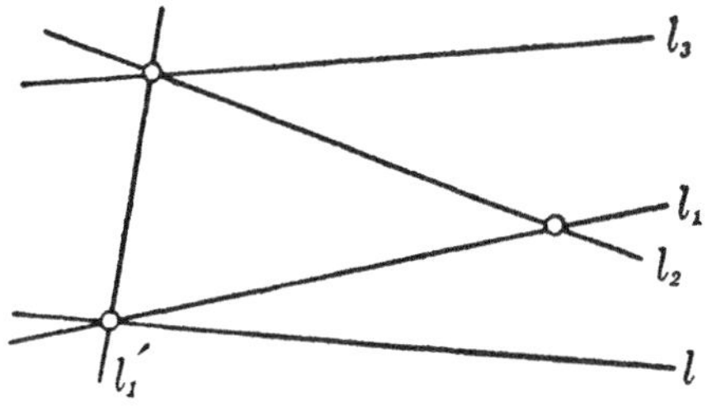

FIG. 26

(Theorem 3). If $S_2$ is on the line joining $ll_1$ and $l_2l_3$, we may replace $l_1$ by any line $l_1''$ through the intersection of $l_1l_2$ which meets $l$ and does not contain the point $S_1$ (Theorem 4). The line joining $l_2l_3$ to $ll_1''$ does not contain the point $S_2''$ which replaces $S_2$. For, since $S_2$ is on the line joining $l_3l_2$ to $ll_1$, the points $l_3l_2$ and $ll_1$ are homologous points of the pencils $[P_3]$ and $[P]$; and if $S_2''$ were on the line joining $l_3l_2$ to $ll_1''$, the point $l_3l_2$ would also be homologous to $ll_1''$. We may then proceed as before. By repeated application of this process we can reduce the number of perspectivities one by one, until finally we obtain the pencil of points $[Q]$ and the perspectivities

$$[P] \overset{S}{\underset{\wedge}{=}} [Q] \overset{S'}{\underset{\wedge}{=}} [P'].$$

As a consequence we have the important theorem:

THEOREM 6. *Any two projective pencils of points on skew lines are axially perspective.* (A, E)

*Proof.* The axis of the perspectivity is the line $SS'$ of the last theorem.

**24. General theory of correspondence. Symbolic treatment.** In preparation for a more detailed study of projective (and other) correspondences, we will now develop certain general ideas applicable to all one-to-one reciprocal correspondences as defined in Chap. II, § 13, p. 35, and show in particular how these ideas may be conveniently represented in symbolic form.* As previously indicated (p. 58), we will represent such correspondences in general by the letters of the Greek alphabet, as A, B, Γ, $\cdots$. The totality of elements affected by the correspondences under consideration forms a *system* which we may denote by S. If, as a result of replacing every element of a system $S_1$ by the element homologous to it in a correspondence A, the system $S_1$ is transformed into a system $S_2$, we express this by the relation $A(S_1) = S_2$. In particular, the element homologous with a given element $P$ is represented by $A(P)$.

I. If two correspondences A, B are applied successively to a system $S_1$, so that we have $A(S_1) = S_2$ and $B(S_2) = S_3$, the single correspondence Γ which transforms $S_1$ into $S_3$ is called the *resultant* or *product* of A by B; in symbols $S_3 = B(S_2) = B(A(S_1)) = BA(S_1)$, or, more briefly, $BA = \Gamma$. Similarly, for a succession of more than two correspondences.

II. Two successions of correspondences $A_m A_{m-1} \cdots A_1$ and $B_q B_{q-1} \cdots B_1$ have the same resultant, or their products are *equal*, provided they transform S into the same S′; in symbols, from the relation

$$A_m A_{m-1} \cdots A_1(S) = B_q B_{q-1} \cdots B_1(S)$$

follows

$$A_m A_{m-1} \cdots A_1 = B_q B_{q-1} \cdots B_1.$$

III. The correspondence which makes every element of the system correspond to itself is called the *identical correspondence* or simply the *identity*, and is denoted by the symbol **1**. It is then readily seen that for any correspondence A we have the relations

$$A\mathbf{1} = \mathbf{1}A = A.$$

IV. If a correspondence A transforms a system $S_1$ into $S_2$, the correspondence which transforms $S_2$ into $S_1$ is called the *inverse* of A and is represented by $A^{-1}$; i.e. if we have $A(S_1) = S_2$, then also $A^{-1}(S_2) = S_1$. The inverse of the inverse of A is then clearly A, and we evidently have also the relations

$$AA^{-1} = A^{-1}A = \mathbf{1}.$$

* In this section we have followed to a considerable extent the treatment given by H. Wiener, Berichte der K. sächsischen Gesellschaft der Wissenschaften, Leipzig, Vol. XLII (1890), pp. 249–252.

Conversely, if A, A′ are two correspondences such that we have $AA' = 1$, then A′ is the inverse of A. Evidently the identity is its own inverse.

V. *The product of three correspondences* A, B, Γ *always satisfies the relation* $(\Gamma B)A = \Gamma(BA)$ (*the associative law*). For from the relations $A(S_1) = S_2$, $B(S_2) = S_3$, $\Gamma(S_3) = S_4$ follows at once $BA(S_1) = S_3$, whence $\Gamma(BA)(S_1) = S_4$; and also $\Gamma B(S_2) = S_4$, and hence $(\Gamma B)A(S_1) = S_4$, which proves the relation in question. More generally, in any product of correspondences any set of successive correspondences may be inclosed in parentheses (provided their order be left unchanged), or any pair of parentheses may be removed; in other words, in a product of correspondences any set of successive correspondences may be replaced by their resultant, or any correspondence may be replaced by a succession of which the given correspondence is the resultant.

VI. In particular, we may conclude from the above that the inverse of the product $M \cdots BA$ is $A^{-1}B^{-1} \cdots M^{-1}$, since we evidently have the relation $M \cdots BAA^{-1}B^{-1} \cdots M^{-1} = 1$ (cf. IV).

VII. Further, it is easy to show that from two relations $A = B$ and $\Gamma = \Delta$ follows $A\Gamma = B\Delta$ and $\Gamma A = \Delta B$. In particular, the relation $A = B$ may also be written $AB^{-1} = 1$, $B^{-1}A = 1$, $BA^{-1} = 1$, or $A^{-1}B = 1$.

VIII. Two correspondences A and B are said to be *commutative* if they satisfy the relation $BA = AB$.

IX. If a correspondence A is repeated $n$ times, the resultant is written $AAA \cdots = A^n$. A correspondence A is said to be of *period* $n$, if $n$ is the smallest positive integer for which the relation $A^n = 1$ is satisfied. When no such integer exists, the correspondence has no period; when it does exist, the correspondence is said to be *periodic* or *cyclic*.

X. The case $n = 2$ is of particular importance. A correspondence of period two is called *involutoric* or *reflexive*.

**25. The notion of a group.** At this point it seems desirable to introduce the notion of a group of correspondences, which is fundamental in any system of geometry. We will give the general abstract definition of a group as follows: *

DEFINITION. A class G of elements, which we denote by $a$, $b$, $c, \cdots$, is said to form a *group with respect to an operation or law of*

* We have used here substantially the definition of a group given by L. E. Dickson, Definitions of a Group and a Field by Independent Postulates, Transactions of the American Mathematical Society, Vol. VI (1905), p. 199.

*combination* $\circ$, acting on pairs of elements of G, provided the following postulates are satisfied:

G 1. *For every pair of (equal or distinct) elements $a$, $b$ of* G, *the result $a \circ b$ of acting with the operation $\circ$ on the pair in the order given* is a uniquely determined element of* G.

G 2. *The relation $(a \circ b) \circ c = a \circ (b \circ c)$ holds for any three (equal or distinct) elements $a$, $b$, $c$ of* G.

G 3. *There occurs in* G *an element $i$, such that the relation $a \circ i = a$ holds for every element $a$* of G.

G 4. *For every element $a$ in* G *there exists an element $a'$ satisfying the relation $a \circ a' = i$.*

From the above set of postulates follow, as theorems, the following:

*The relations $a \circ a' = i$ and $a \circ i = a$ imply respectively the relations $a' \circ a = i$ and $i \circ a = a$.*

An element $i$ of G is called an *identity element*, and an element $a'$ satisfying the relation $a \circ a' = i$ is called an *inverse element* of $a$.

*There is only one identity element in* G.

*For every element $a$ of* G *there is only one inverse.*

We omit the proofs of these theorems.

Definition. A group which satisfies further the following postulate is said to be *commutative* (or *abelian*):

G 5. *The relation $a \circ b = b \circ a$ is satisfied for every pair of elements $a$, $b$ in* G.

**26. Groups of correspondences. Invariant elements and figures.** The developments of the last two sections lead now immediately to the theorem:

*A set of correspondences forms a group provided the set contains the inverse of any correspondence in the set and provided the resultant of any two correspondences is in the set.*

Here the law of combination $\circ$ of the preceding section is simply the formation of the resultant of two successive correspondences.

Definition. If a correspondence A transforms every element of a given figure F into an element of the same figure, the figure F is said to be *invariant under* A, or to be left invariant by A. In particular,

* I.e. $a \circ b$ and $b \circ a$ are not necessarily identical. The operation $\circ$ simply defines a *correspondence*, whereby to every pair of elements $a$, $b$ in G in a given order corresponds a unique element; this element is denoted by $a \circ b$.

an element which is transformed into itself by A is said to be an *invariant element* of A; the latter is also sometimes called a *double element* or a *fixed element* (point, line, plane, etc.).

We now call attention to the following general principle:

*The set of all correspondences in a group* G *which leave a given figure invariant forms a group.*

This follows at once from the fact that if each of two correspondences of G leaves the figure invariant, their product and their inverses will likewise leave it invariant; and these are all in G, since, by hypothesis, G is a group. It may happen, of course, that a group defined in this way consists of the identity only.

These notions are illustrated in the following section:

**27. Group properties of projectivities.** From the definition of a projectivity between one-dimensional forms follows at once

THEOREM 7. *The inverse of any projectivity and the resultant of any two projectivities are projectivities.*

On the other hand, we notice that the resultant of two perspectivities is not, in general, a perspectivity; if, however, two perspectivities connect three concurrent lines, as in Theorem 3, their resultant is a perspectivity. A perspectivity is its own inverse, and is therefore reflexive. As an example of the general principle of § 26, we have the important result:

THEOREM 8. *The set of all projectivities leaving a given pencil of points invariant form a group.*

If the number of points in such a pencil is unlimited, this group contains an unlimited number of projectivities. It is called the *general projective group on the line.* Likewise, the set of all projectivities on a line leaving the figure formed by three distinct points invariant forms a *subgroup* of the general group on the line. If we assume that each permutation (cf. Theorem 1′) of the three points gives rise to only a single projectivity (the proof of which requires an additional assumption), this subgroup consists of six projectivities (including, of course, the identity). Again, the set of all projectivities on a line leaving each of two given distinct points invariant forms a subgroup of the general group.

We will close this section with two examples illustrative of the principles now under discussion, in which the projectivities in question are given by explicit constructions.

EXAMPLE 1. *A group of projectivities leaving each of two given points invariant.* Let $M$, $N$ be two distinct points on a line $l$, and let $m$, $n$ be any two lines through $M$, $N$ respectively and coplanar with $l$ (fig. 27). On $m$ let there be an arbitrary given point $S$. If $S_1$ is any other point on $m$ and not on $l$ or $n$, the points $S$, $S_1$ together with the line $n$ define a projectivity $\pi_1$ on $l$ as follows: The point $\pi_1(A) = A'$ homologous to any point $A$ of $l$ is obtained by the two perspectivities $[A] \overset{S}{\underset{\wedge}{=}} [A_1] \overset{S_1}{\underset{\wedge}{=}} [A']$, where $[A_1]$ is the pencil of points on $n$. Every point $S_i$ then, if not on $l$ or $n$, defines a unique projectivity $\pi_i$; we are to show that the set of all these projectivities $\pi_i$ forms a group. We show first that the product of any two $\pi_1$, $\pi_2$ is a uniquely determined projectivity $\pi_3$ of the set (fig. 27). In the figure, $A' = \pi_1(A)$ and $A'' = \pi_2(A')$ have been

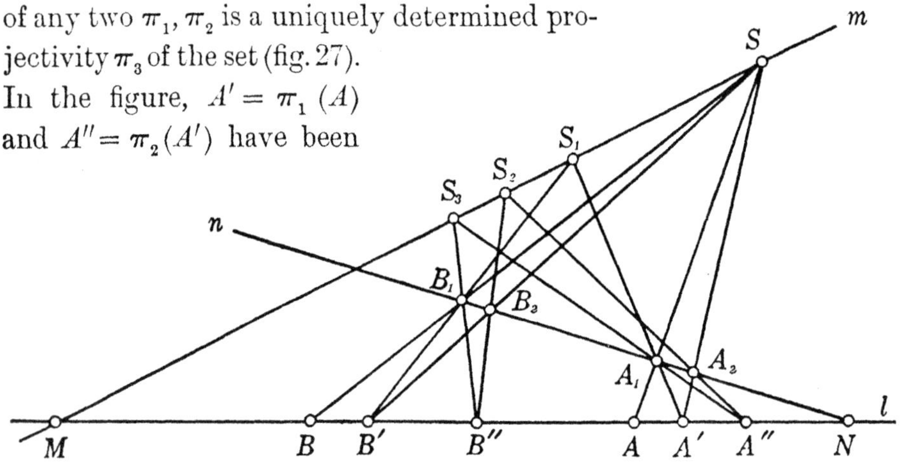


FIG. 27

constructed. The point $S_3$ giving $A''$ directly from $A$ by a similar construction is then uniquely determined as the intersection of the lines $A''A_1$, $m$. Let $B$ be any other point of $l$ distinct from $M$, $N$, and let $B' = \pi_1(B)$ and $B'' = \pi_2(B')$ be constructed; we must show that we have $B'' = \pi_3(B)$. We recognize the quadrangular set $Q(MB'A', NA''B'')$ as defined by the quadrangle $SS_2B_2A_2$. But of this quadrangular set all points except $B''$ are also obtained from the quadrangle $S_1S_3B_1A_1$; whence the line $S_3B_1$ determines the point $B''$ (Theorem 3, Chap. II). It is necessary further to show that the inverse of any projectivity in the set is in the set. For this purpose we need simply determine $S_2$ as the intersection of the line $AA_2$ with $m$ and repeat the former argument. This is left as an exercise. Finally, the identity is in the set, since it is $\pi_1$, when $S_1 = S$.

It is to be noted that in this example the points $M$ and $N$ are *double points* of each projectivity in the group; and also that if $P, P'$ and $Q, Q'$ are any two pairs of homologous points of a projectivity we have Q$(MPQ, NQ'P')$. Moreover, it is clear that any projectivity of the group is uniquely determined by a pair of homologous elements, and that there exists a projectivity which will transform any point $A$ of $l$ into any other point $B$ of $l$, provided only that $A$ and $B$ are distinct from $M$ and $N$. By virtue of the latter property the group is said to be *transitive.*

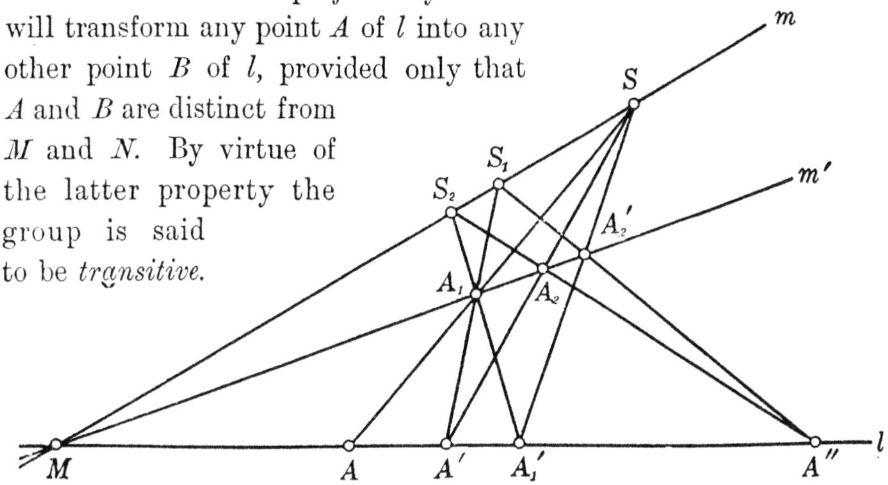


FIG. 28

EXAMPLE 2. *Commutative projectivities.* Let $M$ be a point of a line $l$, and let $m, m'$ be any two lines through $M$ distinct from $l$, but in the same plane with $l$ (fig. 28.) Let $S$ be a given point of $m$, and let a projectivity $\pi_1$ be defined by another point $S_1$ of $m$ which determines the perspectivities $[A] \overset{S}{\underset{\wedge}{=}} [A_1] \overset{S_1}{\underset{\wedge}{=}} [A']$, where $[A_1]$ is the pencil of points on $m'$. *Any two projectivities defined in this way by points $S_i$ are commutative.* Let $\pi_2$ be another such projectivity, and construct the points $A' = \pi_1(A)$, $A'' = \pi_2(A')$, and $A_1' = \pi_2(A)$. The quadrangle $SS_2A_1A_2$ gives Q$(MAA', MA''A_1')$; and the quadrangular set determined on $l$ by the quadrangle $SS_1A_1A_2'$ has the first five points of the former in the same positions in the symbols. Hence we have $\pi_1(A_1') = A''$, and therefore $\pi_1\pi_2 = \pi_2\pi_1$.

## EXERCISES

**1.** Show that the set of all projectivities $\pi_i$ of Example 2 above forms a group, which is then a *commutative group.*

**2.** Show that the projectivity $\pi_1$ of Example 1 above is identical with the projectivity obtained by choosing any other two points of $m$ as centers of perspectivity, provided only that the two projectivities have one homologous pair (distinct from $M$ or $N$) in common. Investigate the general question as to how far the construction may be modified so as still to preserve the proposition that the projectivities are determined by the double points $M$, $N$ and one pair of homologous elements.

**3.** Discuss the same general question for the projectivities of Example 2.

**4.** Apply the method of Example 2 to the projectivities of Example 1. Why does it fail to show that any two of the latter are commutative? State the space and plane duals of the two examples.

**5.** $ABCD$ is a tetrahedron and $\alpha, \beta, \gamma, \delta$ the faces not containing $A, B, C, D$ respectively, and $l$ is any line not meeting an edge. The planes $(lA, lB, lC, lD)$ are projective with the points $(l\alpha, l\beta, l\gamma, l\delta)$.

**6.** On each of the ten sides of a complete 5-point in a plane there are three diagonal points and two vertices. Write down the projectivities among these ten sets of five points each.

## 28. Projective transformations of two-dimensional forms.

DEFINITION. A *projective transformation* between the elements of two two-dimensional or two three-dimensional forms is any one-to-one reciprocal correspondence between the elements of the two forms, such that to every one-dimensional form of one there corresponds a projective one-dimensional form of the other.

DEFINITION. A *collineation* is any (1, 1) correspondence between two two-dimensional or two three-dimensional forms in which to every element of one of the forms corresponds an element of the same kind in the other form, and in which to every one-dimensional form of one corresponds a one-dimensional form of the other. A *projective collineation* is one in which this correspondence is projective. Unless otherwise specified, the term *collineation* will, in the future, always denote a projective collineation.*

In the present chapter we shall confine ourselves to the discussion of some of the fundamental properties of collineations. In this section we discuss the collineations between two-dimensional forms, and shall take the plane (planar field) as typical; the corresponding theorems for the other two-dimensional forms will then follow from duality.

The simplest correspondence between the elements of two distinct planes $\pi, \pi'$ is a *perspective correspondence*, whereby any two homologous elements are on the same element of a bundle whose center $O$ is on neither of the planes $\pi, \pi'$. The simplest collineation in a plane,

* In how far a collineation must be projective will appear later.

i.e. which transforms every element of a plane into an element of the same plane, is the following:

DEFINITION. A *perspective collineation in a plane* is a projective collineation leaving invariant every point on a given line $o$ and every line on a given point $O$. The line $o$ and the point $O$ are called the *axis* and *center* respectively of the perspective collineation. If the center and axis are not united, the collineation is called a *planar homology;* if they are united, a *planar elation.*

A perspective collineation in a plane $\pi$ may be constructed as follows: Let any line $o$ and any point $O$ of $\pi$ be chosen as axis and center respectively, and let $\pi_1$ be any plane through $o$ distinct from $\pi$. Let $O_1$, $O_2$ be any two points collinear with $O$ and in neither of the planes $\pi$, $\pi_1$. The perspective collineation is then obtained by the two perspectivities $[P] \overset{O_1}{\underset{\wedge}{=}} [P_1] \overset{O_2}{\underset{\wedge}{=}} [P']$, where $P$ is any point of $\pi$ and $P_1$, $P'$ are points of $\pi_1$ and $\pi$ respectively. Every point of the line $o$ and every line through the point $O$ clearly remain fixed by the transformation, so that the conditions of the definition are satisfied, if only the transformation is projective. But it is readily seen that every pencil of points is transformed by this process into a perspective pencil of points, the center of perspectivity being the point $O$; and every pencil of lines is transformed into a perspective pencil, the axis of perspectivity being $o$. The above discussion applies whether or not the point $O$ is on the line $o$.

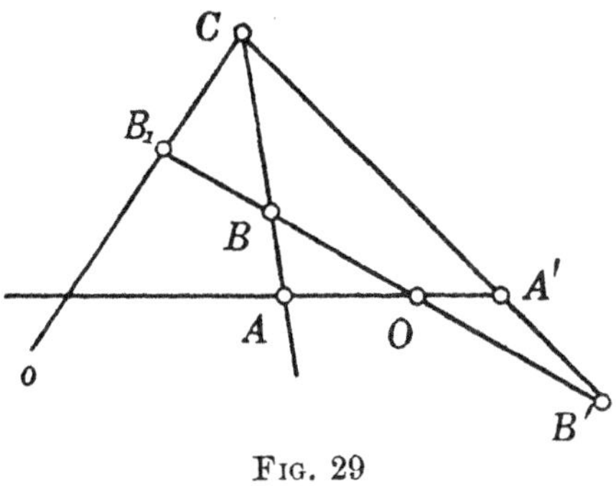


FIG. 29

THEOREM 9. *A perspective collineation in a plane is uniquely defined if the center, axis, and any two homologous points (not on the axis or center) are given, with the single restriction that the homologous points must be collinear with $O$.* (A, E)

*Proof.* Let $O$, $o$ be the center and axis respectively (fig. 29). It is clear from the definition that any two homologous points must be collinear with $O$, since every line through $O$ is invariant; similarly (dually) any two homologous lines must be concurrent with $o$. Let $A$, $A'$ be the given pair of homologous points collinear with $O$. The point $B'$ homologous to any point $B$ of the plane is then determined. We may assume $B$ to be distinct from $O$, $A$ and not to be on $o$. $B'$ is on the line $OB$, and if the line $AB$ meets $o$ in $C$, then, since $C$ is invariant by definition, the line $AB = AC$ is transformed into $A'C$. $B'$ is then determined as the intersection of the lines $OB$ and $A'C$. This applies unless $B$ is on the line $AA'$; in this case we determine as above a pair of homologous points not on $AA'$, and then use the two points thus determined to construct $B'$. This shows that there can be no more than one perspective collineation in the plane with the given elements.

To show that there is one we may proceed as follows: Let $\pi_1$ be any plane through $o$ distinct from $\pi$, the plane of the perspectivity, and let $O_1$ be any point on neither of the planes $\pi$, $\pi_1$. If the line $AO_1$ meets $\pi_1$ in $A_1$, the line $A'A_1$ meets $OO_1$ in a point $O_2$. The perspective collineation determined by the two centers of perspectivity $O_1$, $O_2$ and the plane $\pi_1$ then has $O$, $o$ as center and axis respectively and $A$, $A'$ as a pair of homologous points.

COROLLARY 1. *A perspective collineation in a plane transforms every one-dimensional form into a perspective one-dimensional form.* (A, E)

COROLLARY 2. *A perspective collineation with center $O$ and axis $o$ transforms any triangle none of whose vertices or sides are on $o$ or $O$ into a perspective triangle, the center of perspectivity of the triangles being the center of the collineation and the axis of perspectivity being the axis of the collineation.* (A, E)

COROLLARY 3. *The only planar collineations (whether required to be projective or not) which leave invariant the points of a line $o$ and the lines through a point $O$ are homologies if $O$ is not on $o$, and elations if $O$ is on $o$.* (A, E)

*Proof.* This will be evident on observing that in the first paragraph of the proof of the theorem no use is made of the hypothesis that the collineation is projective.

COROLLARY 4. *If* H *is a perspective collineation such that* $\mathrm{H}(O) = O$, $\mathrm{H}(o) = o$, $\mathrm{H}(A) = A'$, $\mathrm{H}(B) = B'$ *where $A$, $A'$, $B$, $B'$ are collinear with a point $K$ of $o$, then we have* $Q(OAB, KB'A')$. (A, E)

*Proof.* If $C$ is any point not on $AA'$ and $\mathrm{H}(C) = C'$, the lines $AC$ and $A'C'$ meet in a point $L$ of $o$, and $BC$ and $B'C'$ meet in a point $M$ of $o$; and the required quadrangle is $CC'LM$ (cf. fig. 32, p. 77).

THEOREM 10. *Any complete quadrangle of a plane can be transformed into any complete quadrangle of the same or a different plane by a projective collineation which, if the quadrangles are in the same plane, is the resultant of a finite number of perspective collineations.* (A, E)

*Proof.* Let the quadrangles be in the same plane and let their vertices be $A, B, C, D$ and $A', B', C', D'$ respectively. We show first that there exists a collineation leaving any three vertices, say $A'$, $B'$, $C'$, of

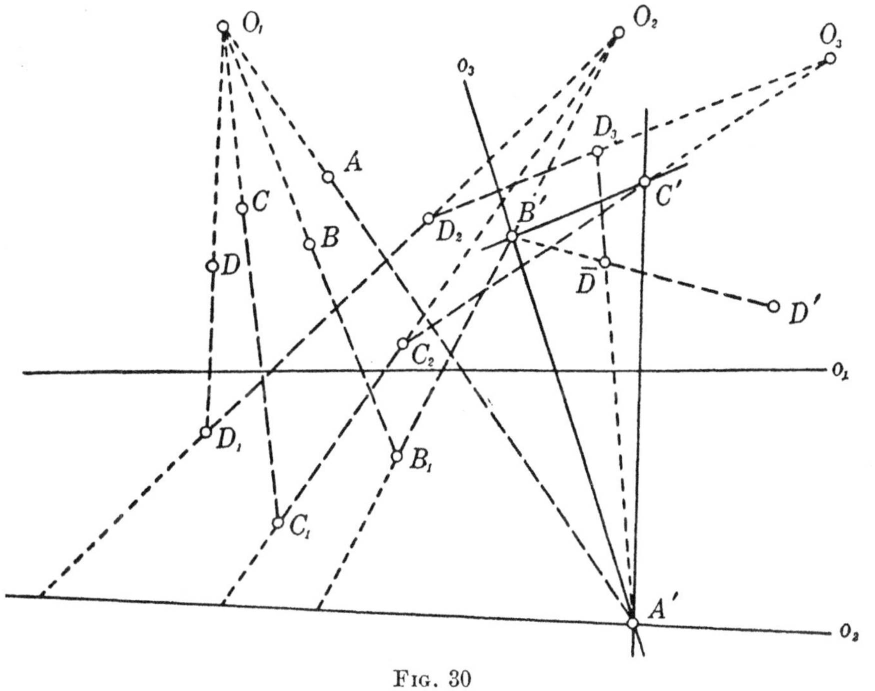


FIG. 30

the quadrangle $A'B'C'D'$ invariant and transforming into the fourth, $D'$, any other point $D_3$ not on a side of the triangle $A'B'C'$ (fig. 30). Let $\bar{D}$ be the intersection of $A'D_3$, $B'D'$ and consider the homology with center $A'$ and axis $B'C'$ transforming $D_3$ into $\bar{D}$. Next consider the homology with center $B'$ and axis $C'A'$ transforming $\bar{D}$ into $D'$. Both these homologies exist by Theorem 9. The resultant of these two homologies is a collineation leaving fixed $A'$, $B'$, $C'$ and transforming $D_3$ into $D'$. (It should be noticed that one or both of the homologies may be the identity.)

Let $O_1$ be any point on the line containing $A$ and $A'$ and let $o_1$ be any line not passing through $A$ or $A'$. By Theorem 9 there exists a perspective collineation $\pi_1$ transforming $A$ to $A'$ and having $O_1$ and $o_1$ as center and axis. Let $B_1$, $C_1$, $D_1$ be points such that

$$\pi_1(ABCD) = A'B_1C_1D_1.$$

In like manner, let $o_2$ be any line through $A'$ not containing $B_1$ or $B'$ and let $O_2$ be any point on the line $B_1B'$. Let $\pi_2$ be the perspective collineation with axis $o_2$, center $O_2$, and transforming $B_1$ to $B'$. Let $C_2 = \pi_2(C_1)$ and $D_2 = \pi_2(D_1)$. Here

$$\pi_2(A'B_1C_1D_1) = A'B'C_2D_2.$$

Now let $O_3$ be any point on the line $C_2C'$ and let $\pi_3$ be the perspective collineation which has $A'B' = o_3$ for axis, $O_3$ for center, and transforms $C_2$ to $C'$. The existence of $\pi_3$ follows from Theorem 9 as soon as we observe that $C'$ is not on the line $A'B'$, by hypothesis, and $C_2$ is not on $A'B'$; because if so, $C_1$ would be on $A'B_1$ and therefore $C$ would be on $AB$. Let $\pi_3(D_2) = D_3$. It follows that

$$\pi_3(A'B'C_2D_2) = A'B'C'D_3.$$

The point $D_3$ cannot be on a side of the triangle $A'B'C'$ because then $D_2$ would be on a side of $A'B'C_2$, and hence $D_1$ on a side of $A'B_1C_1$, and, finally, $D$ on a side of $ABC$. Hence, by the first paragraph of this proof, there exists a projectivity $\pi_4$ such that

$$\pi_4(A'B'C'D_3) = A'B'C'D'.$$

The resultant $\pi_4\pi_3\pi_2\pi_1$ of these four collineations clearly transforms $A$, $B$, $C$, $D$ into $A'$, $B'$, $C'$, $D'$ respectively. If the quadrangles are in different planes, we need only add a perspective transformation between the two planes.

COROLLARY. *There exist projective collineations in a plane which will effect any one of the possible 24 permutations of the vertices of a complete quadrangle in the plane.* (A, E)

**29. Projective collineations of three-dimensional forms.** Projective collineations in a three-dimensional form have been defined at the beginning of § 28.

DEFINITION. A projective collineation in space which leaves invariant every point of a plane $\omega$ and every plane on a point $O$ is called a *perspective collineation.* The plane $\omega$ is called the *plane of perspectivity;* the point $O$ is called the *center.* If $O$ is on $\omega$, the collineation is said to be an *elation* in space; otherwise, a *homology* in space.

THEOREM 11. *If $O$ is any point and $\omega$ any plane, there exists one and only one perspective collineation in space having $O$, $\omega$ for center and plane of perspectivity respectively, which transforms any point $A$ (distinct from $O$ and not on $\omega$) into any other point $A'$ (distinct from $O$ and not on $\omega$) collinear with $AO$.* (A, E)

*Proof.* We show first that there cannot be more than one perspective collineation satisfying the conditions of the theorem, by showing that the point $B'$ homologous to any point $B$ is uniquely

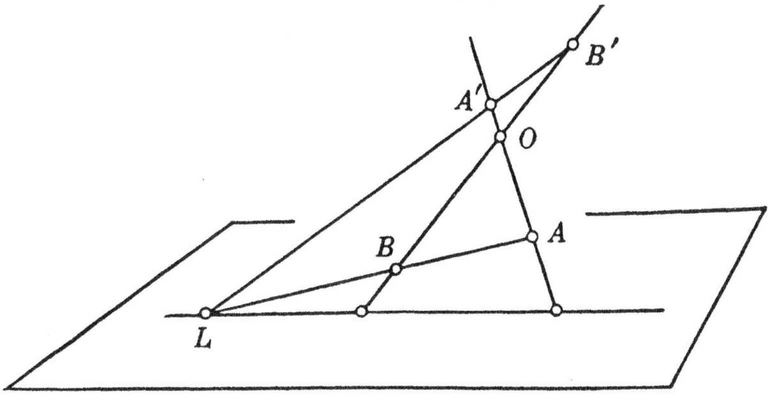


FIG. 31

determined by the given conditions. We may assume $B$ not on $\omega$ and distinct from $O$ and $A$. Suppose first that $B$ is not on the line $AO$ (fig. 31): Since $BO$ is an invariant line, $B'$ is on $BO$; and if the line $AB$ meets $\omega$ in $L$, the line $AB = AL$ is transformed into the line $A'L$. Hence $B'$ is determined as the intersection of $BO$ and $A'L$. There remains the case where $B$ is on $AO$ and distinct from $A$ and $O$ (fig. 32). Let $C$, $C'$ be any pair of homologous points not on $AO$, and let $AC$ and $BC$ meet $\omega$ in $L$ and $M$ respectively. The line $MB = MC$ is transformed into $MC'$, and the point $B'$ is then determined as the intersection of the lines $BO$ and $MC'$. That this point is independent of the choice of the pair $C$, $C'$ now follows from the fact that the quadrangle $MLCC'$ gives the quadrangular set $Q(KAA', OB'B)$, where $K$ is the point in which $AO$ meets $\omega$ ($K$ may coincide with $O$ without affecting the argument). The point $B'$ is then uniquely determined by the five points $O$, $K$, $A$, $A'$, $B$.

The correspondence defined by the construction in the paragraph above has been proved to be one-to-one throughout. On the line $AO$ it is projective because of the perspectivities (fig. 32)

$$[B] \overset{C}{\overline{\wedge}} [M] \overset{C'}{\overline{\wedge}} [B'].$$

On $OB$, any other line through $O$, it is projective because of the perspectivities (fig. 31)

$$[B] \overset{A}{\overline{\wedge}} [L] \overset{A'}{\overline{\wedge}} [B'].$$

That any pencil of points not through $O$ is transformed into a perspective pencil, the center of perspectivity being $O$, is now easily seen and is left as an exercise for the reader. From this it follows

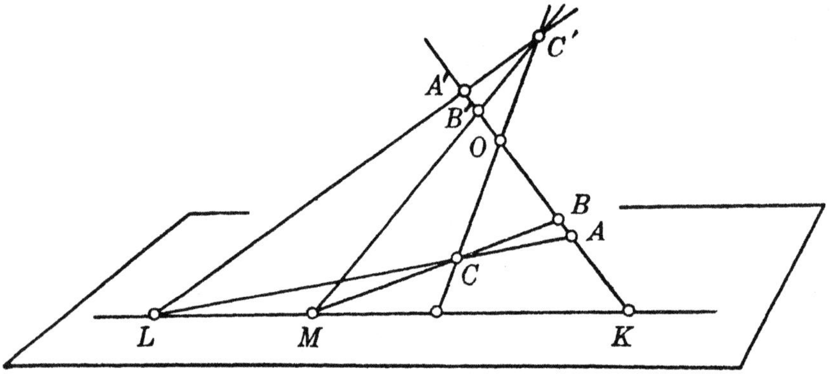


FIG. 32

that any one-dimensional form is transformed into a projective form, so that the correspondence which has been constructed satisfies the definition of a projective collineation.

THEOREM 12. *Any complete five-point in space can be transformed into any other complete five-point in space by a projective collineation which is the resultant of a finite number of perspective collineations.* (A,E)

*Proof.* Let the five-points be $ABCDE$ and $A'B'C'D'E'$ respectively. We will show first that there exists a collineation leaving $A'B'C'D'$ invariant and transforming into $E'$ any point $E_0$ not coplanar with three of the points $A'B'C'D'$. Consider a homology having $A'B'C'$ as plane of perspectivity and $D'$ as center. Any such homology transforms $E_0$ into a point on the line $E_0D'$. Similarly, a homology with plane $A'B'D'$ and center $C'$ transforms $E'$ into a point on the line $E'C'$. If $E_0D'$ and $E'C'$ intersect in a point $E_1$, the resultant of two homologies of the kind described, of which the first transforms $E_0$ into $E_1$ and the second transforms $E_1$ into $E'$, leaves $A'B'C'D'$ invariant and transforms $E_0$ into $E'$. If the lines $E_0D'$ and $E'C'$ are skew, there is a line through $B'$ meeting the lines $E_0D'$ and $E'C'$ respectively in two points $E_1$ and $E_2$. The resultant of the three homologies, of which the first has the plane $A'B'C'$ and center $D'$ and transforms $E_0$ to $E_1$, of which the second has the plane $A'C'D'$ and center $B'$ and transforms $E_1$ to $E_2$, and of which the third has the plane $A'B'D'$ and center $C'$ and transforms $E_2$ to $E'$, is a collineation leaving $A'B'C'D'$ invariant and transforming $E_0$ to $E'$. The remainder of the proof is now entirely analogous to the proof of Theorem 10. The details are left as an exercise.

COROLLARY. *There exist projective collineations which will effect any one of the possible 120 permutations of the vertices of a complete five-point in space.* (A, E)

## EXERCISES

**1.** Prove the existence of perspective collineations in a plane without making use of any points outside the plane.

**2.** Discuss the figure formed by two triangles which are homologous under an elation. How is this special form of the Desargues configuration obtained as a section of a complete five-point in space?

**3.** Given an elation in a plane with center $O$ and axis $o$ and two homologous pairs $A$, $A'$ and $B$, $B'$ on any line through $O$, show that we always have Q$(OAA', OB'B)$.

**4.** What permutations of the vertices of a complete quadrangle leave a given diagonal point invariant? every diagonal point?

**5.** Write down the permutations of the six sides of a complete quadrangle brought about by all possible permutations of the vertices.

**6.** The set of all homologies (elations) in a plane with the same center and axis form a group.

**7.** Prove that two elations in a plane having a common axis and center are commutative. Will this method apply to prove that two homologies with common axis and center are commutative?

**8.** Prove that two elations in a plane having a common axis are commutative. Dualize. Prove the corresponding theorem in space.

**9.** Prove that the resultant of two elations having a common axis is an elation. Dualize. Prove the corresponding theorem in space. What groups of elations are defined by these theorems?

**10.** Discuss the effect of a perspective collineation of space on: (1) a pencil of lines; (2) any plane; (3) any bundle of lines; (4) a tetrahedron; (5) a complete five-point in space.

**11.** The set of all collineations in space (in a plane) form a group.

**12.** The set of all projective collineations in space (in a plane) form a group.

**13.** Show that under certain conditions the configuration of two perspective tetrahedra is left invariant by 120 collineations (cf. Ex. 3, p. 47).

# CHAPTER IV

## HARMONIC CONSTRUCTIONS AND THE FUNDAMENTAL THEOREM OF PROJECTIVE GEOMETRY

**30. The projectivity of quadrangular sets.** We return now to a more detailed discussion of the notion of quadrangular sets introduced at the end of Chap. II. We there defined a *quadrangular set of points* as the section by a transversal of the sides of a complete quadrangle; the plane dual of this figure we call a *quadrangular set of lines;** it consists of the projection of the vertices of a complete quadrilateral from a point which is in the plane of the quadrilateral, but not on any of its sides; the space dual of a quadrangular set of points we call a *quadrangular set of planes;* it is the figure formed by the projection from a point of the figure of a quadrangular set of lines. We may now prove the following important theorem:

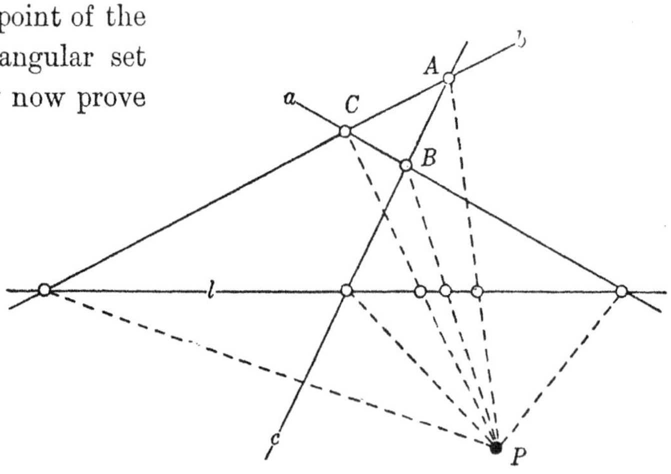


FIG. 33

THEOREM 1. *The section by a transversal of a quadrangular set of lines is a quadrangular set of points.* (A, E)

*Proof.* By Theorem 3′, Chap. II, p. 49, and the dual of Note 2, on p. 48, we may take the transversal $l$ to be one of the sides of a complete quadrilateral the projection of whose vertices from a point $P$ forms the set of lines in question (fig. 33). Let the remaining three sides of such a quadrilateral be $a$, $b$, $c$. Let the points $bc$, $ca$, and $ab$

* It would be more natural at this stage to call such a set a quadrilateral set of lines; the next theorem, however, justifies the term we have chosen, which has the advantage of uniformity.

be denoted by $A$, $B$, and $C$ respectively. The sides of the quadrangle $PABC$ meet $l$ in the same points as the lines of the quadrangular set of lines.

COROLLARY. *A set of collinear points which is projective with a quadrangular set is a quadrangular set.* (A, E)

THEOREM 1′. *The projection from a point of a quadrangular set of points is a quadrangular set of lines.* (A, E)

This is the plane dual of the preceding; the space dual is:

THEOREM 1″. *The section by a plane of a quadrangular set of planes is a quadrangular set of lines.* (A, E)

COROLLARY. *If a set of elements of a primitive one-dimensional form is projective with a quadrangular set, it is itself a quadrangular set.* (A, E)

**31. Harmonic sets.** DEFINITION. A quadrangular set Q(123, 124) is called a *harmonic set* and is denoted by H(12, 34). The elements 3, 4 are called *harmonic conjugates with respect to the elements* 1, 2; and 3 (or 4) is called the *harmonic conjugate of* 4 (*or* 3) *with respect to* 1 *and* 2.

From this definition we see that in a harmonic set of points H($AC$, $BD$), the points $A$ and $C$ are diagonal points of a complete

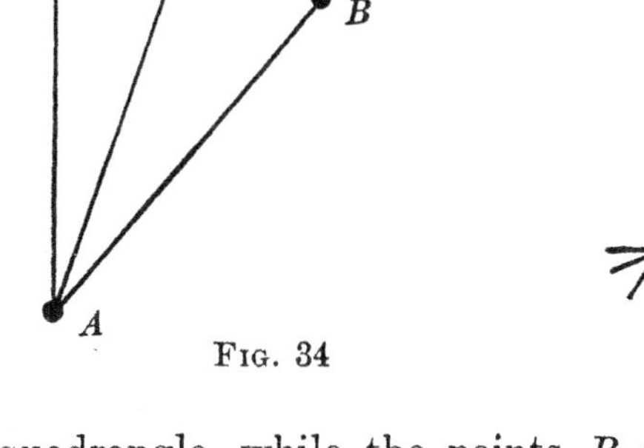


FIG. 34

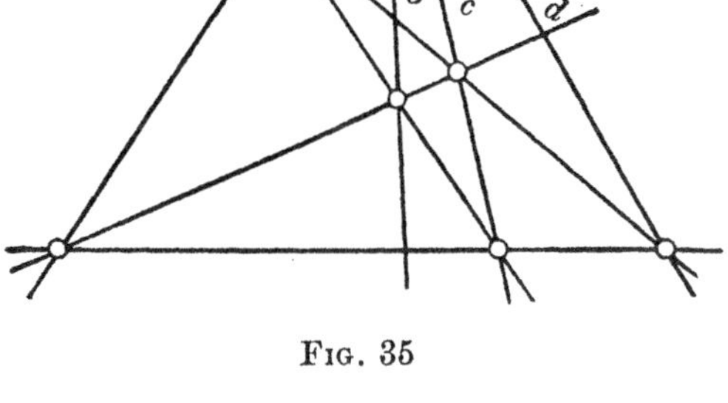


FIG. 35

quadrangle, while the points $B$ and $D$ are the intersections of the remaining two opposite sides of the quadrangle with the line $AC$ (fig. 34). Likewise, in a harmonic set of lines H($ac$, $bd$), the lines $a$ and $c$ are two diagonal lines of a complete quadrilateral, while the lines $b$ and $d$ are the lines joining the remaining pair of opposite vertices of the quadrilateral to the point of intersection $ac$ of the lines $a$ and $c$ (fig. 35). A harmonic set of planes is the space dual of a harmonic set of points, and is therefore the projection from a point of a harmonic set of lines.

In case the diagonal points of a complete quadrangle are collinear, any three points of a line form a harmonic set and any point is its own harmonic conjugate with regard to any two points collinear with it. Theorems on harmonic sets are therefore trivial in those spaces for which Assumption $H_0$ is not true. We shall therefore base our reasoning, in this and the following two sections, on Assumption $H_0$; though most of the theorems are obviously true also in case $H_0$ is false. This is why some of the theorems are labeled as dependent on Assumptions $A$ and $E$, whereas the proofs given involve $H_0$ also.

The corollary of Theorem 3, Chap. II, when applied to harmonic sets yields the following:

THEOREM 2. *The harmonic conjugate of an element with respect to two other elements of a one-dimensional primitive form is a unique element of the form.* (A, E)

Theorem 1 applied to the special case of harmonic sets gives

THEOREM 3. *Any section or projection of a harmonic set is a harmonic set.* (A, E)

COROLLARY. *If a set of four elements of any one-dimensional primitive form is projective with a harmonic set, it is itself a harmonic set.* (A, E)

THEOREM 4. *If* 1 *and* 2 *are harmonic conjugates with respect to* 3 *and* 4, 3 *and* 4 *are harmonic conjugates with respect to* 1 *and* 2. (A, E, $H_0$)

*Proof.* By Theorem 2, Chap. III, there exists a projectivity

$$1234 \,\overline{\wedge}\, 3412.$$

But by hypothesis we have H(34, 12). Hence by the corollary of Theorem 3 we have H(12, 34).

By virtue of this theorem the pairs 1, 2 and 3, 4 in the expression H(12, 34) play the same rôle and may be interchanged.*

* The corresponding theorem for the more general expression Q(123, 456) cannot be derived without the use of an additional assumption (cf. Theorem 24, Chap. IV).

THEOREM 5. *Given two harmonic sets* H (12, 34) *and* H (1′2′, 3′4′), *there exists a projectivity such that* $1234 \;\overline{\wedge}\; 1'2'3'4'$. (A, E)

*Proof.* Any projectivity $123 \;\overline{\wedge}\; 1'2'3'$ (Theorem 1, Chap. III) must transform 4 into 4′ by virtue of Theorem 3, Cor., and the fact that the harmonic conjugate of 3 with respect to 1 and 2 is unique (Theorem 2). This is the converse of Theorem 3, Cor.

COROLLARY 1. *If* H (12, 34) *and* H (12′, 3′4′) *are two harmonic sets of different one-dimensional forms having the element* 1 *in common, we have* $1234 \;\overset{=}{\wedge}\; 12'3'4'$. (A, E)

For under the hypotheses of the corollary the projectivity $123 \;\overline{\wedge}\; 1'2'3'$ of the preceding proof may be replaced by the perspectivity $123 \;\overset{=}{\wedge}\; 12'3'$.

COROLLARY 2. *If* H (12, 34) *is a harmonic set, there exists a projectivity* $1234 \;\overline{\wedge}\; 1243$. (A, E)

This follows directly from the last theorem and the evident fact that if H (12, 34) we have also H (12, 43). The converse of this corollary is likewise valid; the proof, however, is given later in this chapter (cf. Theorem 27, Cor. 5).

We see as a result of the last corollary and Theorem 2, Chap. III, that if we have H (12, 34), there exist projectivities which will transform 1234 into any one of the eight permutations

1234, 1243, 2134, 2143, 3412, 3421, 4312, 4321.*

In other words, if we have H (12, 34), we have likewise H (12, 43), H (21, 34), H (21, 43), H (34, 12), H (34, 21), H (43, 12), H (43, 21).

THEOREM 6. *The two sides of a complete quadrangle which meet in a diagonal point are harmonic conjugates with respect to the two sides of the diagonal triangle which meet in this point.* (A, E)

*Proof.* The four sides of the complete quadrangle which do not pass through the diagonal point in question form a quadrilateral which defines the set of four lines mentioned as harmonic in the way indicated (fig. 36).

It is sometimes convenient to speak of a pair of elements of a form as harmonic with a pair of elements of a form of different kind. For example, we may say that two points are harmonic with two lines in a plane with the points, if the points determine two

* These transformations form the so-called *eight-group*.

lines through the intersection of the given lines which are harmonic with the latter; or, what is the same thing, if the line joining the points meets the lines in two points harmonic with the given points. With this understanding we may restate the last theorem as follows: *The sides of a complete quadrangle which meet in a diagonal point are harmonic with the other two diagonal points.* In like manner, we may say that two points are harmonic with two planes, if the line joining the points meets the planes in a pair of points harmonic with the given points; and a pair of lines is harmonic with a pair of planes, if they intersect on the intersection of the two planes, and if they determine with this intersection two planes harmonic with the given planes.

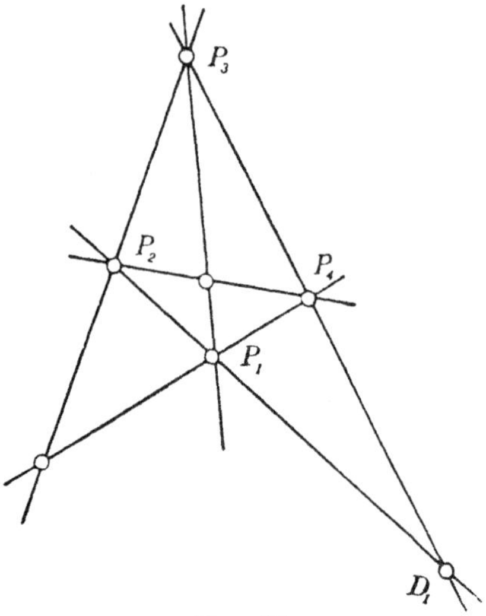

FIG. 36

## EXERCISES

**1.** Prove Theorem 4 directly from a figure without using Theorem 2, Chap. III.

**2.** Prove Theorem 5, Cor. 2, directly from a figure.

**3.** Through a given point in a plane construct a line which passes through the point of intersection of two given lines in the plane, without making use of the latter point.

**4.** A line meets the sides of a triangle $ABC$ in the points $A_1$, $B_1$, $C_1$, and the harmonic conjugates $A_2$, $B_2$, $C_2$ of these points with respect to the two vertices on the same side are determined, so that we have $\text{H}(AB, C_1C_2)$, $\text{H}(BC, A_1A_2)$, and $\text{H}(CA, B_1B_2)$. Show that $A_1, B_2, C_2$; $B_1, C_2, A_2$; $C_1, A_2, B_2$ are collinear; that $AA_2$, $BB_2$, $CC_2$ are concurrent; and that $AA_2$, $BB_1$, $CC_1$; $AA_1$, $BB_2$, $CC_1$; $AA_1$, $BB_1$, $CC_2$ are also concurrent.

**5.** If each of two sides $AB$, $BC$ of a triangle $ABC$ meets a pair of opposite edges of a tetrahedron in two points which are harmonic conjugates with respect to $A$, $B$ and $B$, $C$ respectively, the third side $CA$ will meet the third pair of opposite edges in two points which are harmonic conjugates with respect to $C$, $A$.

**6.** $A$, $B$, $C$, $D$ are the vertices of a quadrangle the sides of which meet a given transversal $l$ in the six points $P_1, P_2, P_3, P_4, P_5, P_6$; the harmonic conjugate of each of these points with respect to the two corresponding vertices of the quadrangle is constructed and these six points are denoted by $P'_1$, $P'_2$, $P'_3$, $P'_4$, $P'_5$, $P'_6$ respectively. The three lines joining the pairs of the latter points which lie on opposite sides of the quadrangle meet in a point $P$, which is the harmonic conjugate of each of the points in which these three lines meet $l$ with respect to the pairs of points $P'$ defining the lines.

**7.** Defining the polar line of a point with respect to a pair of lines as the harmonic conjugate line of the point with regard to the pair of lines, prove that the three polar lines of a point as to the pairs of lines of a triangle form a triangle (called the cogredient triangle) perspective to the given triangle.

**8.** Show that the polar line defined in Ex. 7 is the same as the polar line defined in Ex. 3, p. 52.

**9.** Show that any line through a point $O$ and meeting two intersecting lines $l$, $l'$ meets the polar of $O$ with respect to $l$, $l'$ in a point which is the harmonic conjugate of $O$ with respect to the points in which the line through $O$ meets $l$, $l'$.

**10.** The axis of perspectivity of a triangle and its cogredient triangle is the polar line (cf. p. 46) of the triangle as to the given point.

**11.** If two triangles are perspective, the two polar lines of a point on their axis of perspectivity meet on the axis of perspectivity.

**12.** If the lines joining corresponding vertices of two $n$-lines meet in a point, the points of intersection of corresponding sides meet on a line.

**13.** (Generalization of Exs. 7, 10.) The $n$ polar lines of a point $P$ as to the $n$ $(n-1)$-lines of an $n$-line in a plane form an $n$-line (the cogredient $n$-line) whose sides meet the corresponding sides of the given $n$-line in the points of a line $p$. The line $p$ is called the polar of $P$ as to the $n$-line.*

**14.** (Generalization of Ex. 11.) If two $n$-lines are perspective, the two polar lines of a point on their axis of perspectivity meet on this axis.

**15.** Obtain the plane duals of the last two problems. Generalize them to three- and $n$-dimensional space. These theorems are fundamental for the construction of polars of algebraic curves and surfaces of the $n$-th degree.

**32. Nets of rationality on a line.** DEFINITION. A point $P$ of a line is said to be *harmonically related to three given distinct points* $A$, $B$, $C$ of the line, provided $P$ is one of a sequence of points $A, B, C, H_1, H_2, H_3, \cdots$ of the line, finite in number, such that $H_1$ is the harmonic conjugate of one of the points $A$, $B$, $C$ with respect to the other two, and such that every other point $H_i$ is harmonic with three of the set $A, B, C, H_1, H_2, \cdots, H_{i-1}$. The class of all points harmonically related to three distinct points $A$, $B$, $C$ on a line is called the *one-dimensional net of rationality* defined by $A$, $B$, $C$; it is denoted by $\mathsf{R}(ABC)$. A net of rationality on a line is also called a *linear net*.

* This is a definition by induction of the polar line of a point with respect to an $n$-line.

THEOREM 7. *If $A, B, C, D$ and $A', B', C', D'$ are respectively points of two lines such that $ABCD \barwedge A'B'C'D'$, and if $D$ is harmonically related to $A, B, C$, then $D'$ is harmonically related to $A', B', C'$.* (A, E)

This follows directly from the fact that the projectivity of the theorem makes the set of points $H_j$ which defines $D$ as harmonically related to $A, B, C$ projective with a set of points $H_j'$ such that every harmonic set of points of the sequence $A, B, C, H_1, H_2, \cdots, D$ is homologous with a harmonic set of the sequence $A', B', C', H_1', H_2', \cdots, D'$ (Theorem 3, Cor.).

COROLLARY. *If a class of points on a line is projective with a net of rationality on a line, it is itself a net of rationality.*

THEOREM 8. *If $K, L, M$ are three distinct points of $\mathsf{R}(ABC), A, B, C$ are points of $\mathsf{R}(KLM)$.* (A, E)

*Proof.* From the projectivity $ABCK \barwedge BAKC$ follows, by Theorem 7, that $C$ is a point of $\mathsf{R}(ABK)$. Hence all points harmonically related to $A, B, C$ are, by definition, harmonically related to $A, B, K$. Since $K$ is, by hypothesis, in the net $\mathsf{R}(ABC)$, the definition also requires that all points of $\mathsf{R}(ABK)$ shall be points of $\mathsf{R}(ABC)$. Hence the nets $\mathsf{R}(ABC)$ and $\mathsf{R}(ABK)$ are identical; and so $\mathsf{R}(ABC) = \mathsf{R}(ABK) = \mathsf{R}(AMK) = \mathsf{R}(KLM)$.

COROLLARY. *A net of rationality on a line is determined by any distinct three of its points.*

THEOREM 9. *If all but one of the six (or five, or four) points of a quadrangular set are points of the same net of rationality* $\mathsf{R}$, *this one point is also a point of* $\mathsf{R}$. (A, E)

*Proof.* Let the sides of the quadrangle $PQRS$ (fig. 37) meet the line $l$ as indicated in the points $A, A_1$; $B, B_1$; $C, C_1$, so that $B \neq B_1$; and suppose that the first five of these are points of a net of rationality

$$\mathsf{R} = \mathsf{R}(AA_1B_1) = \mathsf{R}(BCB_1) = \cdots.$$

We must prove that $C_1$ is a point of $\mathsf{R}$. Let the pair of lines $RS$ and $PQ$ meet in $B'$. We then have

$$BCB_1A \overset{S}{\underset{\wedge}{=}} BQB'P \overset{R}{\underset{\wedge}{=}} BA_1B_1C_1.$$

Since $A$ is in $\mathsf{R}(BCB_1)$, it follows from this projectivity, in view of Theorem 7, that $C_1$ is in $\mathsf{R}(BA_1B_1) = \mathsf{R}$.

DEFINITION. A point $P$ of a line is said to be *quadrangularly related to three given distinct points $A, B, C$* of the line, provided

$P$ is one of a sequence of points $A, B, C, H_1, H_2, H_3, \cdots$ of the line, finite in number, such that $H_1$ is the harmonic conjugate of one of the points $A, B, C$ with respect to the other two, and such that every other point $H_i$ is one of a quadrangular set of which the other five belong to the set $A, B, C, H_1, H_2, \cdots, H_{i-1}$.

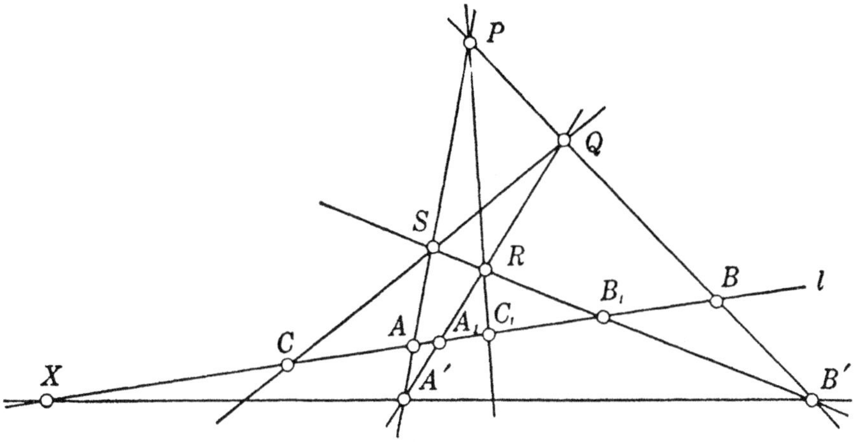


FIG. 37

COROLLARY. *The class of all points quadrangularly related to three distinct collinear points $A, B, C$ is* R$(ABC)$. (A, E)

From the last corollary it is plain that R$(ABC)$ consists of all points that can be constructed from $A$, $B$, $C$ by means of points and lines alone; that is to say, all points whose existence can be inferred from Assumptions A, E, $H_0$. The existence or nonexistence of further points on the line $ABC$ is undetermined as yet. The analogous class of points in a plane is the system of all points constructible, by means of points and lines, out of four points $A$, $B$, $C$, $D$, no three of which are collinear. This class of points is studied by an indirect method in the next section.

**33. Nets of rationality in the plane.** DEFINITION. A point is said to be *rationally related* to two noncollinear nets of rationality $R_1$, $R_2$ having a point in common, provided it is the intersection of two lines each of which joins a point of $R_1$ to a distinct point of $R_2$. A line is said to be *rationally related* to $R_1$ and $R_2$, provided it joins two points that are rationally related to them. The set of all points and lines rationally related to $R_1$, $R_2$ is called the *net of rationality in a plane (or of two dimensions)* determined by $R_1$, $R_2$; it is also called the *planar net* defined by $R_1$, $R_2$.

From this definition it follows directly that all the points of $R_1$ and $R_2$ are points of the planar net defined by $R_1$, $R_2$.

THEOREM 10. *Any line of the planar net* $\mathsf{R}^2$ *defined by* $\mathsf{R}_1$, $\mathsf{R}_2$ *meets* $\mathsf{R}_1$ *and* $\mathsf{R}_2$. (A, E)

*Proof*. We prove first that if a line of the planar net $\mathsf{R}^2$ meets $\mathsf{R}_1$, it meets $\mathsf{R}_2$. Suppose a line $l$ meets $\mathsf{R}_1$ in $A_1$; it then contains a second point $P$ of $\mathsf{R}^2$. By definition, through $P$ pass two lines, each of which joins a point of $\mathsf{R}_1$ to a distinct point of $\mathsf{R}_2$. If $l$ is one of these lines, the proposition is proved; if these lines are distinct from $l$, let them meet $\mathsf{R}_1$ and $\mathsf{R}_2$ respectively in the points $B_1$, $B_2$ and $P_1$, $P_2$ (fig. 38). If $O$ is the common point of $\mathsf{R}_1$, $\mathsf{R}_2$, we then have

$$OA_1B_1P_1 \overset{P}{\underset{\wedge}{=}} OA_2B_2P_2,$$

where $A_2$ is the point in which $l$ meets the line of $\mathsf{R}_2$. Hence $A_2$ is a point of $\mathsf{R}_2$ (Theorem 7).

Now let $l$ be any line of the net $\mathsf{R}^2$, and let $P$, $Q$ be two points of the net and on $l$ (def.). If one of these points is a point of $\mathsf{R}_1$ or $\mathsf{R}_2$, the theorem is proved by the case just considered. If not, two lines, each joining a point of $\mathsf{R}_1$ to a distinct point of $\mathsf{R}_2$, pass through $P$; let them meet $\mathsf{R}_1$ in $A_1$, $B_1$, and $\mathsf{R}_2$ in $A_2$, $B_2$ respectively (fig. 38). Let the lines $QA_1$ and $QB_1$ meet $\mathsf{R}_2$ in $A_2'$ and $B_2'$ respectively (first case).

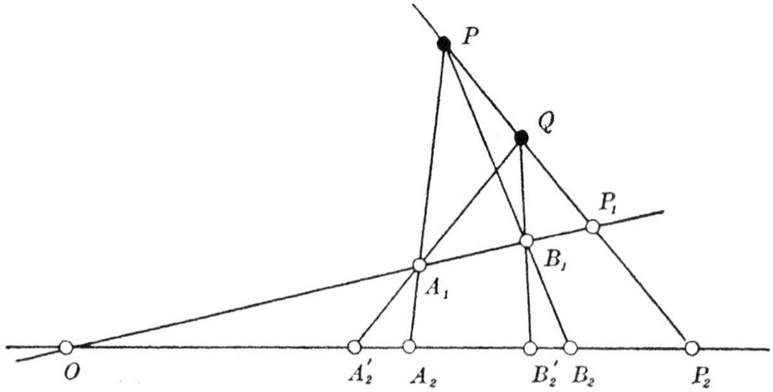


FIG. 38

Then if $l$ meets the lines of $\mathsf{R}_1$ and $\mathsf{R}_2$ in $P_1$ and $P_2$ respectively, the quadrangle $PQA_1B_1$ gives rise to the quadrangular set $\mathsf{Q}(P_2A_2B_2, OB_2'A_2')$ of which five points are points of $\mathsf{R}_2$; hence $P_2$ is a point of $\mathsf{R}_2$ (Theorem 9). $P_1$ is then a point of $\mathsf{R}_1$ by the first case of this proof.

THEOREM 11. *The intersection of any two lines of a planar net is a point of the planar net.* (A, E)

*Proof.* This follows directly from the definition and the last theorem, except when one of the lines passes through $O$, the point common to the two linear nets $R_1$, $R_2$ defining the planar net. In the latter case let the two lines of the planar net be $l_1$, $l_2$ and suppose $l_2$ passes through $O$, while $l_1$ meets $R_1$, $R_2$ in $A_1$, $A_2$ respectively (fig. 39). If the point of intersection $P$ of $l_1 l_2$ were not a point of the planar net, $l_2$ would, by definition, contain a point $Q$ of the planar net, distinct from $O$ and $P$. The lines $QA_1$ and $QA_2$ would meet $R_2$ and $R_1$ in two points $B_2$ and $B_1$ respectively. The point $C_2$ in which the line $PB_1$ met the line of $R_2$ would then be the harmonic conjugate of $B_2$ with respect to $O$ and $A_2$ (through the quadrangle $PQA_1B_1$); $C_2$ would therefore be a point of $R_2$, and hence $P$ would be a point of the planar net, being the intersection of the lines $A_1A_2$ and $B_1C_2$.

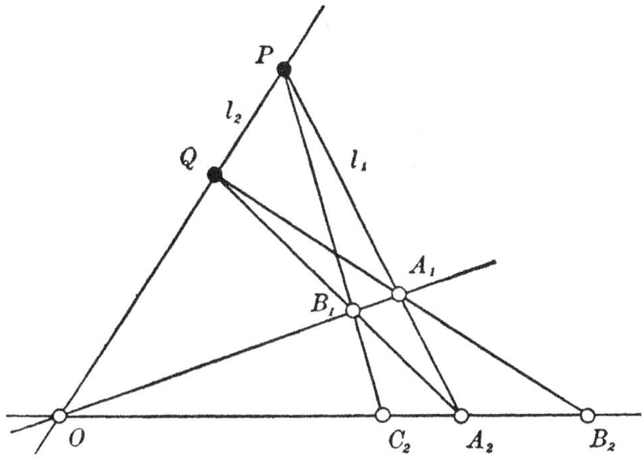


FIG. 39

THEOREM 12. *The points of a planar net* $R^2$ *on a line of the planar net form a linear net.* (A, E)

*Proof.* Let the planar net be defined by the linear nets $R_1$, $R_2$ and let $l$ be any line of the planar net. Let $P$ be any point of the planar net not on $l$ or $R_1$ or $R_2$. The lines joining $P$ to the points of $R^2$ on $l$ meet $R_1$ and $R_2$ by Theorems 10 and 11. Hence $P$ is the center of a perspectivity which makes the points of $R^2$ on $l$ perspective with points of $R_1$ or $R_2$. Hence the points of $l$ belonging to the planar net form a linear net. (Theorem 7, Cor.)

COROLLARY. *The planar net* $R_1^2$ *defined by two linear nets* $R_1$, $R_2$ *is identical with the planar net* $R_2^2$ *defined by two linear nets* $R_3$, $R_4$, *provided* $R_3$, $R_4$ *are linear nets in* $R_1^2$. (A, E)

For every point of $R_1^2$ is a point of $R_2^2$ by the above theorem, and every point of $R_2^2$ is a point of $R_1^2$ by Theorem 10.

**EXERCISE**

If $A$, $B$, $C$, $D$ are the vertices of a complete quadrangle, there is one and only one planar net of rationality containing them; and a point $P$ belongs to this net if and only if $P$ is one of a sequence of points $ABCDD_1D_2 \cdots$, finite in number, such that $D_1$ is the intersection of two sides of the original quadrangle and such that every other point $D_i$ is the intersection of two lines joining pairs of points of the set $ABCDD_1 \cdots D_{i-1}$.

**34. Nets of rationality in space.** DEFINITION. A point is said to be *rationally related* to two planar nets $\mathsf{R}_1^2$, $\mathsf{R}_2^2$ in different planes but having a linear net in common, provided it is the intersection of two lines each of which joins a point of $\mathsf{R}_1^2$ to a distinct point of $\mathsf{R}_2^2$. A line is said to be *rationally related* to $\mathsf{R}_1^2$, $\mathsf{R}_2^2$, if it joins two, a plane if it joins three, points which are rationally related to them. The set of all points, lines, and planes rationally related to $\mathsf{R}_1^2$, $\mathsf{R}_2^2$ is called the *net of rationality in space (or of three dimensions)* determined by $\mathsf{R}_1^2$, $\mathsf{R}_2^2$; it is also called the *spatial net* defined by $\mathsf{R}_1^2$, $\mathsf{R}_2^2$.

Theorems analogous to those derived for planar nets may now be derived for nets of rationality in space. We note first that every point of $\mathsf{R}_1^2$ and of $\mathsf{R}_2^2$ is a point of the spatial net $\mathsf{R}^3$ defined by $\mathsf{R}_1^2$, $\mathsf{R}_2^2$ (the definition applies equally well to the points of the linear net common to $\mathsf{R}_1^2$, $\mathsf{R}_2^2$); and that no other points of the planes of these planar nets are points of $\mathsf{R}^3$. The proofs of the fundamental theorems of alignment, etc., for spatial nets can, for the most part, be readily reduced to theorems concerning planar nets. We note first:

LEMMA. *Any line joining a point $A_1$ of $\mathsf{R}_1^2$ to a distinct point $P$ of $\mathsf{R}^3$ meets $\mathsf{R}_2^2$.* (A, E)

*Proof.* By hypothesis, through $P$ pass two lines, each of which joins a point of $\mathsf{R}_1^2$ to a distinct point of $\mathsf{R}_2^2$. We may assume these lines distinct from the line $PA_1$, since otherwise the lemma is proved. Let the two lines through $P$ meet $\mathsf{R}_1^2$, $\mathsf{R}_2^2$ in $B_1$, $B_2$ and $C_1$, $C_2$ respectively (fig. 40). If $A_1$, $B_1$, $C_1$ are not collinear, the planes $PA_1B_1$ and $PA_1C_1$ meet $\mathsf{R}_1^2$ in the lines $A_1B_1$ and $A_1C_1$ respectively, which meet the linear net common to $\mathsf{R}_1^2$, $\mathsf{R}_2^2$ in two points $S$, $T$ respectively (Theorems 11, 12). The same planes meet the plane of $\mathsf{R}_2^2$ in the lines $SB_2$ and $TC_2$ respectively, which are lines of $\mathsf{R}_2^2$, since $S$, $T$ are points of $\mathsf{R}_2^2$. These lines meet in a point $A_2$ of $\mathsf{R}_2^2$ (Theorem 11), which is evidently the point in which the line $PA_1$ meets the plane of $\mathsf{R}_2^2$. If $A_1$, $B_1$, $C_1$ are collinear, let $A_2$ be the intersection of $PA_1$ with the plane of $\mathsf{R}_2^2$, and $S$ the intersection of $A_1B_1$ with the linear net common to $\mathsf{R}_1^2$ and $\mathsf{R}_2^2$. Since $A_1$ is in $\mathsf{R}(SB_1C_1)$, the perspectivity $SC_1B_1A_1 \overset{P}{\underset{\wedge}{=}} SC_2B_2A_2$ implies that $A_2$ is in $\mathsf{R}(SB_2C_2)$ and hence in $\mathsf{R}_2^2$.

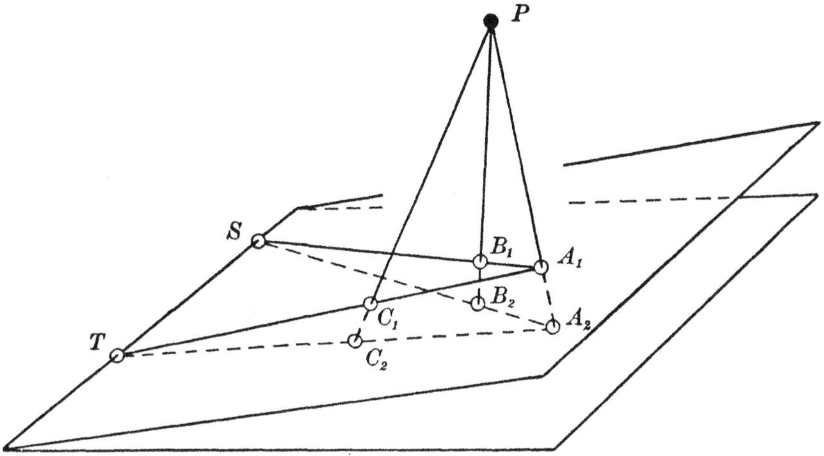


FIG. 40

THEOREM 13. *Any line of the spatial net* $\mathsf{R}^3$ *defined by* $\mathsf{R}_1^2$, $\mathsf{R}_2^2$ *meets* $\mathsf{R}_1^2$ *and* $\mathsf{R}_2^2$. (A, E)

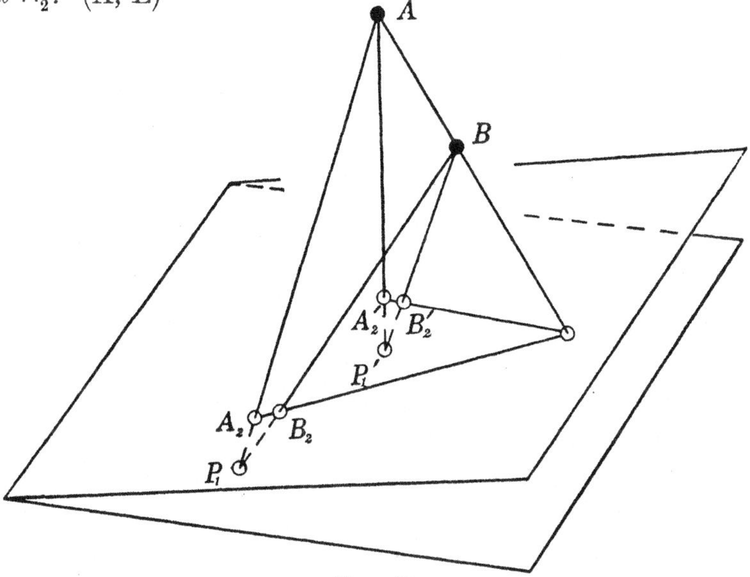


FIG. 41

*Proof.* By definition the given line $l$ contains two points $A$ and $B$ of the net $\mathsf{R}^3$ (fig. 41). If $A$ or $B$ is on $\mathsf{R}_1^2$ or $\mathsf{R}_2^2$, the theorem reduces to the lemma. If not, let $P_1$ be a point of $\mathsf{R}_1^2$, and $A_2$ and $B_2$ the points in which, by the lemma, $P_1A$ and $P_1B$ meet $\mathsf{R}_2^2$; also let $P_1'$ be any point of $\mathsf{R}_1^2$ not in the plane $P_1AB$, and let $P_1'A$ and $P_1'B$ meet $\mathsf{R}_2^2$ in $A_2'$ and $B_2'$. The lines $A_2B_2$ and $A_2'B_2'$ meet in a point of $\mathsf{R}_2^2$ (Theorem 11), and this point is the point of intersection of $l$ with the plane of $\mathsf{R}_2^2$. The argument is now reduced to the case considered in the lemma.

THEOREM 14. *The points of a spatial net lying on a line of the spatial net form a linear net.* (A, E)

*Proof.* Let $l$ be the given line, $\mathsf{R}_1^2$ and $\mathsf{R}_2^2$ the planar nets defining the spatial net $\mathsf{R}^3$, and $L_1$ and $L_2$ the points in which (Theorem 13) $l$ meets $\mathsf{R}_1^2$ and $\mathsf{R}_2^2$ ($L_1$ and $L_2$ may coincide). Let $A_1$ be any point of $\mathsf{R}_1^2$ not on $l$ or on $\mathsf{R}_2^2$, and $S$ the point in which $A_1L_1$ meets the linear net common to $\mathsf{R}_1^2$ and $\mathsf{R}_2^2$ (fig. 42). If $L_1$ and $L_2$ are distinct, the lines

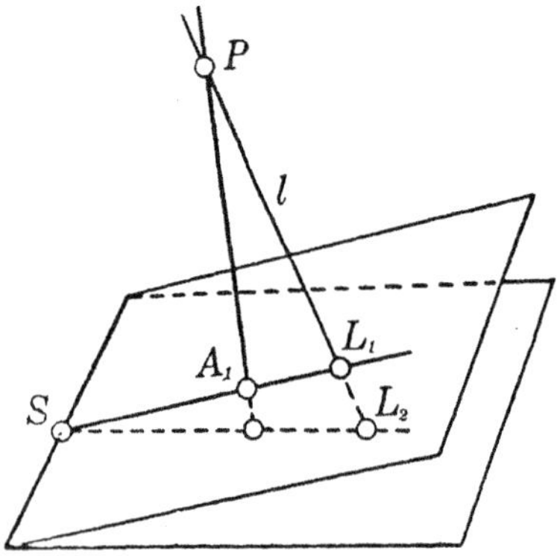


FIG. 42

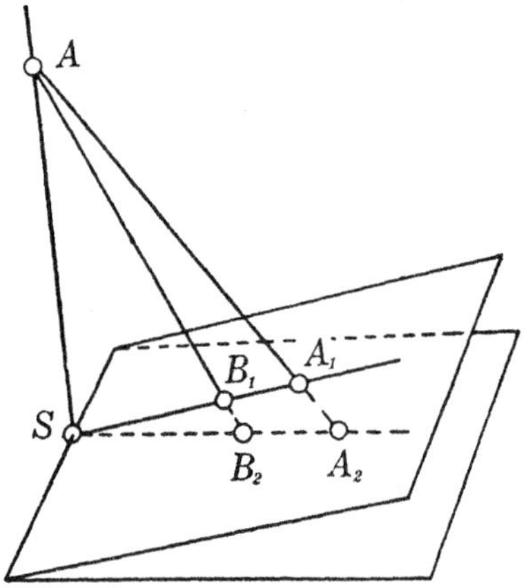


FIG. 43

$SL_1$ and $SL_2$ meet $\mathsf{R}_1^2$ and $\mathsf{R}_2^2$ in linear nets (Theorem 12); and, by Theorem 13, a line joining any point $P$ of $\mathsf{R}^3$ on $l$ to $A_1$ meets each of these linear nets. Hence all points of $\mathsf{R}^3$ on $l$ are in the planar net determined by these two linear nets. Moreover, by the definition of $\mathsf{R}^3$, all the points of the projection from $A_1$ of the linear net on $SL_2$ upon $l$ are points of $\mathsf{R}^3$. Hence the points of $\mathsf{R}^3$ on $l$ are a linear net.

If $L_1 = L_2 = S$, then, by definition, there is on $l$ a point $A$ of $\mathsf{R}^3$, and the line $AA_1$ meets $\mathsf{R}_2^2$ in a point $A_2$ (fig. 43). The lines $SA_1$ and $SA_2$ meet $\mathsf{R}_1^2$ and $\mathsf{R}_2^2$ in linear nets $\mathsf{R}_1$ and $\mathsf{R}_2$ by Theorem 12. If $B_1$ is any point of $\mathsf{R}_1$ other than $A_1$, the line $AB_1$ meets $\mathsf{R}_2^2$ in a point $B_2$ by Theorem 13. By Theorem 12 all points of $l$ in the planar net determined by $\mathsf{R}_1$ and $\mathsf{R}_2$ form a linear net, and they obviously belong to $\mathsf{R}^3$. Moreover, any point of $\mathsf{R}^3$ on $l$, when joined to $A_1$, meets $\mathsf{R}_2^2$ by Theorem 13, and hence belongs to the planar net determined by $\mathsf{R}_1$ and $\mathsf{R}_2$. Hence, in this case also, the points of $\mathsf{R}^3$ on $l$ constitute a linear net.

THEOREM 15. *The points and lines of a spatial net* $R^3$ *which lie on a plane* $\alpha$ *of the net form a planar net.* (A, E)

*Proof.* By definition $\alpha$ contains three noncollinear points $A, B, C$ of $R^3$, and the three lines $AB$, $BC$, $CA$ meet the planar nets $R_1^2$ and $R_2^2$, which determine $R^3$, in points of two linear nets $R_1$ and $R_2$, consisting entirely of points of $R^3$. These linear nets, if distinct, determine a planar net $R^2$ in $\alpha$, which, by Theorem 10, consists entirely of points and lines of $R^3$. Moreover, any line joining a point of $R^3$ in $\alpha$ to $A$ or $B$ or $C$ must, by Theorem 13, meet $R_1$ and $R_2$ and hence be in $R^2$. Hence all points and lines of $R^3$ on $\alpha$ are points and lines of $R^2$. This completes the proof except in case $R_1 = R_2$, which case is left as an exercise.

COROLLARY 1. *A net of rationality in space is a space satisfying Assumptions* A *and* E, *if "line" be interpreted as "linear net" and "plane" as "planar net."* (A, E)

For all assumptions A and E, except A 3, are evidently satisfied; and A 3 is satisfied because there is a planar net of points through any three points of a spatial net $R^3$, and any two linear nets of this planar net have a point in common.

This corollary establishes at once all the theorems of alignment in a net of rationality in space, which are proved in Chap. I, as also the principle of duality. We conclude then, for example, that two planes of a spatial net meet in a line of the net, and that three planes of a spatial net meet in a point of the net (if they do not meet in a line), etc. Moreover, we have at once the following corollary:

COROLLARY 2. *A spatial net is determined by any two of its planar nets.* (A, E)

## EXERCISES

**1.** If $A, B, C, D, E$ are the vertices of a complete space five-point, there is one and only one net of rationality containing them all. A point $P$ belongs to this net if and only if $P$ is one of a sequence of points $ABCDEI_1I_2 \cdots$, finite in number, such that $I_1$ is the point of intersection of three faces of the original five-point and every other point $I_i$ is the intersection of three distinct planes through triples of points of the set $ABCDEI_1 \cdots I_{i-1}$.

**2.** Show that a planar net is determined if three noncollinear points and a line not passing through any of these points are given.

**3.** Under what condition is a planar net determined by a linear net and two points not in this net? Show that two distinct planar nets in the same plane can have at most a linear net and one other point in common.

**4**. Show that a set of points and lines which is projective with a planar net is a planar net.

**5**. A line joining a point $P$ of a planar net to any point not in the net, but on a line of the net not containing $P$, has no other point than $P$ in common with the net.

**6**. Two points and two lines in the same plane do not in general belong to the same planar net.

**7**. Discuss the determination of spatial nets by points and planes, similarly to Exs. 2, 3, and 6.

**8**. Any class of points projective with a spatial net is itself a spatial net.

**9**. If a perspective collineation (homology or elation) in a plane with center $A$ and axis $l$ leaves a net of rationality in the plane invariant, the net contains $A$ and $l$.

**10**. Prove the corresponding proposition for a net of rationality in space invariant under a perspective transformation.

**11**. Show that two linear nets on skew lines always belong to some spatial net; in fact, that the number of spatial nets containing two given linear nets on skew lines is the same as the number of linear nets through two given points.

**12**. Three mutually skew lines and three distinct points on one of them determine one and only one spatial net in which they lie.

**13**. Give further examples of the determination of spatial nets by lines.

**35. The fundamental theorem of projectivity.** It has been shown (Chap. III) that any three distinct elements of a one-dimensional form may be made to correspond to any three distinct points of a line by a projective transformation. Likewise any four elements of a two-dimensional form, no three of which belong to the same one-dimensional form, may be made to correspond to the vertices of a complete planar quadrangle by a projective transformation; and any five elements of a three-dimensional form, no four of which belong to the same two-dimensional form, may be made to correspond to the five vertices of a complete spatial five-point by a projective transformation.

These transformations are of the utmost importance. Indeed, it is the principal object of projective geometry to discover those properties of figures which remain invariant when the figures are subjected to projective transformations. The question now naturally arises, Is it possible to transform any four elements of a one-dimensional form into any four elements of another one-dimensional form? This question must be answered in the negative, since a harmonic set must always correspond to a harmonic set. The question then arises whether or not a projective correspondence between one-dimensional forms is completely determined when three pairs of homologous elements are given. A partial answer to this fundamental question is given in the next theorem.

LEMMA 1. *If a projectivity leaves three distinct points of a line fixed, it leaves fixed every point of the linear net defined by these points.*

This follows at once from the fact that if three points are left invariant by a projectivity, the harmonic conjugate of any one of these points with respect to the other two must also be left invariant by the projectivity (Theorems 2 and 3, Cor.). The projectivity in question must therefore leave invariant every point harmonically related to the three given points.

THEOREM 16. THE FUNDAMENTAL THEOREM OF PROJECTIVITY FOR A NET OF RATIONALITY ON A LINE. *If $A$, $B$, $C$, $D$ are distinct points of a linear net of rationality, and $A'$, $B'$, $C'$ are any three distinct points of another or the same linear net, then for any projectivities giving* $ABCD \barwedge A'B'C'D'$ *and* $ABCD \barwedge A'B'C'D'_1$, *we have* $D' = D'_1$. (A, E)

*Proof.* If $\pi$, $\pi_1$ are respectively the two projectivities of the theorem, the projectivity $\pi_1\pi^{-1}$ leaves $A'B'C'$ fixed and transforms $D'$ into $D'_1$. Since $D'$ is harmonically related to $A'$, $B'$, $C'$ (Theorem 7), the theorem follows from the lemma.

This theorem gives the answer to the question proposed in its relation to the transformation of the points of a linear net. The corresponding proposition for all the points of a line, i.e. the proposition obtained from the last theorem by replacing "linear net" by "line," cannot be proved without the use of one or more additional assumptions (cf. § 50, Chap. VI). We have seen that it is equivalent to the proposition: If a projectivity leaves three points of a line invariant, it leaves every point of the line invariant. Later, by means of a discussion of order and continuity (terms as yet undefined), we shall prove this proposition. This discussion of order and continuity is, however, somewhat tedious and more difficult than the rest of our subject; and, besides, the theorem in question is true in spaces,* where order and continuity do not exist. It has

* Different, of course, from ordinary space; "rational spaces" (cf. p. 98 and the next footnote) are examples in which continuity does not exist; "finite spaces," of which examples are given in the introduction (§ 2), are spaces in which neither order nor continuity exists.

therefore seemed desirable to give some of the results of this theorem before giving its proof in terms of order and continuity. To this end we introduce here the following *provisional assumption of projectivity*, which will later be proved a consequence of the order and continuity assumptions which will replace it. This provisional assumption may take any one of several forms. We choose the following as leading most directly to the desired theorem:

AN ASSUMPTION OF PROJECTIVITY:

P. *If a projectivity leaves each of three distinct points of a line invariant, it leaves every point of the line invariant.**

We should note first that the plane and space duals of this assumption are immediate consequences of the assumption. The principle of duality, therefore, is still valid after our set of assumptions has been enlarged by the addition of Assumption P.

We now have:

THEOREM 17. THE FUNDAMENTAL THEOREM OF PROJECTIVE GEOMETRY.† *If 1, 2, 3, 4 are any four elements of a one-dimensional primitive form, and 1′, 2′, 3′ are any three elements of another or the same one-dimensional primitive form, then for any projectivities giving* $1234 \barwedge 1'2'3'4'$ *and* $1234 \barwedge 1'2'3'4_1'$, *we have* $4' = 4_1'$. (A, E, P)

*Proof.* The proof is the same under the principle of duality as that of Theorem 16, Assumption P replacing the previous lemma.

This theorem may also be stated as follows:

*A projectivity between one-dimensional primitive forms is uniquely determined when three pairs of homologous elements are given.* (A, E, P)

COROLLARY. *If two pencils of points on different lines are projective and have a self-corresponding point, they are perspective.* (A, E, P)

* We have seen in the lemma of the preceding theorem that the projectivity described in this assumption leaves invariant every point of the net of rationality defined by the three given points. The assumption simply states that if all the points of a linear net remain invariant under a projective transformation, then all the points of the line containing this net must also remain invariant. It will be shown later that in the ordinary geometry the points of a linear net of rationality on a line correspond to the points of the line whose coördinates, when represented analytically, are rational numbers. This consideration should make the last assumption almost, if not quite, as intuitionally acceptable as the previous Assumptions A and E.

† On this theorem and related questions there is an extensive literature to which references can be found in the Encyklopädie articles on Projective Geometry and Foundations of Geometry. It is associated with the names of von Staudt, Klein, Zeuthen, Lüroth, Darboux, F. Schur, Pieri, Wiener, Hilbert. Cf. also § 50, Chap. VI.

*Proof.* For if $O$ is the self-corresponding point, and $AA'$ and $BB'$ are any two pairs of homologous points distinct from $O$, the perspectivity whose center is the intersection of the lines $AA'$, $BB'$ is a projectivity between the two lines which has the three pairs of homologous points $OO$, $AA'$, $BB'$, which must be the projectivity of the corollary by virtue of the last theorem.

The corresponding theorems for two- and three-dimensional forms are now readily derived. We note first, as a lemma, the propositions in a plane and in space corresponding to Assumption P.

LEMMA 2. *A projective transformation which leaves invariant each of a set of* $\genfrac{}{}{0pt}{}{\textit{four}}{\textit{five}}$ *points of* $\genfrac{}{}{0pt}{}{\textit{a plane}}{\textit{space}}$ *no* $\genfrac{}{}{0pt}{}{\textit{three}}{\textit{four}}$ *of which belong to the same* $\genfrac{}{}{0pt}{}{\textit{line}}{\textit{plane}}$ *leaves invariant every point of* $\genfrac{}{}{0pt}{}{\textit{the plane.}}{\textit{space.}}$ (A, E, P)

*Proof.* If $A$, $B$, $C$, $D$ are four points of a plane no three of which are collinear, a projective transformation leaving each of them invariant must also leave the intersection $O$ of the lines $AB$, $CD$ invariant. By Assumption P it then leaves every point of each of the lines $AB$, $CD$ invariant. Any line of the plane which meets the lines $AB$ and $CD$ in two distinct points is therefore invariant, as well as the intersection of any two such lines. But any point of the plane may be determined as the intersection of two such lines. The proof for the case of a projective transformation leaving invariant five points no four of which are in the same plane is entirely similar. The existence of perspective collineations shows that the condition that no three (four) of the points shall be on the same line (plane) is essential.

THEOREM 18. *A projective collineation* * *between two planes (or within a single plane) is uniquely determined when four pairs of homologous points are given, provided no three of either set of four points are collinear.* (A, E, P)

*Proof.* Suppose there were two collineations $\pi$, $\pi_1$ having the given pairs of homologous points. The collineation $\pi_1\pi^{-1}$ is then, by the lemma, the identical collineation in one of the planes. This gives at once $\pi_1 = \pi$, contrary to the hypothesis.

* We confine the statement to the case of the collineation for the sake of simplicity of enunciation. Projective transformations which are not collineations will be discussed in detail later, at which time attention will be called explicitly to the fundamental theorem.

By precisely similar reasoning we have:

THEOREM 19. *A projective collineation in space is uniquely determined when five pairs of homologous points are given, provided no four of either set of five points are in the same plane.* (A, E, P)

The fundamental theorem deserves its name not only because so large a part of projective geometry is logically connected with it, but also because it is used explicitly in so many arguments. It is indeed possible to announce a general course of procedure that appears in the solution of most "linear" problems, i.e. problems which depend on constructions involving points, lines, and planes only. If it is desired to prove that certain three lines $l_1, l_2, l_3$ pass through a point, find two other lines $m_1, m_2$ such that the four points $m_1l_1, m_1l_2, m_1l_3, m_1m_2$ may be shown to be projective with the four points $m_2l_1, m_2l_2, m_2l_3, m_2m_1$ respectively. Then, since in this projectivity the point $m_1m_2$ is self-corresponding, the three lines $l_1, l_2, l_3$ joining corresponding points are concurrent (Theorem 17, Cor.). The dual of this method appears when three points are to be shown collinear. This method may be called the *principle of projectivity*, and takes its place beside the principle of duality as one of the most powerful instruments of projective geometry. The theorems of the next section may be regarded as illustrations of this principle. They are all propositions from which the principle of projectivity could be derived, i.e. they are propositions which might be chosen to replace Assumption P.

We have already said that ordinary real (or complex) space is a space in which Assumption P is valid. Any such space we call a *properly projective space.* It will appear in Chap. VI that there exist spaces in which this assumption is not valid. Such a space, i.e. a space satisfying Assumptions A and E but not P, we will call an *improperly projective space.*

From Theorem 15, Cor. 1 and Lemma 1, we then have

THEOREM 20. *A net of rationality in space is a properly projective space.* (A, E)

It should here be noted that if we added to our list of Assumptions A and E another assumption of closure, to the effect that all points of space belong to the same net of rationality, we should obtain a space in which all our previous theorems are valid, including the fundamental theorem (without using Assumption P).

Such a space may be called a *rational space*. In general, it is clear that any complete five-point in any properly or improperly projective space determines a subspace which is rational and therefore properly projective.

**36. The configuration of Pappus. Mutually inscribed and circumscribed triangles.**

THEOREM 21. *If $A$, $B$, $C$ are any three distinct points of a line $l$, and $A'$, $B'$, $C'$ any three distinct points of another line $l'$ meeting $l$, the three points of intersection of the pairs of lines $AB'$ and $A'B$, $BC'$ and $B'C$, $CA'$ and $C'A$ are collinear.* (A, E, P)

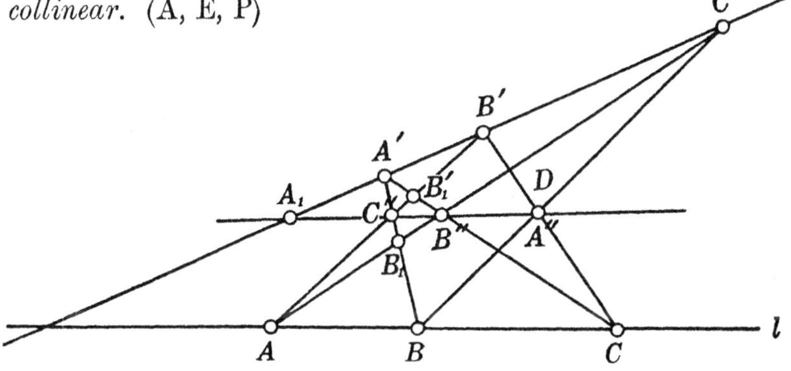


FIG. 44

*Proof.* Let the three points of intersection referred to in the theorem be denoted by $C''$, $A''$, $B''$ respectively (fig. 44). Let the line $B''C''$ meet the line $B'C$ in a point $D$ (to be proved identical with $A''$); also let $B''C''$ meet $l'$ in $A_1$, the line $A'B$ meet $AC'$ in $B_1$, the line $AB'$ meet $A'C$ in $B_1'$. We then have the following perspectivities:

$$A'C''B_1B \overset{A}{\underset{\wedge}{=}} A'B_1'B''C \overset{B'}{\underset{\wedge}{=}} A_1C''B''D.$$

By the principle of projectivity then, since in the projectivity thus established $C''$ is self-corresponding, we conclude that the three lines $A_1A'$, $B''B_1$, $DB$ meet in the point $C'$. Hence $D$ is identical with $A''$, and $A''$, $B''$, $C''$ are collinear.

It should be noted that the figure of the last theorem is a configuration of the symbol

| | |
|---|---|
| 9 | 3 |
| 3 | 9 |

It is known as the *configuration of Pappus*.* It should also be noted that this configuration may be considered as a simple plane hexagon (six-point) inscribed in two intersecting lines. If the sides of such a hexagon be denoted in order by 1, 2, 3, 4, 5, 6, and if we call the sides 1 and 4 opposite, likewise the sides 2 and 5, and the sides 3 and 6 (cf. Chap. II, § 14), the last theorem may be stated in the following form:

COROLLARY. *If a simple hexagon be inscribed in two intersecting lines, the three pairs of opposite sides will intersect in collinear points.*†

Finally, we may note that the nine points of the configuration of Pappus may be arranged in sets of three, the sets forming three triangles, 1, 2, 3, such that 2 is inscribed in 1, 3 in 2, and 1 in 3. This observation leads to another theorem connected with the Pappus configuration.

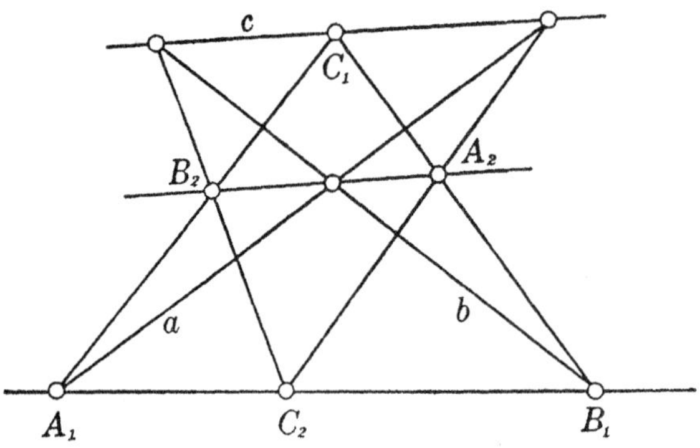

FIG. 45

THEOREM 22. *If $A_2B_2C_2$ be a triangle inscribed in a triangle $A_1B_1C_1$, there exists a certain set of triangles each of which is inscribed in the former and circumscribed about the latter.* (A, E, P)

*Proof.* Let $[a]$ be the pencil of lines with center $A_1$; $[b]$ the pencil with center $B_1$; and $[c]$ the pencil with center $C_1$ (fig. 45). Consider the perspectivities $[a] \overset{B_2A_2}{\overline{\overline{\wedge}}} [b] \overset{B_2C_2}{\overline{\overline{\wedge}}} [c]$. In the projectivity thus established between $[a]$ and $[c]$ the line $A_1C_1$ is self-corresponding; the pencils of lines $[a]$, $[c]$ are therefore perspective (Theorem 17, Cor. (dual)). Moreover, the axis of this perspectivity is $C_2A_2$; for the lines $A_1C_2$ and $C_1C_2$ are clearly homologous, as also the lines $A_1A_2$ and $C_1A_2$. Any three homologous lines of the perspective pencils $[a]$, $[b]$, $[c]$ then form a triangle which is circumscribed about $A_1B_1C_1$ and inscribed in $A_2B_2C_2$.

* Pappus, of Alexandria, lived about 340 A.D. A special case of this theorem may be proved without the use of the fundamental theorem (cf. Ex. 3, p. 52).

† In this form it is a special case of Pascal's theorem on conic sections (cf. Theorem 3, Chap. V).

## EXERCISES

**1.** Given a triangle $ABC$ and two distinct points $A'$, $B'$; determine a point $C'$ such that the lines $AA'$, $BB'$, $CC'$ are concurrent, and also the lines $AB'$, $BC'$, $CA'$ are concurrent, i.e. such that the two triangles are perspective from two different points. The two triangles are then said to be doubly perspective.

**2.** If two triangles $ABC$ and $A'B'C'$ are doubly perspective in such a way that the vertices $A$, $B$, $C$ are homologous with $A'$, $B'$, $C'$ respectively in one perspectivity and with $B'$, $C'$, $A'$ respectively in the other, they will also be perspective from a third point in such a way that $A$, $B$, $C$ are homologous respectively with $C'$, $A'$, $B'$; i.e. they will be triply perspective.

**3.** Show that if $A''$, $B''$, $C''$ are the centers of perspectivity for the triangles in Ex. 2, the three triangles $ABC$, $A'B'C'$, $A''B''C''$ are so related that any two are triply perspective, the centers of perspectivity being in each case the vertices of the remaining triangle. The nine vertices of the three triangles form the points of one configuration of Pappus, and the nine sides form the lines of another configuration of Pappus.

### 37. Construction of projectivities on one-dimensional forms.

THEOREM 23. *A necessary and sufficient condition for the projectivity on a line* $MNAB \barwedge MNA'B'$ $(M \neq N)$ *is* $Q(MAB, NB'A')$. (A, E, P)

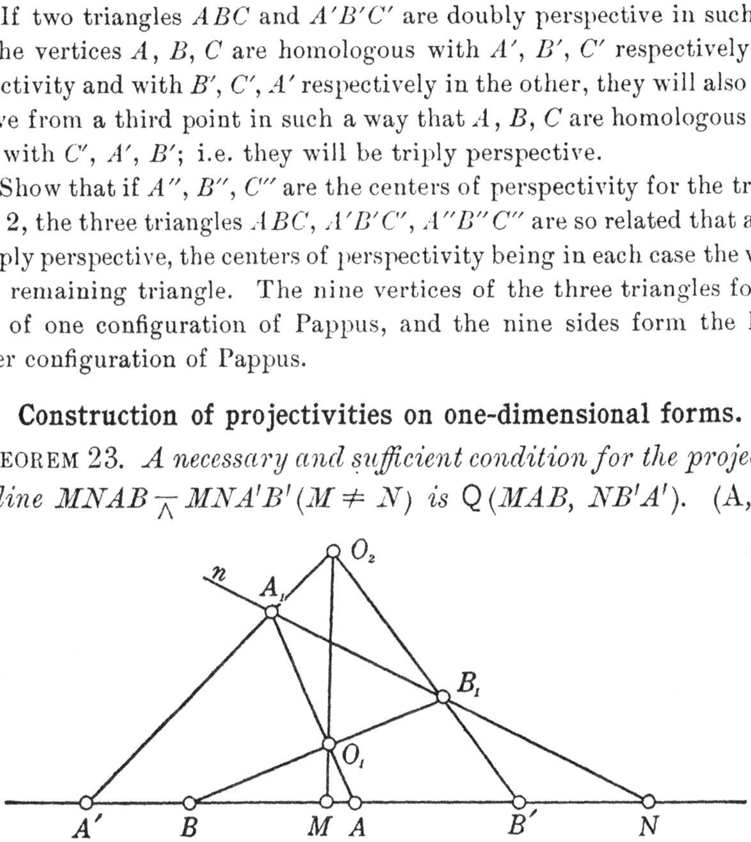


FIG. 46

*Proof.* Let $n$ be any line on $N$ not passing through $A$ (fig. 46). Let $O_1$ be any point not on $n$ or on $MA$, and let $A_1$ and $B_1$ be the intersections respectively of $O_1A$ and $O_1B$ with $n$. Let $O_2$ be the intersection of $A'A_1$ and $B'B_1$. Then

$$NAB \overset{O_1}{\underset{\wedge}{=}} NA_1B_1 \overset{O_2}{\underset{\wedge}{=}} NA'B'.$$

By Theorem 17 the projectivity so determined on the line $AM$ is the same as

$$MNAB \barwedge MNA'B'.$$

The only possible double points of the projectivity are $N$ and the intersection of $AN$ with $O_1O_2$. Hence $O_1O_2$ passes through $M$, and $Q(MAB, NB'A')$ is determined by the quadrangle $O_1O_2A_1B_1$.

Conversely, if $Q(MAB, NB'A')$ we have a quadrangle $O_1O_2A_1B_1$, and hence

$$NAB \overset{O_1}{\underset{\wedge}{=}} NA_1B_1 \overset{O_2}{\underset{\wedge}{=}} NA'B',$$

and by this construction $M$ is self-corresponding, so that

$$MNAB \barwedge MNA'B'.$$

If in the above construction we have $M = N$, we obtain a projectivity with the single double point $M = N$.

Definition. A projectivity on a one-dimensional primitive form with a single double element is called *parabolic*. If the double element is $M$, and $AA'$, $BB'$ are any two homologous pairs, the projectivity is completely determined and is conveniently represented by $MMAB \barwedge MMA'B'$.

Corollary. *A necessary and sufficient condition for a parabolic projectivity* $MMAB \barwedge MMA'B'$ *is* $Q(MAB, MB'A')$. (A, E, P)

Theorem 24. *If we have*

$$Q(ABC, A'B'C'),$$

*we have also* $\quad Q(A'B'C', ABC)$.

*Proof.* By the theorem above,

$$Q(ABC, A'B'C')$$

implies $\quad AA'BC \barwedge AA'C'B'$,

which is the inverse of $\quad A'AB'C' \barwedge A'ACB$,

which, by the theorem above, implies

$$Q(A'B'C', ABC).$$

The notation $Q(ABC, A'B'C')$ implies that $A, B, C$ are the traces of a point triple of sides of the quadrangle determining the quadrangular set. The theorem just proved states the existence of another quadrangle for which $A', B', C'$ are a point triple, and consequently $A, B, C$ are a triangle triple. This theorem therefore establishes the existence of oppositely placed quadrangles, as stated in § 19, p. 50. This result can also be propounded as follows:

Theorem 25. *If two quadrangles* $P_1P_2P_3P_4$ *and* $Q_1Q_2Q_3Q_4$ *are so related — $P_1$ to $Q_1$, $P_2$ to $Q_2$, etc. — that five of the sides* $P_iP_j$ $(i, j = 1, 2, 3, 4;$ $i \neq j)$ *meet the five sides of the second which are opposite to* $Q_iQ_j$ *in points of a line $l$, the remaining sides of the two quadrangles meet on $l$.* (A, E, P)

*Proof.* The sides of the first quadrangle meet $l$ in a quadrangular set $Q(P_{12}P_{13}P_{14}, P_{34}P_{24}P_{23})$; hence $Q(P_{34}P_{24}P_{23}, P_{12}P_{13}P_{14})$. But, by hypothesis, five of the sides of the second quadrangle pass through these points as follows: $Q_1Q_2$ through $P_{34}$, $Q_1Q_3$ through $P_{24}$, $Q_1Q_4$ through $P_{23}$, $Q_3Q_4$ through $P_{12}$, $Q_4Q_2$ through $P_{13}$, $Q_3Q_2$ through $P_{14}$. As five of these conditions are satisfied, by Theorem 3, Chap. II, they must all be satisfied.

### EXERCISES

**1.** Given one double point of a projectivity on a line and two pairs of homologous points, construct the other double point.

**2.** If $a, b, c$ are three nonconcurrent lines and $A', B', C'$ are three collinear points, give a construction for a triangle whose vertices $A, B, C$ are respectively on the given lines and whose sides $BC, CA, AB$ pass respectively through the given points. What happens when the three lines $a, b, c$ are concurrent? Dualize.

**38. Involutions.** DEFINITION. If a projectivity in a one-dimensional form is of period two, it is called an *involution*. Any pair of homologous points of an involution is called a *conjugate pair* of the involution or a *pair of conjugates*.

It is clear that if an involution transforms a point $A$ into a point $A'$, then it also transforms $A'$ into $A$; this is expressed by the phrase that the points $A, A'$ *correspond to each other doubly*. The effect of an involution is then simply a pairing of the elements of a one-dimensional form such that each element of a pair corresponds to the other element of the pair. This justifies the expression "a conjugate pair" applied to an involution.

THEOREM 26. *If for a single point $A$ of a line which is not a double point of a projectivity $\pi$ on the line we have the relations $\pi(A) = A'$ and $\pi(A') = A$, the projectivity is an involution.* (A, E, P)

*Proof.* For suppose $P$ is any other point on the line (not a double point of $\pi$), and suppose $\pi(P) = P'$. There then exists a projectivity giving

$$AA'PP' \barwedge A'AP'P$$

(Theorem 2, Chap. III). By Theorem 17 this projectivity is $\pi$, since it has the three pairs of homologous points $A, A'$; $A', A$; $P, P'$. But in this projectivity $P'$ is transformed into $P$. Thus every pair of homologous points corresponds doubly.

COROLLARY. *An involution is completely determined when two pairs of conjugate points are given.* (A, E, P)

THEOREM 27. *A necessary and sufficient condition that three pairs of points $A, A'$; $B, B'$; $C, C'$ be conjugate pairs of an involution is* $\mathsf{Q}(ABC, A'B'C')$. (A, E, P)

*Proof.* By hypothesis we have

$$AA'BC \,\overline{\wedge}\, A'AB'C'.$$

By Theorem 2, Chap. III, we also have

$$A'AB'C' \,\overline{\wedge}\, AA'C'B',$$

which, with the first projectivity, gives

$$AA'BC \,\overline{\wedge}\, AA'C'B'.$$

A necessary and sufficient condition that the latter projectivity hold is $\mathsf{Q}(ABC, A'B'C')$ (Theorem 23).

COROLLARY 1. *If an involution has double points, they are harmonic conjugates with respect to every pair of the involution.* (A, E, P)

For the hypothesis $A = A'$, $B = B'$ gives at once $\mathsf{H}(AB, CC')$ as the condition of the theorem.

COROLLARY 2. *An involution is completely determined when two double points are given, or when one double point and one pair of conjugates are given.* (A, E, P)

COROLLARY 3. *If $M, N$ are distinct double points of a projectivity on a line, and $A, A'$; $B, B'$ are any two pairs of homologous elements, the pairs $M, N$; $A, B'$; $A', B$ are conjugate pairs of an involution.** (A, E, P)

COROLLARY 4. *If an involution has one double element, it has another distinct from the first.* (A, E, $H_0$, P)

COROLLARY 5. *The projectivity $ABCD \,\overline{\wedge}\, ABDC$ between four distinct points of a line implies the relation* $\mathsf{H}(AB, CD)$. (A, E, P)

For the projectivity is an involution (Theorem 26) of which $A, B$ are double points. The result then follows from Cor. 1.

**39. Axis and center of homology.**

THEOREM 28. *If $[A]$ and $[B]$ are any two projective pencils of points in the same plane on*

THEOREM 28′. *If $[l]$ and $[m]$ are any two projective pencils of lines in the same plane on distinct*

* This relation is sometimes expressed by saying, "The pairs of points are *in involution.*" From what precedes it is clear that any two pairs of elements of a one-dimensional form are in involution, but in general three pairs are not.

*distinct lines* $l_1$, $l_2$, *there exists a line* $l$ *such that if* $A_1, B_1$ *and* $A_2, B_2$ *are any two pairs of homologous points of the two pencils, the lines* $A_1B_2$ *and* $A_2B_1$ *intersect on* $l$. (A, E, P)

DEFINITION. The line $l$ is called the *axis of homology* of the two pencils of points.

*points* $S_1$, $S_2$, *there exists a point* $S$ *such that if* $a_1$, $b_1$ *and* $a_2$, $b_2$ *are any two pairs of homologous lines of the two pencils, the points* $a_1b_2$ *and* $a_2b_1$ *are collinear with* $S$. (A, E, P)

DEFINITION. The point $S$ is called the *center of homology* of the pencils of lines.

*Proof.* The two theorems being plane duals of each other, we may confine ourselves to the proof of the theorem on the left. From the projectivity $[B] \overline{\wedge} [A]$ follows $A_1[B] \overline{\wedge} B_1[A]$ (fig. 47). But in this projectivity the line $A_1B_1$ is self-corresponding, so that (Theorem 17, Cor.)

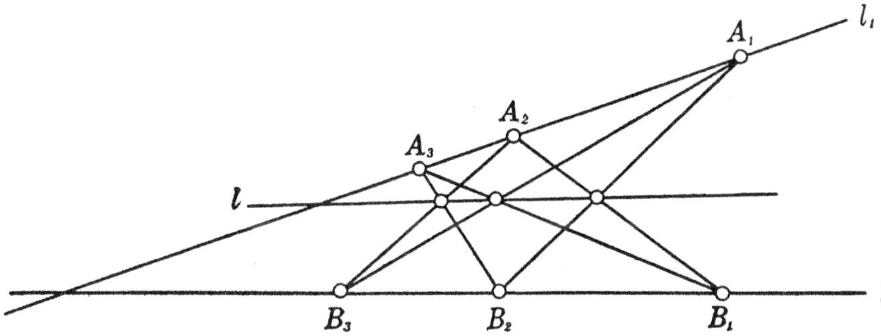


FIG. 47

the two pencils are perspective. Hence pairs of corresponding lines meet on a line $l$; e.g. the lines $A_1B_3$ and $B_1A_3$ meet on $l$ as well as $A_1B_2$ and $B_1A_2$. To prove our theorem it remains only to show that $B_2A_3$ and $A_2B_3$ also meet on $l$. But the latter follows at once from Theorem 21, since the figure before us is the configuration of Pappus.

COROLLARY. *If* $[A]$, $[B]$ *are not perspective, the axis of homology is the line joining the points homologous with the point* $l_1l_2$ *regarded first as a point of* $l_1$ *and then as a point of* $l_2$.

COROLLARY. *If* $[l]$, $[m]$ *are not perspective, the center of homology is the point of intersection of the lines homologous with the line* $S_1S_2$ *regarded first as a line of* $[l]$ *and then as a line of* $[m]$.

For in the perspectivity $A_1[B] \overset{l}{\underset{\wedge}{=}} B_1[A]$ the line $l_1$ corresponds to $B_1(ll_1)$, and hence the point $l_1l_2$ corresponds to $ll_1$ in the projectivity $[B] \overline{\wedge} [A]$. Similarly, $ll_2$ corresponds to $l_1l_2$.

### EXERCISES

**1**. There is one and only one projectivity of a one-dimensional form leaving invariant one and only one element $O$, and transforming a given other element $A$ to an element $B$.

**2**. Two projective ranges on skew lines are always perspective.

**3**. Prove Cor. 5, Theorem 27, without using the notion of involution.

**4**. If $MNAB \barwedge MNA'B'$, then $MNAA' \barwedge MNBB'$.

**5**. If $P$ is any point of the axis of homology of two projective ranges $[A] \barwedge [B]$, then the projectivity $P[A] \barwedge P[B]$ is an involution. Dualize.

**6**. Call the faces of one tetrahedron $\alpha_1, \alpha_2, \alpha_3, \alpha_4$ and the opposite vertices $A_1, A_2, A_3, A_4$ respectively, and similarly the faces and vertices of another tetrahedron $\beta_1, \beta_2, \beta_3, \beta_4$ and $B_1, B_2, B_3, B_4$. If $A_1, A_2, A_3, A_4$ lie on $\beta_1, \beta_2, \beta_3, \beta_4$ respectively, and $B_1$ lies on $\alpha_1$, $B_2$ on $\alpha_2$, $B_3$ on $\alpha_3$, then $B_4$ lies on $\alpha_4$. Thus each of the two tetrahedra related in this fashion is both inscribed and circumscribed to the other.

**7**. Prove the theorem of Desargues (Chap. II) by the principle of projectivity.

**8**. Given a triangle $ABC$ and a point $A'$, show how to construct two points $B'$, $C'$ such that the triangles $ABC$ and $A'B'C'$ are perspective from four different centers.

**9**. If two triangles $A_1B_1C_1$ and $A_2B_2C_2$ are perspective, the three points

$$(A_1B_2,\ A_2B_1) = C_3,\ (A_1C_2,\ A_2C_1) = B_3,\ (B_1C_2,\ B_2C_1) = A_3,$$

if not collinear, form a triangle perspective with the first two, and the three centers of perspectivity are collinear.

* **10**. (*a*) If $\pi$ is a projectivity in a pencil of points $[A]$ on a line $a$ with invariant points $A_1, A_2$, and if $[L]$, $[M]$ are the pencils of points on two lines $l$, $m$ through $A_1, A_2$ respectively, show by the methods of Chap. III that there exist three points $S_1, S_2, S_3$ such that we have

$$[A] \overset{S_1}{\doublebarwedge} [L] \overset{S_2}{\doublebarwedge} [M] \overset{S_3}{\doublebarwedge} [A'],$$

where $\pi(A) = A'$; that $S_1, S_2, A_2$ are collinear; and that $S_2, S_3, A_1$ are collinear.

(*b*) Using the fundamental theorem, show that there exists on the line $S_1A_2$ a point $S$ such that we have

$$[A] \overset{S_1}{\doublebarwedge} [L] \overset{S}{\doublebarwedge} [A'].$$

(*c*) Show that (*b*) could be used as an assumption of projectivity instead of Assumption P; i.e. $P$ could be replaced by: If $\pi$ is a projectivity with fixed points $A_1, A_2$, giving $\pi(A) = A'$ in a pencil of points $[A]$, and $[L]$ is a pencil of points on a line $l$ through $A_1$, there exist two points $S_1, S_2$ such that

$$[A] \overset{S_1}{\doublebarwedge} [L] \overset{S_2}{\doublebarwedge} [A'].$$

* **11**. Show that Assumption P could be replaced by the corollary of Theorem 17.

* **12**. Show that Assumption P could be replaced by the following: If we have a projectivity in a pencil of points defined by the perspectivities

$$[X] \overset{S_1}{\underset{\wedge}{=}} [L] \overset{S_2}{\underset{\wedge}{=}} [X'],$$

and $[M]$ is the pencil of points on the line $S_1S_2$, there exist on the base of $[L]$ two points $S_1'$, $S_2'$ such that we have also

$$[X] \overset{S_1'}{\underset{\wedge}{=}} [M] \overset{S_2'}{\underset{\wedge}{=}} [X'].$$

**40. Types of collineations in the plane.** We have seen in the proof of Theorem 10, Chap. III, that if $O_1O_2O_3$ is any triangle, there exists a collineation $\Pi$ leaving $O_1$, $O_2$, and $O_3$ invariant, and transforming any point not on a side of the triangle into any other such

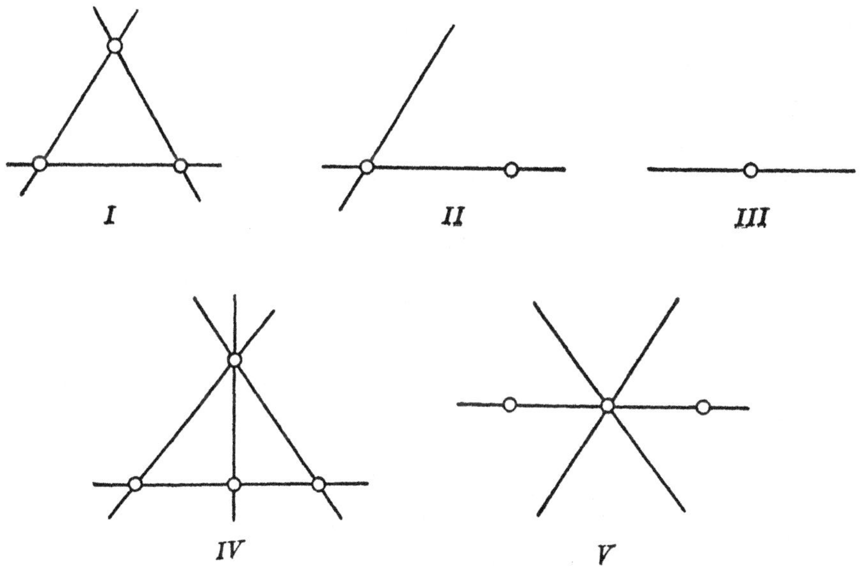


FIG. 48

point. By Theorem 18 there is only one such collineation $\Pi$. By the same theorem it is clear that $\Pi$ is fully determined by the projectivity it determines on two of the sides of the invariant triangle, say $O_2O_3$ and $O_1O_3$. Hence, if $H_1$ is a homology with center $O_1$ and axis $O_2O_3$, which determines the same projectivity as $\Pi$ on the line $O_1O_3$, and if $H_2$ is a homology with center $O_2$ and axis $O_1O_3$, which determines the same projectivity as $\Pi$ on the line $O_2O_3$, then it is evident that

$$\Pi = H_1H_2 = H_2H_1.$$

It is also evident that no point not a vertex of the invariant triangle can be fixed unless $\Pi$ reduces to a homology or to the identity. Such a transformation $\Pi$ when it is not a homology is said to be of *Type I*, and is denoted by Diagram *I* (fig. 48).

**EXERCISE**

Prove that two homologies with the same center and axis are commutative, and hence that two projectivities of Type *I* with the same invariant figure are commutative.

Consider the figure of two points $O_1$, $O_2$ and two lines $o_1$, $o_2$, such that $O_1$ and $O_2$ are on $o_1$, and $o_1$ and $o_2$ are on $O_1$. A collineation $\Pi$ which is the product of a homology H, leaving $O_2$ and $o_2$ invariant, and an elation E, leaving $O_1$ and $o_1$ invariant, evidently leaves this figure invariant and also leaves invariant no other point or line. If $A$ and $B$ are two points not on the lines of the invariant figure, and we require that

$$\Pi(A) = B,$$

this fixes the transformation (with two distinct double lines) among the lines at $O_1$, and the parabolic transformation among the lines at $O_2$, and thus determines $\Pi$ completely. Clearly if $\Pi$ is not to reduce to a homology or an elation, the line $AB$ must not pass through $O_1$ or $O_2$. Such a transformation $\Pi$, when it does not reduce to a homology or an elation or the identity, is said to be of *Type II* and is denoted by Diagram *II* (fig. 48).

**EXERCISE**

Two projective collineations of Type *II*, having the same invariant figure, are commutative.

Definition. The figure of a point $O$ and a line $o$ on $O$ is called a *lineal element* $Oo$.

A collineation having a lineal element as invariant figure must effect a parabolic transformation both on the points of the line and on the lines through the point. Suppose $Aa$ and $Bb$ are any two lineal elements whose points are not on $o$ or collinear with $O$, and whose lines are not on $O$ or concurrent with $o$. Let $E_1$ be an elation with center $O$ and axis $OA$, which transforms the point $(oa)$ to the point $(ob)$. Let $E_2$ be an elation of center $(AB, o)$ and axis $o$, which transforms $A$ to $B$. Then $\Pi = E_2E_1$ has evidently no other invariant elements than $O$ and $o$ and transforms $Aa$ to $Bb$.

Suppose that another projectivity $\Pi'$ would transfer $Aa$ to $Bb$ with $Oo$ as only invariant elements. The transformation $\Pi'$ would evidently have the same effect on the lines of $O$ and points of $o$ as $\Pi$. Hence $\Pi'\Pi^{-1}$ would be the identity or an elation. But as $\Pi'\Pi^{-1}(B) = B$ it would be the identity. Hence $\Pi$ is the only projectivity which transforms $Aa$ to $Bb$ with $Oo$ as only invariant.

A transformation having as invariant figure a lineal element and no other invariant point or line is said to be of *Type III*, and is denoted by Diagram *III* (fig. 48).

A homology is said to be of *Type IV* and is denoted by Diagram *IV*.

An elation is said to be of *Type V* and is denoted by Diagram *V*.

It will be shown later that any collineation can be regarded as belonging to one of these five types. The results so far obtained may be summarized as follows:

THEOREM 29. *A projective collineation with given invariant figure* F, *if of Type I or II will transform any point P not on a line of* F *into any other such point not on a line joining P to a point of* F; *if of Type III will transform any lineal element Pp such that p is not on a point, or P on a line, of* F *into any other such element Qq; if of Type IV or V, will transform any point P into any other point on the line joining P to the center of the collineation.*

The rôle of Assumption P is well illustrated by this theorem. In case of each of the first three types the *existence* of the required collineation was proved by means of Assumptions A and E, together with the existence of a sufficient number of points to effect the construction. But its *uniqueness* was established only by means of Assumption P. In case of Types *IV* and *V*, both existence and uniqueness follow from Assumptions A and E.

### EXERCISES

**1.** State the dual of Theorem 29.

**2.** If the number of points on a line is $p + 1$, the number of collineations with a given invariant figure is as follows:

Type *I*, $(p - 2)(p - 3)$.
Type *II*, $(p - 2)(p - 1)$.
Type *III*, $p(p - 1)^2$.
Type *IV*, $p - 2$.
Type *V*, $p - 1$.

In accordance with the results of this exercise, when the number of points on a line is infinite it is said that there are $\infty^2$ transformations of Type *I* or *II*; $\infty^3$ of Type *III*; and $\infty^1$ of Types *IV* and *V*.

# CHAPTER V *

## CONIC SECTIONS

### 41. Definitions. Pascal's and Brianchon's theorems.

DEFINITION. The set of all points of intersection of homologous lines of two projective, nonperspective flat pencils which are on the same plane but not on the same point is called a *point conic* (fig. 49). The plane dual of a point conic is called a *line conic* (fig. 50). The space dual of a point conic is called a *cone of planes;* the space dual

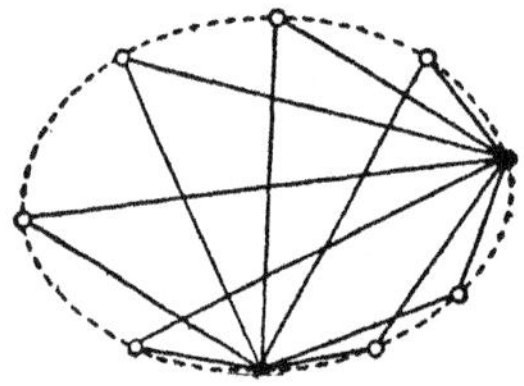

FIG. 49

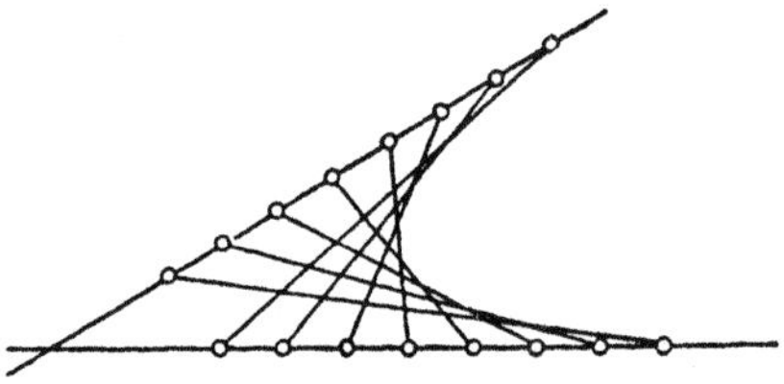

FIG. 50

of a line conic is called a *cone of lines.* The point through which pass all the lines (or planes) of a cone of lines (or planes) is called the *vertex* of the cone. The point conic, line conic, cone of planes, and cone of lines are called *one-dimensional forms of the second degree.*†

The following theorem is an immediate consequence of this definition.

THEOREM 1. *The section of a cone of lines by a plane not on the vertex of the cone is a point conic. The section of a cone of planes by a plane not on the vertex is a line conic.*

Now let $A_1$ and $B_1$ be the centers of two flat pencils defining a point conic. They are themselves, evidently, points of the conic, for the line $A_1B_1$ regarded as a line of the pencil on $A_1$ corresponds to some other line through $B_1$ (since the pencils are, by hypothesis, projective

* All the developments of this chapter are on the basis of Assumptions A, E, P, and $H_0$.

† A fifth one-dimensional form — a self-dual form of lines in space called the *regulus* — will be defined in Chap. XI. This definition of the first four one-dimensional forms of the second degree is due to Jacob Steiner (1796–1863). Attention will be called to other methods of definition in the sequel.

but not perspective), and the intersection of these homologous lines is $B_1$. The conic is clearly determined by any other three of its points, say $A_2$, $B_2$, $C_2$, because the projectivity of the pencils is then determined by

$$A_1(A_2B_2C_2) \,\overline{\wedge}\, B_1(A_2B_2C_2)$$

(Theorem 17, Chap. IV).

Let us now see how to determine a sixth point of the conic on a line through one of the given points, say on a line $l$ through $B_2$. If the line $l$ is met by the lines $A_1A_2$, $A_1C_2$, $B_1A_2$, $B_1C_2$ in the points $S$, $T$, $U$, $A$

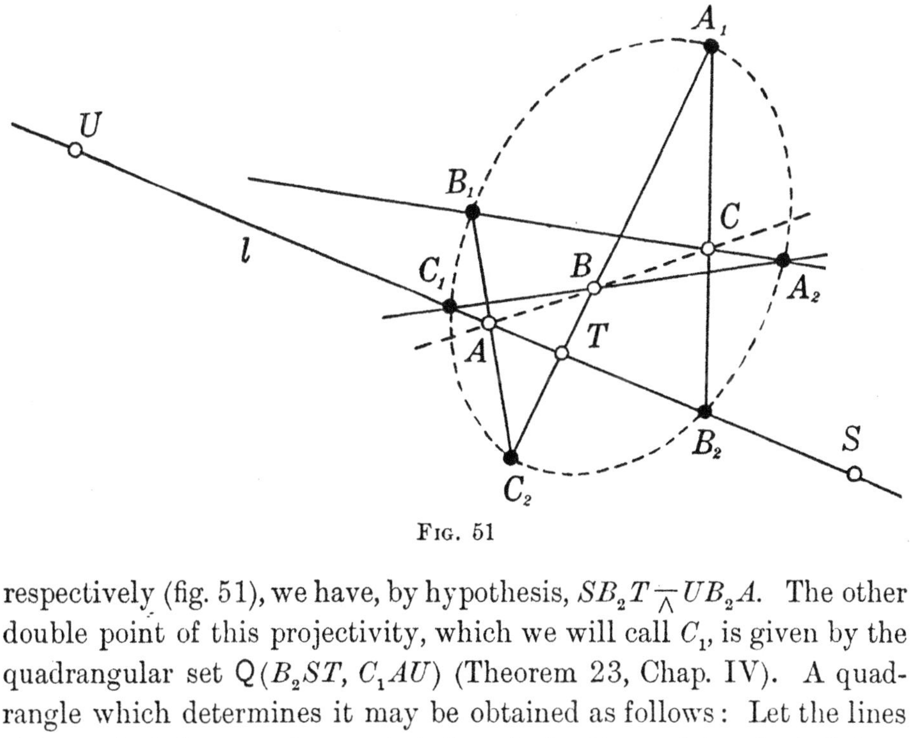


FIG. 51

respectively (fig. 51), we have, by hypothesis, $SB_2T \,\overline{\wedge}\, UB_2A$. The other double point of this projectivity, which we will call $C_1$, is given by the quadrangular set $Q(B_2ST, C_1AU)$ (Theorem 23, Chap. IV). A quadrangle which determines it may be obtained as follows: Let the lines $A_2B_1$ and $A_1B_2$ meet in a point $C$, and the lines $AC$ and $A_1C_2$ in a point $B$; then the required quadrangle is $A_1A_2CB$, and $C_1$ is determined as the intersection of $A_2B$ with $l$.

$C_1$ will coincide with $B_2$, if and only if $B$ is on $A_2B_2$ (fig. 52). This means that $AC$, $A_1C_2$, and $A_2B_2$ are concurrent in $B$. In other words, $A$ must be the point of intersection of $B_1C_2$ with the line joining $C=(A_2B_1)(A_1B_2)$ and $B=(A_1C_2)(A_2B_2)$, and $l$ must be the line joining $B_2$ and $A$. This gives, then, *a construction for a line which meets a given conic in only one point.*

The result of the preceding discussion may be summarized as follows: *The four points $A_2$, $B_2$, $C_2$, $C_1$ are points of a point conic* *determined by two projective pencils on* $A_1$ *and* $B_1$, *if and only if the three points* $C = (A_1B_2)(A_2B_1)$, $B = (A_1C_2)(A_2C_1)$, $A = (B_1C_2)(B_2C_1)$ *are collinear.* The three points in question are clearly the intersections of pairs of opposite sides of the simple hexagon $A_1B_2C_1A_2B_1C_2$.

Since $A_1$, $B_1$, $C_1$ may be interchanged with $A_2$, $B_2$, $C_2$ respectively in the above statement, it follows that $A_1$, $B_1$, $C_1$, $C_2$ are points of a conic determined by projective pencils on $A_2$ and $B_2$. Thus, if $C_1$ is any point of the first conic, it is also a point of the second conic, and vice versa. Hence we have established the following theorem:

THEOREM 2. STEINER'S THEOREM. *If $A$ and $B$ are any two given points of a conic, and $P$ is a variable point of this conic, we have* $A[P] \overline{\wedge} B[P]$.

In view of this theorem the six points in the discussion may be regarded as any six points of a conic, and hence we have

THEOREM 3. PASCAL'S THEOREM.* *The necessary and sufficient condition that six points, no three of which are collinear, be points of the same conic is that the three pairs of opposite sides of a simple hexagon of which they are vertices shall meet in collinear points.*†

The plane dual of this theorem is

THEOREM 3'. BRIANCHON'S THEOREM. *The necessary and sufficient condition that six lines, no three of which are concurrent, be lines of a line conic is that the lines joining the three pairs of opposite vertices of any simple hexagon of which the given lines are sides, shall be concurrent.*†

As corollaries of these theorems we have

COROLLARY 1. *A line in the plane of a point conic cannot have more than two points in common with the conic.*

COROLLARY 1'. *A point in the plane of a line conic cannot be on more than two lines of the conic.*

* Theorem 3 was proved by B. Pascal in 1640 when only sixteen years of age. He proved it first for the circle and then obtained it for any conic by projection and section. This is one of the earliest applications of this method. Theorem 3' was first given by C. J. Brianchon in 1806 (Journal de l'École Polytechnique, Vol. VI, p. 301).

† The line thus determined by the intersections of the pairs of opposite sides of any simple hexagon whose vertices are points of a point conic is called the *Pascal line* of the hexagon. The dual construction gives rise to the *Brianchon point* of a hexagon whose sides belong to a line conic.

Also as immediate corollaries of these theorems we have

THEOREM 4. *There is one and only one point conic containing five given points of a plane no three of which are collinear.*

THEOREM 4'. *There is one and only one line conic containing five given lines of a plane no three of which are concurrent.*

### EXERCISES

**1.** What are the space duals of the above theorems?

**2.** Prove Brianchon's theorem without making use of the principle of duality.

**3.** A necessary and sufficient condition that six points, no three of which are collinear, be points of a point conic, is that they be the points of intersection $(ab')$, $(bc')$, $(ca')$, $(ba')$, $(cb')$, $(ac')$ of the sides $a$, $b$, $c$ and $a'$, $b'$, $c'$ of two perspective triangles, in which $a$ and $a'$, $b$ and $b'$, $c$ and $c'$ are homologous.

**42. Tangents. Points of contact.** DEFINITION. A line $p$ in the plane of a point conic which meets the point conic in one and only one point $P$ is called a *tangent* to the point conic at $P$. A point $P$ in the plane of a line conic through which passes one and only one line $p$ of the line conic is called a *point of contact* of the line conic on $p$.

THEOREM 5. *Through any point of a point conic there is one and only one tangent to the point conic.*

*Proof.* If $P_0$ is the given point of the point conic and $P_1$ is any other point of the point conic, while $P$ is a variable point of this conic, we have, by Theorem 2,

$$P_0[P] \barwedge P_1[P].$$

Any line through $P_0$ meets its homologous line of the pencil on $P_1$ in a point distinct from $P_0$, except when its homologous line is $P_1P_0$. Since a projectivity is a one-to-one correspondence, there is only one line on $P_0$ which has $P_1P_0$ as its homologous line.

THEOREM 5'. *On any line of a line conic there is one and only one point of contact of the line conic.*

This is the plane dual of the preceding theorem.

### EXERCISE

Give the space duals of the preceding definitions and theorems.

Returning now to the construction in the preceding section for the points of a point conic containing five given points, we recall that the point of intersection $C_1$ of a line $l$ through $B_2$ was determined by the quadrangular set Q$(B_2ST,\ C_1AU)$. The points $B_2$ and $C_1$ can, by the preceding theorem, coincide on one and only one of the lines through $B_2$.* For this particular line $l$, $A$ becomes the intersection

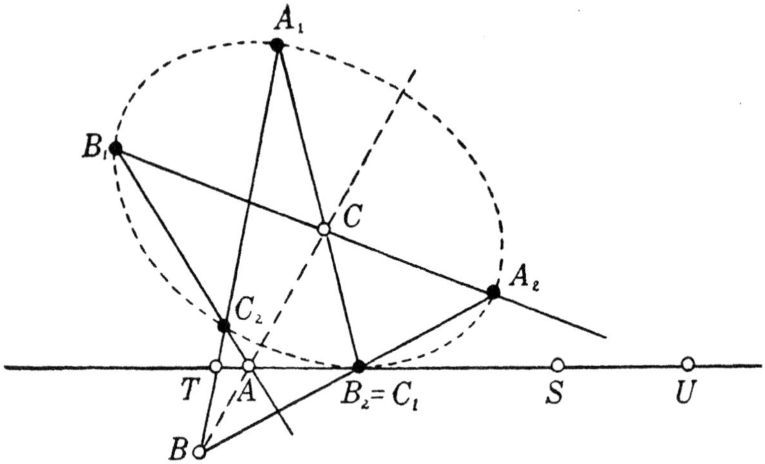


FIG. 52

of the tangent at $B_2$ with $B_1C_2$, and the collinearity of the points $A, B, C$ may be stated as follows:

THEOREM 6. *If the vertices of a simple plane five-point are points of a point conic, the tangent to the point conic at one of the vertices meets the opposite side in a point collinear with the points of intersection of the other two pairs of nonadjacent sides.*

This theorem, by its derivation, is a degenerate case of Pascal's theorem. It may also be regarded as a degenerate case in its statement, if the tangent be thought of as taking the place of one side of the simple hexagon.

It should be clearly understood that the theorem has been obtained by specializing the figure of Theorem 3, and not by a continuity argument. The latter would be clearly impossible, since our assumptions do not require the conic to contain more than a finite number of points.

Theorem 6 may be applied to the construction of a tangent to a point conic at any one of five given points $P_1, P_2, P_3, P_4, P_5$ of the point conic (fig. 53). By this theorem the tangent $p_1$ at $P_1$ must be

* As explained in the fine print on page 110, this occurs when $l$ passes through the point of intersection of $B_1C_2$ with the line joining $C = (A_1B_2)(A_2B_1)$ and $B = (A_1C_2)(A_2B_2)$.

such that the points $p_1(P_3P_4) = A$, $(P_1P_2)(P_4P_5) = B$, and $(P_2P_3)(P_5P_1) = C$ are collinear. But $B$ and $C$ are determined by $P_1$, $P_2$, $P_3$, $P_4$, $P_5$, and hence $p_1$ is the line joining $P_1$ to the intersection of the lines $BC$ and $P_3P_4$.

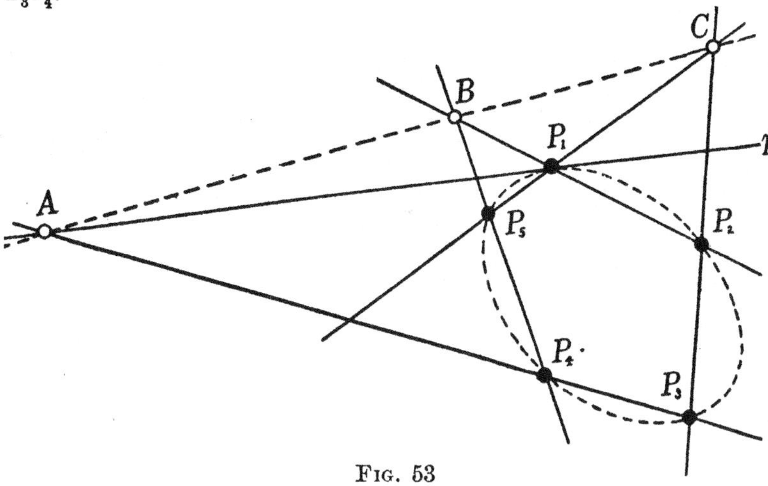


FIG. 53

In like manner, if $P_1$, $P_2$, $P_3$, $P_4$, and $p_1$ are given, to construct the point $P_5$ on any line $l$ through $P_4$ of a point conic containing $P_1, P_2, P_3, P_4$ and of which $p_1$ is the tangent at $P_1$, we need only determine the points $A = p_1(P_3P_4)$, $B = l(P_1P_2)$, and $C = (AB)(P_2P_3)$; then $P_1C$ meets $l$ in $P_5$ (fig. 53).

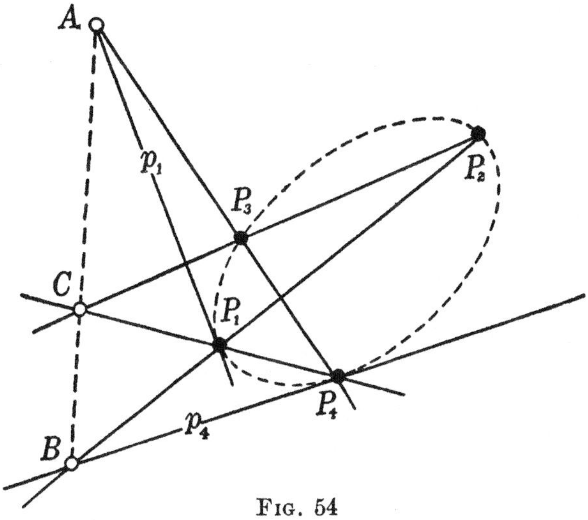


FIG. 54

In case $l$ is the tangent $p_4$ at $P_4$, $P_5$ coincides with $P_4$ and the following points are collinear (fig. 54):

$$A = p_1(P_3P_4),\ B = p_4(P_1P_2),\ C = (P_1P_4)(P_2P_3).$$

Hence we have the following theorem:

THEOREM 7. *If the vertices $P_1$, $P_2$, $P_3$, $P_4$ of a simple quadrangle are points of a point conic, the tangent at $P_1$ and the side $P_3P_4$, the tangent at $P_4$ and the side $P_1P_2$, and the pair of sides $P_1P_4$ and $P_2P_3$ meet in three collinear points.*

If $P_1$, $P_2$, $P_3$, $P_5$ and the tangent $p_1$ at $P_1$ are given, the construction determined by Theorem 6 for a point $P_4$ of the point conic on a line $l$ through $P_3$ is as follows (fig. 53): Determine $C=(P_1P_5)(P_2P_3)$, $A=p_1l$, and $B=(AC)(P_1P_2)$; then $P_5B$ meets $l$ in $P_4$.

In case $l$ is the tangent at $P_3$, $P_4$ coincides with $P_3$ and we have the result that $C=(P_1P_5)(P_2P_3)$, $A=p_1p_3$, $B=(P_1P_2)(P_5P_3)$ are collinear points, which gives

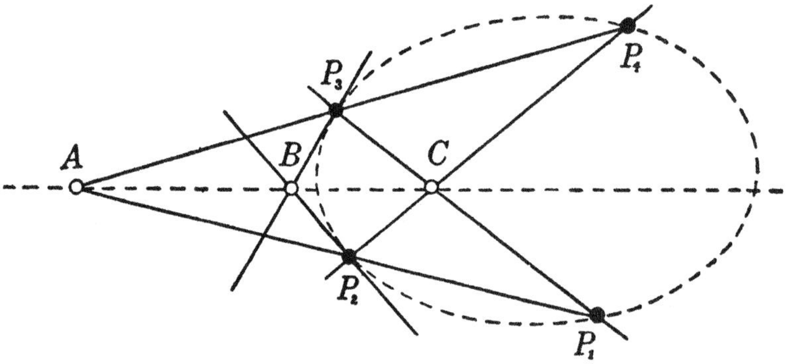


FIG. 55

THEOREM 8. *If the vertices of a complete quadrangle are points of a point conic, the tangents at a pair of vertices meet in a point of the line joining the diagonal points of the quadrangle which are not on the side joining the two vertices* (fig. 55).

The last two theorems lead to the construction for a point conic of which there are given three points and the tangents at two of them. Reverting to the notation of Theorem 7 (fig. 54), let the given points be $P_4$, $P_1$, $P_3$ and the given tangents be $p_4$, $p_1$. Let $l$ be any line through $P_3$. If $P_2$ is the other point in which $l$ meets the point conic, the points $A=p_1(P_3P_4)$, $B=p_4(P_1P_2)$, and $C=(P_2P_3)(P_4P_1)$ are collinear. Hence, if $C=l(P_1P_4)$ and $B=p_4(AC)$, then $P_2$ is the intersection of $l$ with $BP_1$.

In case $l$ is the tangent $p_3$ at $P_3$, the points $P_2$ and $P_3$ coincide, and the points

$$p_1(P_3P_4), \quad p_3(P_1P_4), \quad p_4(P_1P_3)$$

are collinear. Hence the two triangles $P_1P_3P_4$ and $p_1p_3p_4$ are perspective, and we obtain as a last specialization of Pascal's theorem (fig. 56)

THEOREM 9. *A triangle whose vertices are points of a point conic is perspective with the triangle formed by the tangents at these points, the tangent at any vertex being homologous with the side of the first triangle which does not contain this vertex.*

COROLLARY. *If $P_1$, $P_3$, $P_4$ are three points of a point conic, the lines $P_3P_1$, $P_3P_4$ are harmonic with the tangent at $P_3$ and the line joining $P_3$ to the intersection of the tangents at $P_1$ and $P_4$.*

*Proof.* This follows from the definition of a harmonic set of lines, on considering the quadrilateral $P_1A$, $AB$, $BP_4$, $P_4P_1$ (fig. 56).

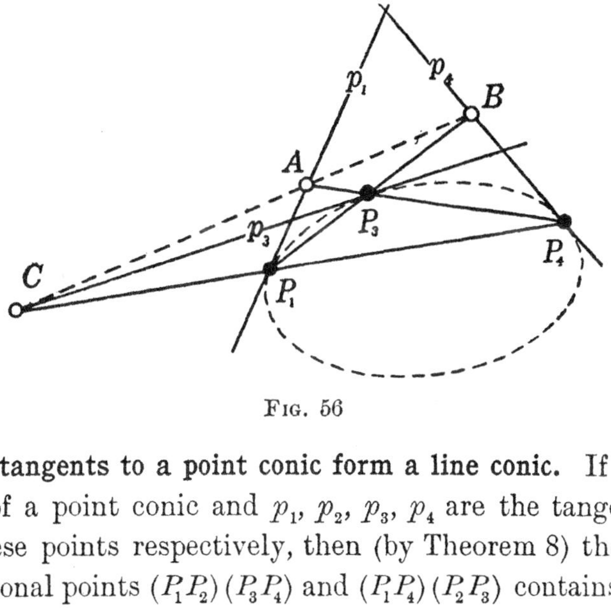


FIG. 56

**43. The tangents to a point conic form a line conic.** If $P_1$, $P_2$, $P_3$, $P_4$ are points of a point conic and $p_1$, $p_2$, $p_3$, $p_4$ are the tangents to the conic at these points respectively, then (by Theorem 8) the line joining the diagonal points $(P_1P_2)(P_3P_4)$ and $(P_1P_4)(P_2P_3)$ contains the intersection of the tangents $p_1$, $p_3$ and also the intersection of $p_2$, $p_4$. This line is a diagonal line not only of the quadrangle $P_1P_2P_3P_4$, but also of the quadrilateral $p_1p_2p_3p_4$. Theorem 8 may therefore be stated in the form:

THEOREM 10. *The complete quadrangle formed by four points of a point conic and the complete quadrilateral of the tangents at these points have the same diagonal triangle.*

Looked at from a slightly different point of view, Theorem 8 gives also

THEOREM 11. *The tangents to a point conic form a line conic.*

*Proof.* Let $P_1$, $P_2$, $P_3$ be any three fixed points on a conic, and let $P$ be a variable point of this conic. Let $p_1$, $p_2$, $p_3$, $p$ be respectively the tangents at these points (fig. 57). By the corollary of Theorem 28, Chap. IV, $P_1P_2$ is the axis of homology of the projectivity between the pencils of points on $p_1$ and $p_2$ defined by

$$P_1(p_1p_2)(p_1p_3) \barwedge (p_2p_1)P_2(p_2p_3).$$

But by Theorem 10, if $Q = (P_1P_2)(P_3P)$, the points $pp_2$, $p_1p_3$, and $Q$ are collinear. For the same reason the points $p_2p_3$, $pp_1$, $Q$ are collinear. It follows, by Theorem 28, Chap. IV, that the homolog of the variable

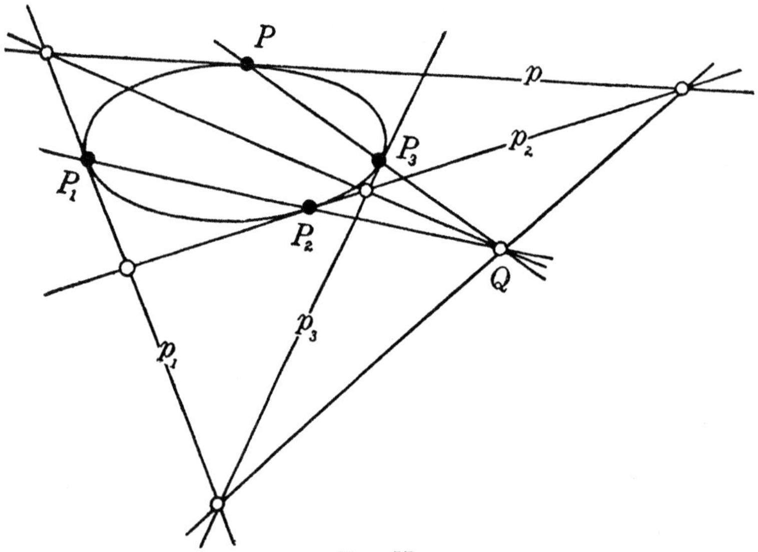


FIG. 57

point $p_1p$ is $p_2p$; i.e. $p$ is the line joining pairs of homologous points on the two lines $p_1$, $p_2$, so that the totality of the lines $p$ satisfies the definition of a line conic.

COROLLARY. *The center of homology of the projectivity* $P_1[P] \barwedge P_2[P]$ *determined by the points* $P$ *of a point conic containing* $P_1$, $P_2$ *is the intersection of the tangents at* $P_1$, $P_2$. *The axis of homology of the projectivity* $p_1[p] \barwedge p_2[p]$ *determined by the lines* $p$ *of a line conic containing the lines* $p_1$, $p_2$ *is the line joining the points of contact of* $p_1$, $p_2$.

THEOREM 12. *If* $P_1$ *is a fixed and* $P$ *a variable point of a point conic, and* $p_1$, $p$ *are the tangents at these two points respectively, then we have* $P_1[P] \barwedge p_1[p]$.

*Proof.* Using the notation of the proof of Theorem 11 (fig. 57), we have

$$P_1[P] \,\overline{\wedge}\, P_3[P] \,\overline{\overline{\wedge}}\, [Q],$$

where $Q$ is always on $P_1P_2$. But we also have

$$[Q] \overset{p_1p_3}{\overline{\overline{\wedge}}} p_2[p],$$

and, by Theorem 11, $$p_2[p] \,\overline{\wedge}\, p_1[p].$$

Combining these projectivities, we have

$$P_1[P] \,\overline{\wedge}\, p_1[p].$$

The plane dual of Theorem 11 states that *the points of contact of a line conic form a point conic.* In view of these two theorems and their space duals we now make the following

DEFINITION. A *conic section* or a *conic* is the figure formed by a point conic and its tangents. A *cone* is the figure formed by a cone of lines and its tangent planes.

The figure formed by a line conic and its points of contact is then likewise a conic as defined above; i.e. a conic (and also a cone) is a *self-dual* figure.

The duals of Pascal's theorem and its special cases now give us a set of theorems of the same consequence for point conics as for line conics. We content ourselves with restating Brianchon's theorem (Theorem 3′) from this point of view.

BRIANCHON'S THEOREM. *If the sides of a simple hexagon are tangents to a conic, the lines joining opposite vertices are concurrent; and conversely.*

It follows from the preceding discussion that in forming the plane duals of theorems concerning conics, the word *conic* is left unchanged, while the words *point* (of a conic) and *tangent* (of a conic) are interchanged. We shall also, in the future, make use of the phrase a conic *passes through* a point $P$, and $P$ *is on* the conic, when $P$ is a point of a conic, etc.

DEFINITION. If the points of a plane figure are on a conic, the figure is said to be *inscribed* in the conic; if the lines of a plane figure are tangent to a conic, the figure is said to be *circumscribed* about the conic.

### EXERCISES

**1.** State the plane and space duals of the special cases of Pascal's theorem.

**2.** Construct a conic, given (1) five tangents, (2) four tangents and the point of contact of one of them, (3) three tangents and the points of contact of two of them.

**3.** $ABX$ is a triangle whose vertices are on a conic, and $a, b, x$ are the tangents at $A$, $B$, $X$ respectively. If $A$, $B$ are given points and $X$ is variable, determine the locus of (1) the center of perspectivity of the triangles $ABX$ and $abx$; (2) the axis of perspectivity.

**4.** $X$, $Y$, $Z$ are the vertices of a variable triangle, such that $X$, $Y$ are always on two given lines $a$, $b$ respectively, while the sides $XY$, $ZX$, $ZY$ always pass through three given points $P$, $A$, $B$ respectively. Show that the locus of the point $Z$ is a point conic containing $A$, $B$, $D=(ab)$, $M=(AP)b$, and $N=(BP)a$ (Maclaurin's theorem). Dualize. (The plane dual of this theorem is known as the theorem of Braikenridge.)

**5.** If a simple plane $n$-point varies in such a way that its sides always pass through $n$ given points, while $n-1$ of its vertices are always on $n-1$ given lines, the $n$th vertex describes a conic (Poncelet).

**6.** If the vertices of two triangles are on a conic, the six sides of these two triangles are tangents of a second conic; and conversely. Corresponding to every point of the first conic there exists a triangle having this point as a vertex, whose other two vertices are also on the first conic and whose sides are tangents to the second conic. Dualize.

**7.** If two triangles in the same plane are perspective, the points in which the sides of one triangle meet the nonhomologous sides of the other are on the same conic; and the lines joining the vertices of one triangle to the nonhomologous vertices of the other are tangents to another conic.

**8.** If $A$, $B$, $C$, $D$ be the vertices of a complete quadrangle, whose sides $AB$, $AC$, $AD$, $BC$, $BD$, $CD$ are cut by a line in the points $P$, $Q$, $R$, $S$, $T$, $V$ respectively, and if $E$, $F$, $G$, $K$, $L$, $M$ are respectively the harmonic conjugates of these points with respect to the pairs of vertices of the quadrangle so that we have $\mathrm{H}(AB, PE)$, $\mathrm{H}(AC, QF)$, etc., then the six points $E$, $F$, $G$, $K$, $L$, $M$ are on a conic which also passes through the diagonal points of the quadrangle (Holgate, Annals of Mathematics, Ser. 1, Vol. VII (1893), p. 73).

**9.** If a plane $\alpha$ cut the six edges of a tetrahedron in six distinct points, and the harmonic conjugates of each of these points with respect to the two vertices of the tetrahedron that lie on the same edge are determined, then the lines joining the latter six points to any point $O$ of the plane $\alpha$ are on a cone, on which are also the lines through $O$ and meeting a pair of opposite edges of the tetrahedron (Holgate, Annals of Mathematics, Ser. 1, Vol. VII (1893), p. 73).

**10.** Given four points of a conic and the tangent at one of them, construct the tangents at the other three points. Dualize.

**11.** $A$, $A'$, $B$, $B'$ are the vertices of a quadrangle, and $m$, $n$ are two lines in the plane of the quadrangle which meet on $AA'$. $M$ is a variable point on $m$, the lines $BM$, $B'M$ meet $n$ in the points $N$, $N'$ respectively; the lines $AN$, $A'N'$ meet in a point $P$. Show that the locus of the lines $PM$ is a line conic, which contains the lines $m$, $p = P(n, BB')$, and also the lines $AA'$, $BB'$, $A'B'$, $AB$ (Amodeo, Lezioni di Geometria Projettiva, Naples (1905), p. 331).

**12.** Use the result of Ex. 11 to give a construction of a line conic determined by five given lines, and show that by means of this construction it is possible to obtain two lines of the conic at the same time (Amodeo, loc. cit.).

**13.** If $a$, $b$, $c$ are the sides of a triangle whose vertices are on a conic, and $m$, $m'$ are two lines meeting on the conic which meet $a$, $b$, $c$ in the points $A$, $B$, $C$ and $A'$, $B'$, $C'$ respectively, and which meet the conic again in $N$, $N'$ respectively, we have $ABCN \barwedge A'B'C'N'$ (cf. Ex. 6).

**14.** If $A$, $B$, $C$, $D$ are points on a conic and $a$, $b$, $c$, $d$ are the tangents to the conic at these points, the four diagonals of the simple quadrangle $ABCD$ and the simple quadrilateral $abcd$ are concurrent.

### 44. The polar system of a conic.

THEOREM 13. *If $P$ is a point in the plane of a conic, but not on the conic, the points of intersection of the tangents to the conic at all the pairs of points which are collinear with $P$ are on a line, which also contains the harmonic conjugates of $P$ with respect to these pairs of points.*

THEOREM 13′. *If $p$ is a line in the plane of a conic, but not tangent to the conic, the lines joining the points of contact of pairs of tangents to the conic which meet on $p$ pass through a point $P$, through which pass also the harmonic conjugates of $p$ with respect to these pairs of tangents.*

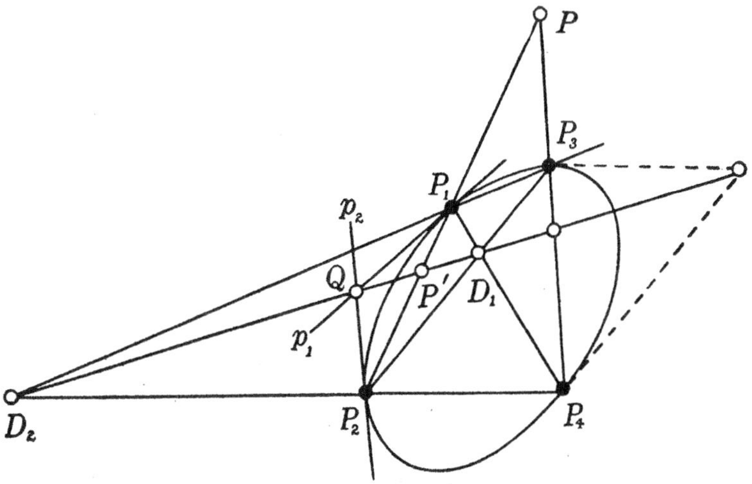


FIG. 58

*Proof.* Let $P_1$, $P_2$ and $P_3$, $P_4$ be two pairs of points on the conic which are collinear with $P$, and let $p_1$, $p_2$ be the tangents to the conic at $P_1$, $P_2$ respectively (fig. 58). If $D_1$, $D_2$ are the points $(P_1P_4)(P_2P_3)$ and $(P_1P_3)(P_2P_4)$

respectively, the line $D_1D_2$ passes through the intersection $Q$ of $p_1$, $p_2$ (Theorem 8). Moreover, the point $P'$ in which $D_1D_2$ meets $P_1P_2$ is the harmonic conjugate of $P$ with respect to $P_1$, $P_2$ (Theorem 6, Chap. IV). This shows that the line $D_1D_2 = QP'$ is completely determined by the pair of points $P_1$, $P_2$. Hence the same line $QP'$ is obtained by replacing $P_3$, $P_4$ by any other pair of points on the conic collinear with $P$, and distinct from $P_1$, $P_2$. This proves Theorem 13. Theorem 13′ is the plane dual of Theorem 13.

DEFINITION. The line thus associated with any point $P$ in the plane of a conic, but not on the conic, is called the *polar* of $P$ with respect to the conic. If $P$ is a point on the conic, the *polar* is defined as the tangent at $P$.

DEFINITION. The point thus associated with any line $p$ in the plane of a conic, but not tangent to the conic, is called the *pole* of $p$ with respect to the conic. If $p$ is a tangent to the conic, the *pole* is defined as the point of contact of $P$.

THEOREM 14. *The line joining two diagonal points of any complete quadrangle whose vertices are points of a conic is the polar of the other diagonal point with respect to the conic.*

THEOREM 14′. *The point of intersection of two diagonal lines of any complete quadrilateral whose sides are tangent to a conic is the pole of the other diagonal line with respect to the conic.*

*Proof.* Theorem 14 follows immediately from the proof of Theorem 13. Theorem 14′ is the plane dual of Theorem 14

THEOREM 15. *The polar of a point $P$ with respect to a conic passes through the points of contact of the tangents to the conic through $P$, if such tangents exist.*

THEOREM 15′. *The pole of a line $p$ with respect to a conic is on the tangents to the conic at the points in which $p$ meets the conic, if such points exist.*

*Proof.* Let $P_1$ be the point of contact of a tangent through $P$, and let $P_2$, $P_3$ be any pair of distinct points of the conic collinear with $P$. The line through $P_1$ and the intersection of the tangents at $P_2$, $P_3$ meets the line $P_3P_2$ in the harmonic conjugate of $P$ with respect to $P_3$, $P_2$ (Theorem 9, Cor.). But the line thus determined is the polar of $P$ (Theorem 13). This proves Theorem 15. Theorem 15′ is its plane dual.

THEOREM 16. *If $p$ is the polar of a point $P$ with respect to a conic, $P$ is the pole of $p$ with respect to the same conic.*

If $P$ is not on the conic, this follows at once by comparing Theorem 13 with Theorem 13′. If $P$ is on the conic, it follows immediately from the definition.

THEOREM 17. *If the polar of a point $P$ passes through a point $Q$, the polar of $Q$ passes through $P$.*

*Proof.* If $P$ or $Q$ is on the conic, the theorem is equivalent to Theorem 15. If neither $P$ nor $Q$ is on the conic, let $PP_1$ be a line

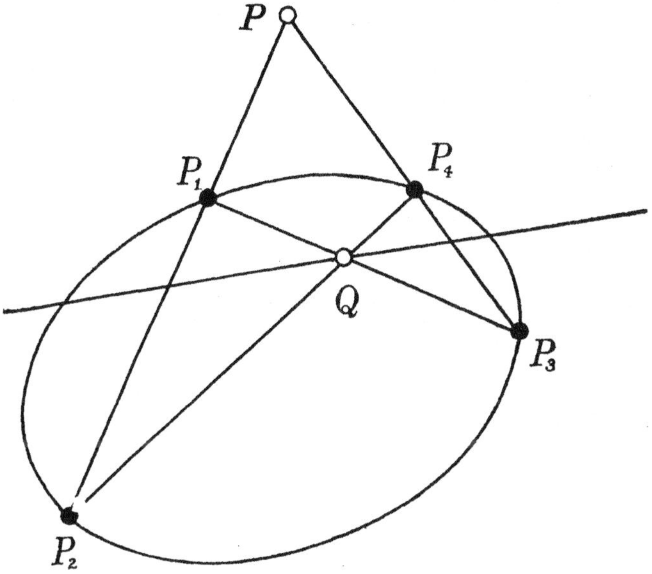


FIG. 59

meeting the conic in two points, $P_1$, $P_2$. If one of the lines $P_1Q$, $P_2Q$ is a tangent to the conic, the other is also a tangent (Theorem 13); the line $P_1P_2 = P_1P$ is then the polar of $Q$, which proves the theorem under this hypothesis. If, on the other hand, the lines $P_1Q$, $P_2Q$ meet the conic again in the points $P_3$, $P_4$ respectively (fig. 59), the point $(P_1P_2)(P_3P_4)$ is on the polar of $Q$ (Theorem 14). By Theorems 13 and 14 the polar of $(P_1P_2)(P_3P_4)$ contains the intersection of the tangents at $P_1$, $P_2$ and the point $Q$. By hypothesis, however, and Theorem 13, the polar of $P$ contains these points also. Hence we have $(P_1P_2)(P_3P_4) = P$, which proves the theorem.

COROLLARY 1. *If two vertices of a triangle are the poles of their opposite sides with respect to a conic, the third vertex is the pole of its opposite side.*

DEFINITION. Any point on the polar of a point $P$ is said to be *conjugate* to $P$ with regard to the conic; and any line on the pole of a line $p$ is said to be conjugate to $p$ with regard to the conic. The figure obtained from a given figure in the plane of a conic by constructing the polar of every point and the pole of every line of the given figure with regard to the conic is called the *polar* or *polar reciprocal* of the given figure with regard to the conic.* A triangle, of which each vertex is the pole of the opposite side, is said to be *self-polar* or *self-conjugate* with regard to the conic.

COROLLARY 2. *The diagonal triangle of a complete quadrangle whose vertices are on a conic, or of a complete quadrilateral whose sides are tangent to a conic, is self-polar with regard to the conic; and, conversely, every self-polar triangle is the diagonal triangle of a complete quadrangle whose points are on the conic, and of a complete quadrilateral whose sides are tangent to the conic. Corresponding to a given self-polar triangle, one vertex or side of such a quadrangle or quadrilateral may be chosen arbitrarily on the conic.*

Theorem 17 may also be stated as follows: If $P$ is a variable point on a line $q$, its polar $p$ is a variable line through the pole $Q$ of $q$. In the special case where $q$ is a tangent to the conic, we have already seen (Theorem 12) that we have

$$[P] \,\overline{\wedge}\, [p].$$

If $Q$ is not on $q$, let $A$ (fig. 60) be a fixed point on the conic, $a$ the tangent at $A$, $X$ the point (distinct from $A$, if $AP$ is not tangent) in which $AP$ meets the conic, and $x$ the tangent at $X$. We then have, by Theorem 12,

$$[P] \,\overline{\overline{\wedge}}\, A[X] \,\overline{\wedge}\, a[x] \,\overline{\overline{\wedge}}\, Q[(ax)].$$

By Theorem 13, $(ax)$ is on $p$, and hence $p = Q\,(ax)$. Hence we have

$$[P] \,\overline{\wedge}\, [p].$$

If $P'$ is the point $pq$, this gives

$$[P] \,\overline{\wedge}\, [P'].$$

But since the polar of $P'$ also passes through $P$, this projectivity is an involution. The result of this discussion may then be stated as follows:

* It was by considering the polar reciprocal of Pascal's theorem that Brianchon derived the theorem named after him. This method was fully developed by Poncelet and Gergonne in the early part of the last century in connection with the principle of duality.

THEOREM 18. *On any line not a tangent to a given conic the pairs of conjugate points are pairs of an involution. If the line meets the conic in two points, these points are the double points of the involution.*

COROLLARY. *As a point $P$ varies over a pencil of points, its polar with respect to any conic varies over a projective pencil of lines.*

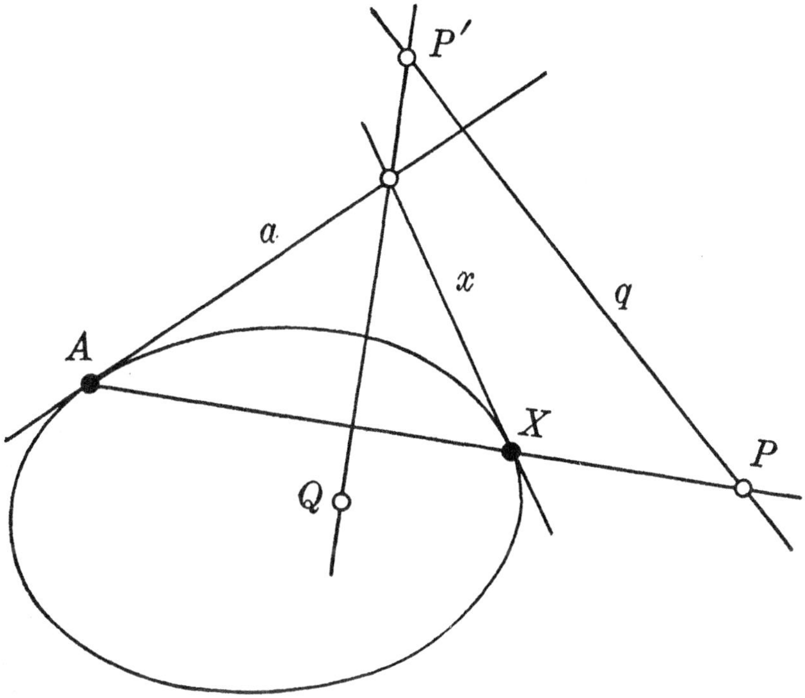


FIG. 60

DEFINITION. The pairing of the points and lines of a plane brought about by associating with every point its polar and with every line its pole with respect to a given conic in the plane is called a *polar system.*

## EXERCISES

**1.** If in a polar system two points are conjugate to a third point $A$, the line joining them is the polar of $A$.

**2.** State the duals of the last two theorems.

**3.** If $a$ and $b$ are two nonconjugate lines in a polar system, every point $A$ of $a$ has a conjugate point $B$ on $b$. The pencils of points $[A]$ and $[B]$ are projective; they are perspective if and only if $a$ and $b$ intersect on the conic of the polar system.

**4.** Let $A$ be a point and $b$ a line not the polar of $A$ with respect to a given conic, but in the plane of the conic. If on any line $l$ through $A$ we determine that point $P$ which is conjugate with the point $lb$, the locus of $P$ is a conic passing through $A$ and the pole $B$ of $b$, unless the line $AB$ is tangent to the conic, in which case the locus of $P$ is a line. If $AB$ is not tangent to the conic, the locus of $P$ also passes through the points in which $b$ meets the given conic (if such points exist), and also through the points of contact of the tangents to the given conic through $A$ (if such tangents exist). Dualize (Reye-Holgate, Geometry of Position, p. 106).

**5.** If the vertices of a triangle are on a given conic, any line conjugate to one side meets the other two sides in a pair of conjugate points. Conversely, a line meeting two sides of the triangle in conjugate points passes through the pole of the third side (von Staudt).

**6.** If two lines conjugate with respect to a conic meet the conic in two pairs of points, these pairs are projected from any point on the conic by a harmonic set of lines, and the tangents at these pairs of points meet any tangent in a harmonic set of points.

**7.** With a given point not on a given conic as center and the polar of this point as axis, the conic is transformed into itself by a homology of period two.

**8.** The Pascal line of any simple hexagon whose vertices are on a conic is the polar with respect to the conic of the Brianchon point of the simple hexagon whose sides are the tangents to the conic at the vertices of the first hexagon.

**9.** If the line joining two points $A$, $B$, conjugate with respect to a conic, meets the conic in two points, these two points are harmonic with $A$, $B$.

**10.** If in a plane there are given two conics $C_1^2$ and $C_2^2$, and the polars of all the points of $C_1^2$ with respect to $C_2^2$ are determined, these polars are the tangents of a third conic.

**11.** If the tangents to a given conic meet a second conic in pairs of points, the tangents at these pairs of points meet on a third conic.

**12.** Given five points of a conic (or four points and the tangent through one of them, or any one of the other conditions determining a conic), show how to construct the polar of a given point with respect to the conic.

**13.** If two pairs of opposite sides of a complete quadrangle are pairs of conjugate lines with respect to a conic, the third pair of opposite sides are conjugate with respect to the conic (von Staudt).

**14.** If each of two triangles in a plane is the polar of the other with respect to a conic, they are perspective, and the axis of perspectivity is the polar of the center of perspectivity (Chasles).

**15.** Two triangles that are self-polar with respect to the same conic have their six vertices on a second conic and their six sides tangent to a third conic (Steiner).

**16.** Regarding the Desargues configuration as composed of a quadrangle and a quadrilateral mutually inscribed (cf. § 18, Chap. II), show that the diagonal triangle of the quadrangle is perspective with the diagonal triangle of the quadrilateral.

**17.** Let $A$, $B$ be any two conjugate points with respect to a conic, and let the lines $AM$, $BM$ joining them to an arbitrary point of the conic meet the latter again in the points $C$, $D$ respectively. The lines $AD$, $BC$ will then meet on the conic, and the lines $CD$ and $AB$ are conjugate. Dualize.

**45. Degenerate conics.** For a variety of reasons it is desirable to regard two coplanar lines or one line (thought of as two coincident lines) as degenerate cases of a point conic; and dually to regard two points or one point (thought of as two coincident points) as degenerate cases of a line conic. This conception makes it possible to leave out the restriction as to the plane of section in Theorem 1. For the section of a cone of lines by a plane through the vertex of the cone consists evidently of two (distinct or coincident) lines, i.e. of a degenerate point conic; and the section of a cone of planes by a plane through the vertex of the cone is the figure formed by some or all the lines of a flat pencil, i.e. a degenerate line conic.

### EXERCISE

Dualize in all possible ways the degenerate and nondegenerate cases of Theorem 1.

Historically, the first definition of a conic section was given by the ancient Greek geometers (e.g. Menæchmus, about 350 B.C.), who defined them as the plane sections of a "right circular cone." In a later chapter we will show that in the "geometry of reals" any nondegenerate point conic is projectively equivalent to a circle, and thus that for the ordinary geometry the modern projective definition given in § 41 is equivalent to the old definition. We are here using one of the modern definitions because it can be applied before developing the Euclidean metric geometry.

Degenerate conics would be included in our definition (p. 109), if we had not imposed the restriction on the generating projective pencils that they be nonperspective; for the locus of the point of intersection of pairs of homologous lines in two perspective flat pencils in the same plane consists of the axis of perspectivity and the line joining the centers of the pencils.

It will be seen, as we progress, that many theorems regarding nondegenerate conics apply also when the conics are degenerate. For example, Pascal's theorem (Theorem 3) becomes, for the case of a degenerate conic consisting of two distinct lines, the theorem of Pappus already proved as Theorem 21, Chap. IV (cf. in particular the corollary). The polar of a point with regard to a degenerate conic consisting of two lines is the harmonic conjugate of the point with respect to the two lines (cf. the definition, p. 84, Ex. 7). Hence the polar system of a degenerate conic of two lines (and dually of two points) determines an involution at a point (on a line).

### EXERCISES

1. State Brianchon's theorem (Theorem 3′) for the case of a degenerate line conic consisting of two points.

2. Examine all the theorems of the preceding sections with reference to their behavior when the conic in question becomes degenerate.

## 46. Desargues's theorem on conics.

THEOREM 19. *If the vertices of a complete quadrangle are on a conic which meets a line in two points, the latter are a pair in the involution determined on the line by the pairs of opposite sides of the quadrangle.**

*Proof.* Reverting to the proof of Theorem 2 (fig. 51), let the line meet the conic in the points $B_2$, $C_1$ and let the vertices of the quadrangle be $A_1$, $A_2$, $B_1$, $C_2$. This quadrangle determines on the line an involution in which $S$, $A$ and $T$, $U$ are conjugate pairs. But in the proof of Theorem 2 we saw that the quadrangle $A_1A_2BC$ determines $Q(B_2ST, C_1AU)$. Hence the two quadrangles determine the same involution on the line, and therefore $B_2$, $C_1$ are a pair of the involution determined by the quadrangle $A_1A_2B_1C_2$.

Since the quadrangles $A_1A_2B_1C_2$ and $A_1A_2BC$ determine the same involution on the line when the latter is a tangent to the conic, we have as a special case of the above theorem:

COROLLARY. *If the vertices of a complete quadrangle are on a conic, the pairs of opposite sides meet the tangent at any other point in pairs of an involution of which the point of contact of the tangent is a double point.*

The Desargues theorem leads to a slightly different form of statement for the construction of a conic through five given points: On any line through one of the points the complete quadrangle of the other four determine an involution; the conjugate in this involution of the given point on the line is a sixth point on the conic.

As the Desargues theorem is related to the theorem of Pascal, so are certain degenerate cases of the Desargues theorem related to the degenerate cases of the theorem of Pascal (Theorems 6, 7, 8, 9). Thus in fig. 53 we see (by Theorem 6) that the quadrangle $BCP_2P_5$ determines on the line $P_3P_4$ an involution in which the points $P_3$, $P_4$ of the conic are one pair, while the points determined by $p_1$, $P_2P_5$ and those

* First given by Desargues in 1639; cf. Œuvres, Paris, Vol. I (1864), p. 188.

determined by $P_1P_2$, $P_1P_5$ are two other pairs. This gives the following special case of the theorem of Desargues:

THEOREM 20. *If the vertices of a triangle are on a conic, and a line $l$ meets the conic in two points, the latter are a pair of the involution determined on $l$ by the pair of points in which two sides of the triangle meet $l$, and the pair in which the third side and the tangent at the opposite vertex meet $l$. In case $l$ is a tangent to the conic, the point of contact is a double point of this involution.*

In terms of this theorem we may state the construction of a conic through four points and tangent to a line through one of them as follows: On any line through one of the points which is not on the tangent an involution is determined in which the tangent and the line passing through the other two points determine one pair, and the lines joining the point of contact to the other two points determine another pair. The conjugate of the given point on the line in this involution is a point of the conic.

A further degenerate case is derived either from Theorem 7 or Theorem 8. In fig. 54 (Theorem 7) let $l$ be the line $P_2P_3$. The quadrangle $ABP_1P_4$ determines on $l$ an involution in which $P_2$, $P_3$ are one pair, in which the tangents at $P_1$, $P_4$ determine another pair, and in which the line $P_1P_4$ determines a double point. Hence we have

THEOREM 21. *If a line $l$ meets a conic in two points and $P_1$, $P_4$ are any other two points on the conic, the points in which $l$ meets the conic are a pair of an involution through a double point of which passes the line $P_1P_4$ and through a pair of conjugate points of which pass the tangents at $P_1$, $P_4$. If $l$ is tangent to the conic, the point of contact is the second double point of this involution.*

The construction of the conic corresponding to this theorem may be stated as follows: Given two tangents and their points of contact and one other point of the conic. On any line $l$ through the latter point is determined an involution of which one double point is the intersection with $l$ of the line joining the two points of contact, and of which one pair is the pair of intersections with $l$ of the two tangents. The conjugate in this involution of the given point of the conic on $l$ is a point of the conic.

## EXERCISE

State the duals of the theorems in this section.

**47. Pencils and ranges of conics. Order of contact.** The theorems of the last section and their plane duals determine the properties of certain systems of conics which we now proceed to discuss briefly.

DEFINITION. The set of all conics through the vertices of a complete quadrangle is called a *pencil of conics of Type I* (fig. 61).

DEFINITION. The set of all conics tangent to the sides of a complete quadrilateral is called a *range of conics of Type I* (fig. 62).

Theorem 19 and its plane dual give at once:

THEOREM 22. *Any line (not through a vertex of the determining quadrangle) is met by the conics of a pencil of Type I in the pairs of an involution.**

THEOREM 22′. *The tangents through any point (not on a side of the determining quadrilateral) to the conics of a range of Type I are the pairs of an involution.*

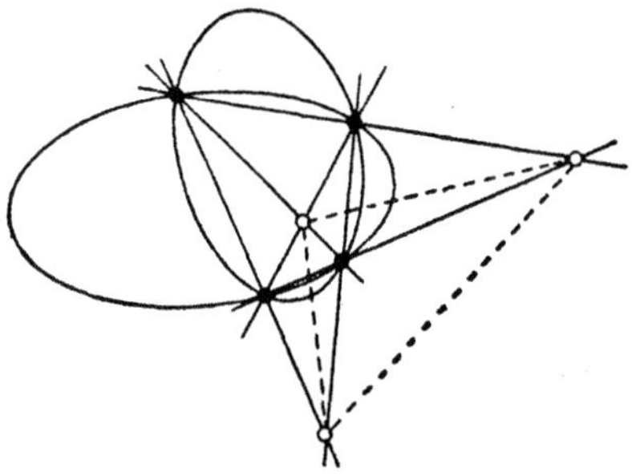

FIG. 61

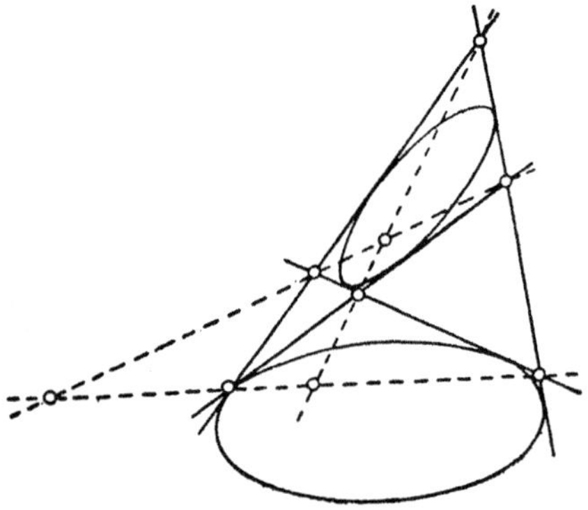

FIG. 62

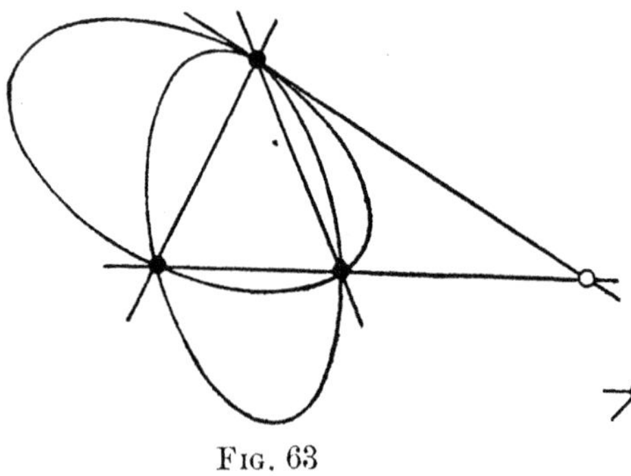

FIG. 63

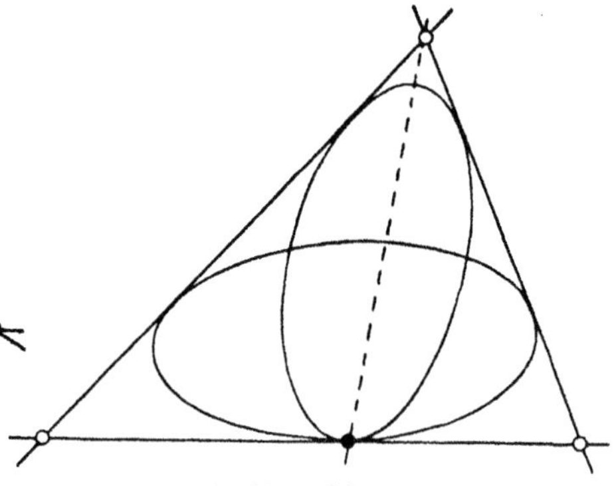

FIG. 64

COROLLARY. *Through a general*† *point in the plane there is one and only one, and tangent to a general line there are two or no conics of a given pencil of Type I.*

COROLLARY. *Tangent to a general line in the plane there is one and only one, and through a general point there are two or no conics of a given range of Type I.*

* This form of Desargues's theorem is due to Ch. Sturm, Annales de Mathématiques, Vol. XVII (1826), p. 180.

† The vertices of the quadrangle are regarded as exceptional points.

DEFINITION. The set of all conics through the vertices of a triangle and tangent to a fixed line through one vertex is called a *pencil of conics of Type II* (fig. 63).

DEFINITION. The set of all conics tangent to the sides of a triangle and passing through a fixed point on one side is called a *range of conics of Type II* (fig. 64).

Theorem 20 and its plane dual then give at once:

THEOREM 23. *Any line in the plane of a pencil of conics of Type II (which does not pass through a vertex of the determining triangle) is met by the conics of the pencil in the pairs of an involution.*

THEOREM 23′. *The tangents through any point in the plane of a range of conics of Type II (which is not on a side of the determining triangle) to the conics of the range are the pairs of an involution.*

COROLLARY. *Through a general point in the plane there is one and only one conic of the pencil; and tangent to a general line in the plane there are two or no conics of the pencil.*

COROLLARY. *Tangent to a general line in the plane there is one and only one conic of the range; and through a general point in the plane there are two or no conics of the range.*

DEFINITION. The set of all conics through two given points and tangent to two given lines through these points respectively is called a *pencil or range of conics of Type IV* * (fig. 65).

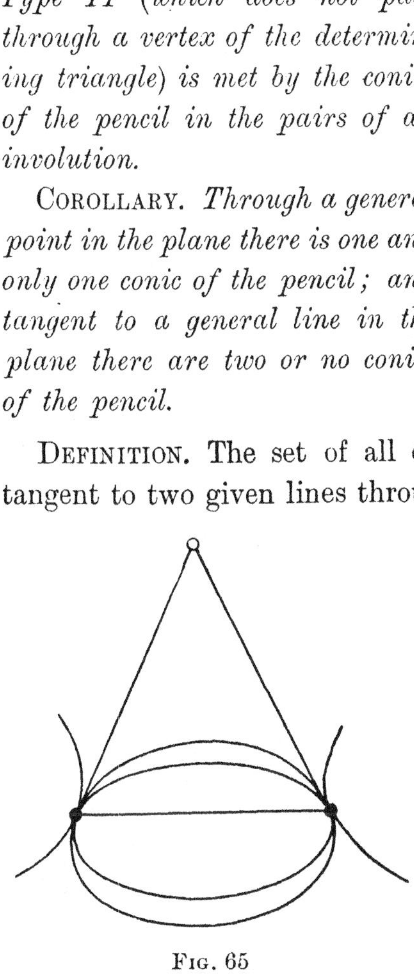

FIG. 65

Theorem 21 now gives at once:

THEOREM 24. *Any line in the plane of a pencil of conics of Type IV (which does not pass through either of the points common to all the conics of the pencil) is met by the conics of the pencil in the pairs of an involution. Through any point in the plane (not on either of the lines that are tangent to all the conics of the pencil) the tangents to the conics of the pencil are the pairs of an involution. The line joining the two points common to all the conics of the pencil meets*

* The classification of pencils and ranges of conics into types corresponds to the classification of the corresponding plane collineations (cf. Exs. 2, 4, 7, below).

*any line in a double point of the involution determined on that line. And the point of intersection of the common tangents is joined to any point by a double line of the involution determined at that point.*

COROLLARY. *Through any general point or tangent to any general line in the plane there is one and only one conic of the pencil.*

**EXERCISES**

**1.** What are the degenerate conics of a pencil or range of Type *I*? The diagonal triangle of the fundamental quadrangle (quadrilateral) of the pencil (range) is the only triangle which is self-polar with respect to two conics of the pencil (range).

**2.** Let $A^2$ and $B^2$ be any two conics of a pencil of Type *I*, and let $P$ be any point in the plane of the pencil. If $p$ is the polar of $P$ with respect to $A^2$, and $P'$ is the pole of $p$ with respect to $B^2$, the correspondence thus established between $[P]$ and $[P']$ is a projective collineation of Type *I*, whose invariant triangle is the diagonal triangle of the fundamental quadrangle. Do all projective collineations thus determined by a pencil of conics of Type *I* form a group? Dualize.

**3.** What are the degenerate conics of a pencil or range of Type *II*?

**4.** Let a pencil of conics of Type *II* be determined by a triangle $ABC$ and a tangent $a$ through $A$. Further, let $a'$ be the harmonic conjugate of $a$ with respect to $AB$ and $AC$, and let $A'$ be the intersection of $a$ and $BC$. Then $A$, $a$ and $A'$, $a'$ are pole and polar with respect to every conic of the pencil; and no pair of conics of the pencil have the same polars with regard to any other points than $A$ and $A'$. Dualize, and show that all the collineations determined as in Ex. 2 are in this case of Type *II*.

**5.** What are the degenerate conics of a pencil or range of Type *IV*?

**6.** Show that any point on the line joining the two points common to all the conics of a pencil of Type *IV* has the same polar with respect to all the conics of the pencil, and that these all pass through the point of intersection of the two common tangents.

**7.** Show that the collineations determined by a pencil of Type *IV* by the method of Ex. 2 are all homologies (i.e. of Type *IV*).

* The pencils and ranges of conics thus far considered have in common the properties (1) that the pencil (range) is completely defined as soon as two conics of the pencil (range) are given; (2) the conics of the pencil (range) determine an involution on any line (point) in the plane (with the exception of the lines (points) on the determining points (lines) of the pencil (range)). Three other systems of conics may be defined which likewise have these properties. These new systems

* The remainder of this section may be omitted on a first reading.

may be regarded as degenerate cases of the pencils and ranges already defined. Their existence is established by the theorems given below, which, together with their corollaries, may be regarded as degenerate cases of the theorem of Desargues. We shall need the following

LEMMA. *Any conic is transformed by a projective collineation in the plane of the conic into a conic such that the tangents at homologous points are homologous.*

*Proof.* This follows almost directly from the definition of a conic. Two projective flat pencils are transformed by a projective collineation into two projective flat pencils. The intersections of pairs of homologous lines of one pencil are therefore transformed into the intersections of the corresponding pairs of homologous lines of the transformed pencils. If any line meets the first conic in a point $P$, the transformed line will meet the transformed conic in the point homologous with $P$. Therefore a tangent at a point of the first conic must be transformed into the tangent at the corresponding point of the second conic.

THEOREM 25. *If a line $p_0$ is a tangent to a conic $A^2$ at a point $P_0$, and $Q$ is any point of $A^2$, then through any point on the plane of $A^2$ but not on $A^2$ or $p_0$, there is one and only one conic $B^2$ through $P_0$ and $Q$, tangent to $p_0$, and such that there is no point of $p_0$, except $P_0$, having the same polar with regard to both $A^2$ and $B^2$.*

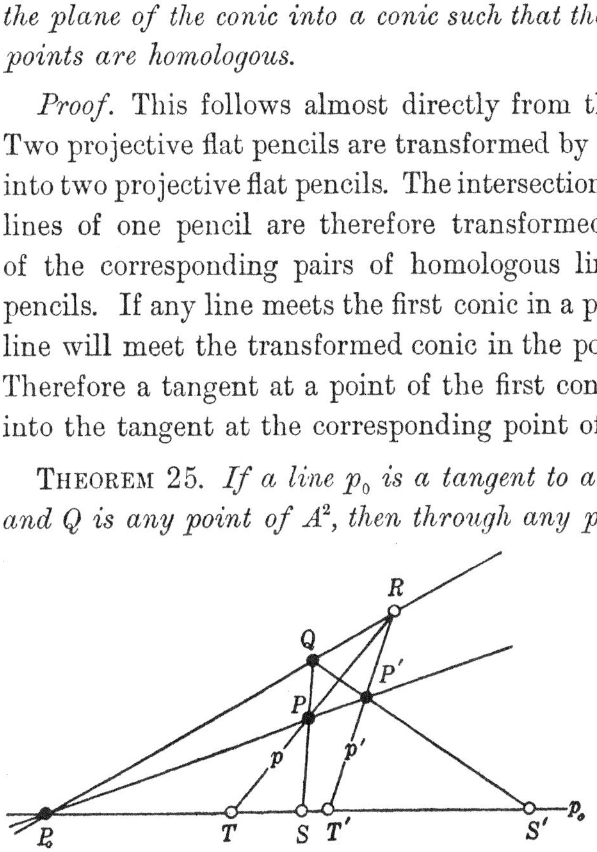

FIG. 66

*Proof.* If $P'$ is any point of the plane not on $p_0$ or $A^2$, let $P$ be the second point in which $P_0P'$ meets $A^2$ (fig. 66). There is one and only one elation with center $P_0$ and axis $P_0Q$ changing $P$ into $P'$ (Theorem 9, Chap. III). This elation (by the lemma above) changes $A^2$ into another conic $B^2$ through the points $P_0$ and $Q$ and tangent to $p_0$. The lines through $P_0$ are unchanged by the elation, whereas their poles (on $p_0$) are subjected to a parabolic projectivity. Hence no point on $p_0$ (distinct from $P_0$) has the same polar with regard to $A^2$ as with regard to $B^2$. Since $A^2$ is transformed into $B^2$ by an elation, the two conics can have no other points in common than $P_0$ and $Q$.

That there is only one conic $B^2$ through $P'$ satisfying the conditions of the theorem is to be seen as follows: Let $QP$ meet $p_0$ in $S$, and $QP'$ meet $p_0$ in $S'$ (fig. 66). The point $S$ has the same polar with regard to $A^2$ as $S'$ with regard to any conic $B^2$, since this polar must be the harmonic conjugate of $p_0$ with regard to $P_0Q$ and $P_0P$. Let $p$ be the tangent to $A^2$ at $P$ and $p'$ be the tangent to $B^2$ at $P'$, and let $p$ and $p'$ meet $p_0$ in $T$ and $T'$ respectively. The points

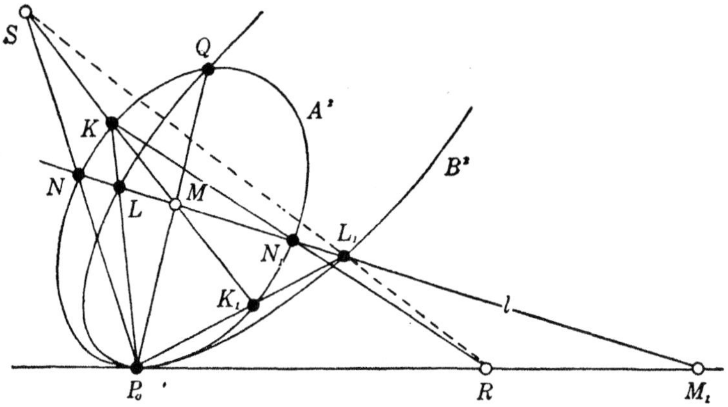


FIG. 67

$T$ and $T'$ have the same polar, namely $P_0P$, with regard to $A^2$ and any conic $B^2$. By the conditions of the theorem the projectivity

$$P_0ST \;\overline{\wedge}\; P_0S'T'$$

must be parabolic. Hence, by Theorem 23, Cor., Chap. IV,

$$Q(P_0ST,\ P_0T'S').$$

Hence $p$ and $p'$ must meet on $P_0Q$ in a point $R$ so as to form the quadrangle $RQPP'$. This determines the elements $P_0$, $Q$, $P'$, $p_0$, $p'$ of $B^2$, and hence there is only one possible conic $B^2$.

COROLLARY 1. *The conics $A^2$ and $B^2$ can have no other points in common than $P_0$ and $Q$.*

COROLLARY 2. *Any line $l$ not on $P_0$ or $Q$ which meets $A^2$ and $B^2$ meets them in pairs of an involution in which the points of intersection of $l$ with $P_0Q$ and $p_0$ are conjugate.*

*Proof.* Let $l$ meet $A^2$ in $N$ and $N_1$, $B^2$ in $L$ and $L_1$, $P_0Q$ in $M$, and $p_0$ in $M_1$ (fig. 67). Let $K$ and $K_1$ be the points of $A^2$ which are transformed by the elation into $L$ and $L_1$ respectively. By the definition of an elation $K$ and $K_1$ are collinear with $M$, while $K$ is on the line $LP_0$ and $K_1$ on $L_1P_0$. Let $KN_1$ meet $p_0$ in $R$, and $NP_0$ meet $KK_1$ in $S$.

Then, since $N, K, N_1, K_1$ are on the conic to which $p_0$ is tangent at $P_0$, we have, by Theorem 6, applied to the degenerate hexagon $P_0P_0K_1KN_1N$, that $S$, $L_1$, and $R$ are collinear. Hence the complete quadrilateral $SR$, $KN_1$, $KK_1$, $l$ has pairs of opposite vertices on $P_0M$ and $P_0M_1$, $P_0N$ and $P_0N_1$, $P_0L$ and $P_0L_1$. Hence $Q(MNL, M_1N_1L_1)$.*

DEFINITION. The set of all conics through a point $Q$ and tangent to a line $p_0$ at a point $P_0$, and such that no point of $p_0$ except $P_0$ has the same polar with regard to two conics of the set, is called a *pencil of conics of Type III* (fig. 68).

DEFINITION. The set of all conics tangent to a line $q$ and tangent to a line $p_0$ at a point $P_0$, and such that no line on $P$ except $p_0$ has the same pole with regard to two conics of the set, is called a *range of conics of Type III* (fig. 69).

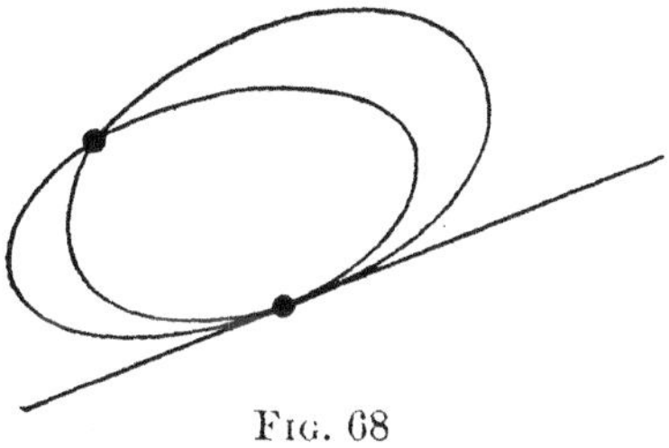

FIG. 68

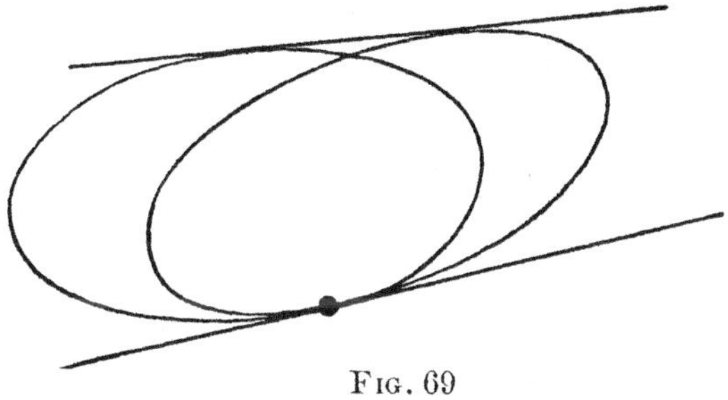

FIG. 69

Two conics of such a pencil (range) are said to have *contact of the second order*, or to *osculate*, at $P_0$.

Corollary 2 of Theorem 25 now gives at once:

THEOREM 26. *Any line in the plane of a pencil of conics of Type III, which is not on either of the common points of the pencil, is met by the conics of the pencil in the pairs of an involution. Through any point in the plane except the common points there is one and only one conic of the pencil; and tangent to any line not through either of the common points there are two or no conics of the pencil.*

THEOREM 26′. *Through any point in the plane of a range of conics of Type III, which is not on either of the common tangents of the range, the tangents to the conics of the pencil are the pairs of an involution. Tangent to any line in the plane except the common tangents there is one and only one conic of the range; and through any point not on either of the common tangents there are two or no conics of the range.*

* This argument has implicitly proved that three pairs of points of a conic, as $KK_1$, $NN_1$, $P_0Q$, such that the lines joining them meet in a point $M$, are projected from any point of the conic by a quadrangular set of lines (Theorem 16, Chap. VIII).

The pencil is determined by the two common points, the common tangent, and one conic of the pencil.

The range is determined by the two common tangents, the common point, and one conic of the range.

**EXERCISES**

**1.** What are the degenerate conics of this pencil and range?

**2.** Show that the collineation obtained by making correspond to any point $P$ the point $P'$ which has the same polar $p$ with regard to one given conic of the pencil (range) that $P$ has with regard to another given conic of the pencil (range) is of Type *III*.

THEOREM 27. *If a line $p_0$ is tangent to a conic $A^2$ at a point $P_0$, there is one and only one conic tangent to $p_0$ at $P_0$ and passing through any other point $P'$ of the plane of $A^2$ not on $p_0$ or $A^2$ which determines for every point of $p_0$ the same polar line as does $A^2$.*

*Proof.* Let $P$ be the second point in which $P_0P'$ meets $A^2$ (fig. 70). There is one and only one elation of which $P_0$ is center and $p_0$ axis, changing $P$ to $P'$. This elation changes $A^2$ into a conic $B^2$ through

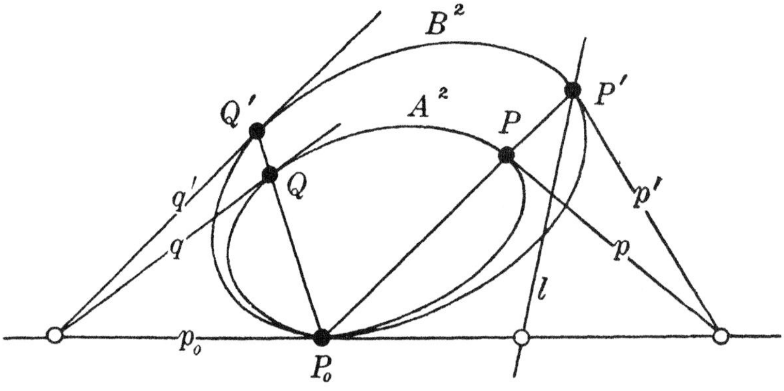


FIG. 70

$P'$, and is such that if $q$ is any tangent to $A^2$ at a point $Q$, then $q$ is transformed to a tangent $q'$ of $B^2$ passing through $qp_0$, and $Q$ is transformed into the point of contact $Q'$ of $q'$, collinear with $Q$ and $P_0$. Hence there is one conic of the required type through $P'$.

That there is only one is evident, because if $l$ is any line through $P'$, any conic $B^2$ must pass through the fourth harmonic of $P'$ with regard to $lp_0$ and the polar of $lp_0$ as to $A^2$ (Theorem 13). By considering two lines $l$ we thus determine enough points to fix $B^2$.

COROLLARY 1. *By duality there is one and only one conic $B^2$ tangent to any line not passing through $P_0$.*

COROLLARY 2. *Any line $l$ not on $P_0$ which meets $A^2$ and $B^2$ meets them in pairs of an involution one double point of which is $lp_0$, and the other the point of $l$ conjugate to $lp_0$ with respect to $A^2$. A dual statement holds for any point $L$ not on $p_0$.*

COROLLARY 3. *The conics $A^2$ and $B^2$ can have no other point in common than $P_0$ and no other tangent in common than $p_0$.*

*Proof.* If they had one other point $P$ in common, they would have in common the conjugate of $P$ in the involution determined on any line through $P$ according to Corollary 2.

DEFINITION. The set of all conics tangent to a given line $p_0$ at a given point $P_0$, and such that each point on $p_0$ has the same polar with regard to all conics of the set, is called a *pencil* or *range of conics of Type V*. Two conics of such a pencil are said to have *contact of the third order*, or to *hyperosculate* at $P_0$.

Theorem 27 and its first two corollaries now give at once:

THEOREM 28. *Any line $l$ not on the common point of a pencil of Type V is met by the conics of the pencil in pairs of an involution one double point of which is the intersection of $l$ with the common tangent. Through any point $L$ not on the common tangent the pairs of tangents to the conics of the pencil form an involution one double line of which is the line joining $L$ to the common point. There is one conic of the set through each point of the plane not on the common tangent, and one conic tangent to each line not on the common point.*

The pencil or range is determined by the common point, the common tangent, and one conic of the set.

## EXERCISES

1. What are the degenerate conics of a pencil of Type $V$?

2. Show that the collineation obtained by making correspond to any point $P$ the point $Q$ which has the same pole $p$ with regard to one conic of a pencil of Type $V$ that $P$ has with regard to another conic of the pencil is an elation.

3. The lines polar to a point $A$ with regard to all the conics of a pencil of any of the five types pass through a point $A'$. The points $A$ and $A'$ are double points of the involution determined by the pencil on the line $AA'$. Construct $A'$. Dualize. Derive a theorem on the complete quadrangle as a special case of this one.

4. Construct the polar line of a point $A$ with regard to a conic $C^2$ being given four points of $C^2$ and a conjugate of $A$ with regard to $C^2$.

**5.** Given an involution I on a line $l$, a pair of points $A$ and $A'$ on $l$ not conjugate in I, and any other point $B$ on $l$, construct a point $B'$ such that $A$ and $A'$ and $B$ and $B'$ are pairs of an involution I′ whose double points are a pair in I. The involution I′ may also be described as one which is commutative with I, or such that the product of I and I′ is an involution.

**6.** There is one and only one conic through three points and having a given point $P$ and line $p$ as pole and polar.

**7.** The conics through three points and having a given pair of points as conjugate points form a pencil of conics.

## MISCELLANEOUS EXERCISES

**1.** If $O$ and $o$ are pole and polar with regard to a conic, and $A$ and $B$ are two points of the conic collinear with $O$, then the conic is generated by the two pencils $A[P]$ and $B[P']$ where $P$ and $P'$ are paired in the involution on $o$ of conjugates with regard to the conic.

**2.** Given a complete plane five-point $ABCDE$. The locus of all points $X$ such that

$$X(BCDE) \barwedge A(BCDE)$$

is a conic.

**3.** Given two projective nonperspective pencils, $[p]$ and $[q]$. Every line $l$ upon which the projectivity $l[p] \barwedge l[q]$ is involutoric passes through a fixed point $O$. The point $O$ is the pole of the line joining the centers of the pencils with respect to the conic generated by them.

**4.** If two complete quadrangles have the same diagonal points, their eight vertices lie on a conic (Cremona, Projective Geometry (Oxford, 1885), Chap. XX).

**5.** If two conics intersect in four points, the eight tangents to them at these points are on the same line conic. Dualize and extend to the cases where the conics are in pencils of Types *II–V*.

**6.** All conics with respect to which a given triangle is self-conjugate, and which pass through a fixed point, also pass through three other fixed points. Dualize.

**7.** Construct a conic through two given points and with a given self-conjugate triangle. Dualize.

**8.** If the sides of a triangle are tangent to a conic, the lines joining two of its vertices to any point conjugate with regard to the conic to the third vertex are conjugate with regard to the conic. Dualize.

**9.** If two points $P$ and $Q$ on a conic are joined to two conjugate points $P'$, $Q'$ on a line conjugate to $PQ$, then $PP'$ and $QQ'$ meet on the conic.

**10.** If a simple quadrilateral is circumscribed to a conic, and if $l$ is any transversal through the intersection of its diagonals, $l$ will meet the conic and the pairs of opposite sides in conjugate pairs of an involution. Dualize.

**11.** Given a conic and three fixed collinear points $A$, $B$, $C$. There is a fourth point $D$ on the line $AB$ such that if three sides of a simple quadrangle inscribed in the conic pass through $A$, $B$, and $C$ respectively, the fourth passes through $D$ (Cremona, Chap. XVII).

**12.** If a variable simple $n$-line ($n$ even) is inscribed in a conic in such a way that $n-1$ of its sides pass through $n-1$ fixed collinear points, then the other side passes through another fixed point of the same line. Dualize this theorem.

**13.** If two conics intersect in two points $A$, $B$ (or are tangent at a point $A$) and two lines through $A$ and $B$ respectively (or through the point of contact $A$) meet the conics again in $O$, $O'$ and $L$, $L'$, then the lines $OL$ and $O'L'$ meet on the line joining the remaining points of intersection (if existent) of the two conics.

**14.** If a conic $C^2$ passes through the vertices of a triangle which is self-polar with respect to another conic $K^2$, there is a triangle inscribed in $C^2$ and self-polar with regard to $K^2$, and having one vertex at any point of $C^2$. The lines which cut $C^2$ and $K^2$ in two pairs of points which are harmonically conjugate to one another constitute a line conic $C_2^2$, which is the polar reciprocal of $C^2$ with regard to $K^2$ (Cremona, Chap. XXII).

**15.** If a variable triangle is such that two of its sides pass respectively through two fixed points $O'$ and $O$ lying on a given conic, and the vertices opposite them lie respectively on two fixed lines $u$ and $u'$, while the third vertex lies always on the given conic, then the third side touches a fixed conic, which touches the lines $u$ and $u'$. Dualize (Cremona, Chap. XXII).

**16.** If $P$ is a variable point on a conic containing $A$, $B$, $C$, and $l$ is a variable line through $P$ such that all throws $\mathsf{T}\,(PA,\ PB;\ PC,\ l)$ are projective, then all lines $l$ meet in a point of the conic (Schröter, Journal für die reine und angewandte Mathematik, Vol. LXII, p. 222).

**17.** Given a fixed conic and a fixed line, and three fixed points $A$, $B$, $C$ on the conic, let $P$ be a variable point on the conic and let $PA$, $PB$, $PC$ meet the fixed line in $A'$, $B'$, $C'$. If $O$ is a fixed point of the plane and $(OA', PB') = K$ and $(KC') = l$, then $K$ describes a conic and $l$ a pencil of lines whose center is on the conic described by $K$ (Schröter, loc. cit.).

**18.** Two triangles $ABC$ and $PQR$ are perspective in four ways. Show that if $ABC$ and the point $P$ are fixed and $Q$, $R$ are variable, the locus of each of the latter points is a conic (cf. Ex. 8, p. 105, and Schröter, Mathematische Annalen, Vol. II (1870), p. 553).

**19.** Given six points on a conic. By taking these in all possible orders 60 different simple hexagons inscribed in the conic are obtained. Each of these simple hexagons gives rise to a Pascal line. The figure thus associated with any six points of a conic is called the *hexagrammum mysticum*.* Prove the following properties of the hexagrammum mysticum:

i. The Pascal lines of the three hexagons $P_1P_2P_3P_4P_5P_6$, $P_1P_4P_3P_6P_5P_2$, and $P_1P_6P_3P_2P_5P_4$ are concurrent. The point thus associated with such a set of three hexagons is called a *Steiner point*.

ii. There are in all 20 Steiner points.

* On the Pascal hexagram cf. Steiner-Schröter, Vorlesungen über Synthetische Geometrie, Vol. II, § 28; Salmon, Conic Sections in the Notes; Christine Ladd, American Journal of Mathematics, Vol. II (1879), p. 1.

iii. From a given simple hexagon five others are obtained by permuting in all possible ways a set of three vertices no two of which are adjacent. The Pascal lines of these six hexagons pass through two Steiner points, which are called *conjugate* Steiner points. The 20 Steiner points fall into ten pairs of conjugates.

iv. The 20 Steiner points lie by fours on 15 lines called *Steiner lines.*

v. What is the symbol of the configuration composed of the 20 Steiner points and the 15 Steiner lines?

**20.** Discuss the problem corresponding to that of Ex. 19 for all the special cases of Pascal's theorem.

**21.** State the duals of the last two exercises.

**22.** If in a plane there are given two conics, any point $A$ has a polar with respect to each of them. If these polars intersect in $A'$, the points $A$, $A'$ are conjugate with respect to both conics. The polars of $A'$ likewise meet in $A$. In this way every point in the plane is paired with a unique other point. By the dual process every line in the plane is paired with a unique line to which it is conjugate with respect to both conics. Show that in this correspondence the points of a line correspond in general to the points of a conic. All such conics which correspond to lines of the plane have in common a set of at most three points. The polars of every such common point coincide, so that to each of them is made to correspond all the points of a line. They form the exceptional elements of the correspondence. Dualize (Reye-Holgate, p. 110).*

**23.** If in the last exercise the two given conics pass through the vertices of the same quadrangle, the diagonal points of this quadrangle are the "common points" mentioned in the preceding exercise (Reye-Holgate, p. 110).

**24.** Given a cone of lines with vertex $O$ and a line $u$ through $O$. Then a one-to-one correspondence may be established among the lines through $O$ by associating with every such line $a$ its conjugate $a'$ with respect to the cone lying in the plane $au$. If, then, $a$ describes a plane $\pi$, $a'$ will describe a cone of lines passing through $u$ and through the polar line of $\pi$, and which has in common with the given cone any lines common to it and to the given cone and the polar plane of $u$ (Reye-Holgate, p. 111).*

**25.** Two conics are determined by the two sets of five points $A$, $B$, $C$, $D$, $E$ and $A$, $B$, $C$, $H$, $K$. Construct the fourth point of intersection of the two conics (Castelnuovo, Lezioni di Geometria, p. 391).

**26.** Apply the result of the preceding Exercise to construct the point $P$ such that the set of lines $P(A, B, C, D, E)$ joining $P$ to the vertices of any given complete plane five-point be projective with any given set of five points on a line (Castelnuovo, loc. cit.).

**27.** Given any plane quadrilateral, construct a line which meets the sides of the quadrilateral in a set of four points projective with any given set of four collinear points.

* The correspondences defined in Exs. 22 and 24 are examples of so-called quadratic correspondences.

**28.** Two sets of five points $A$, $B$, $C$, $D$, $E$ and $A$, $B$, $H$, $K$, $L$ determine two conics which intersect again in two points $X$, $Y$. Construct the line $XY$ and show that the points $X$, $Y$ are the double points of a certain involution (Castelnuovo, loc. cit.).

**29.** If three conics pass through two given points $A$, $B$ and the three pairs of conics cut again in three pairs of points, show that the three lines joining these pairs of points are concurrent (Castelnuovo, loc. cit.).

**30.** Prove the converse of the second theorem of Desargues: The conics passing through three fixed points and meeting a given line in the pairs of an involution pass through a fourth fixed point. This theorem may be used to construct a conic, given three of its points and a pair of points conjugate with respect to the conic. Dualize (Castelnuovo, loc. cit.).

**31.** The poles of a line with respect to all the conics of a pencil of conics of Type $I$ are on a conic which passes through the diagonal points of the quadrangle defining the pencil. This conic cuts the given line in the points in which the latter is tangent to conics of the pencil. Dualize.

**32.** Let $p$ be the polar of a point $P$ with regard to a triangle $ABC$. If $P$ varies on a conic which passes through $A$, $B$, $C$, then $p$ passes through a fixed point $Q$ (Cayley, Collected Works, Vol. I, p. 361).

**33.** If two conics are inscribed in a triangle, the six points of contact are on a third conic.

**34.** Any two vertices of a triangle circumscribed to a conic are separated harmonically by the point of contact of the side containing them and the point where this side meets the line joining the points of contact of the other sides.

# CHAPTER VI

## ALGEBRA OF POINTS AND ONE-DIMENSIONAL COÖRDINATE SYSTEMS

**48. Addition of points.** That analytic methods may be introduced into geometry on a strictly projective basis was first shown by von Staudt.* The point algebra on a line which is defined in this chapter without the use of any further assumptions than A, E, P is essentially equivalent to von Staudt's algebra of throws (p. 60), a brief account of which will be found in § 55. The original method of von Staudt has, however, been considerably clarified and simplified by modern researches on the foundations of geometry.† All the definitions and theorems of this chapter before Theorem 6 are independent of Assumption P. Indeed, if desired, this part of the chapter may be read before taking up Chap. IV.

Given a line $l$, and on $l$ three distinct (arbitrary) fixed points which for convenience and suggestiveness we denote by $P_0$, $P_1$, $P_\infty$, we define two *one-valued operations* ‡ on pairs of points of $l$ with reference to the *fundamental points* $P_0$, $P_1$, $P_\infty$. The fundamental points are said to determine a *scale* on $l$.

Definition. In any plane through $l$ let $l_\infty$ and $l'_\infty$ be any two lines through $P_\infty$, and let $l_0$ be any line through $P_0$ meeting $l_\infty$ and $l'_\infty$ in points $A$ and $A'$ respectively (fig. 71). Let $P_x$ and $P_y$ be any two points of $l$, and let the lines $P_xA$ and $P_yA'$ meet $l'_\infty$ and $l_\infty$ in the points $X$ and $Y$ respectively. The point $P_{x+y}$, in which the line $XY$ meets $l$, is called the *sum* of the points $P_x$ and $P_y$ (in symbols $P_x + P_y = P_{x+y}$) in

* K. G. C. von Staudt (1798–1867), Beiträge zur Geometrie der Lage, Heft 2 (1857), pp. 166 et seq. This book is concerned also with the related question of the interpretation of imaginary elements in geometry.

† Cf., for example, G. Hessenberg, Ueber einen Geometrischen Calcul, Acta Mathematica, Vol. XXIX, p. 1.

‡ By a *one-valued operation* o on a pair of points $A$, $B$ is meant any process whereby with every pair $A$, $B$ is associated a point $C$, which is unique provided the order of $A$, $B$ is given; in symbols $A \circ B = C$. Here "order" has no geometrical significance, but implies merely the formal difference of $A \circ B$ and $B \circ A$. If $A \circ B = B \circ A$, the operation is *commutative;* if $(A \circ B) \circ C = A \circ (B \circ C)$, the operation is *associative.*

the scale $P_0, P_1, P_\infty$. The operation of obtaining the sum of two points is called *addition*.*

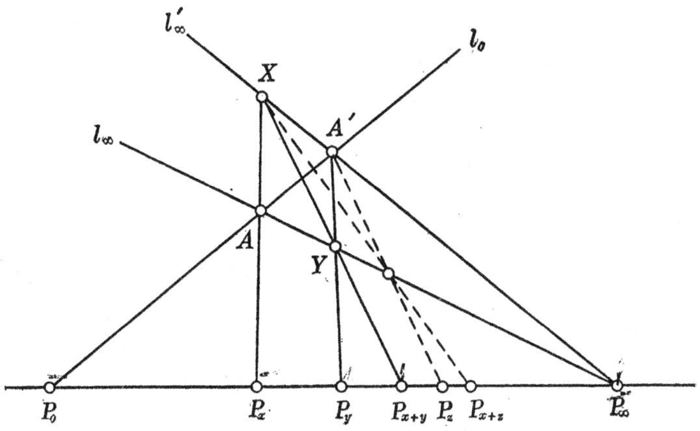


FIG. 71

THEOREM 1. *If $P_x$ and $P_y$ are distinct from $P_0$ and $P_\infty$, $Q(P_\infty P_x P_0, P_\infty P_y P_{x+y})$ is a necessary and sufficient condition for the equality $P_x + P_y = P_{x+y}$.* (A, E)

This follows immediately from the definition, $AXA'Y$ being a quadrangle which determines the given quadrangular set.

COROLLARY 1. *If $P_x$ is any point of $l$, we have $P_x + P_0 = P_0 + P_x = P_x$, and $P_x + P_\infty = P_\infty + P_x = P_\infty$ $(P_x \neq P_\infty)$.* (A, E)

This is also an immediate consequence of the definition.

COROLLARY 2. *The operation of addition is one-valued for every pair of points $P_x$, $P_y$ of $l$, except for the pair $P_\infty$, $P_\infty$.* (A, E)

This follows from the theorem above and the corollary of

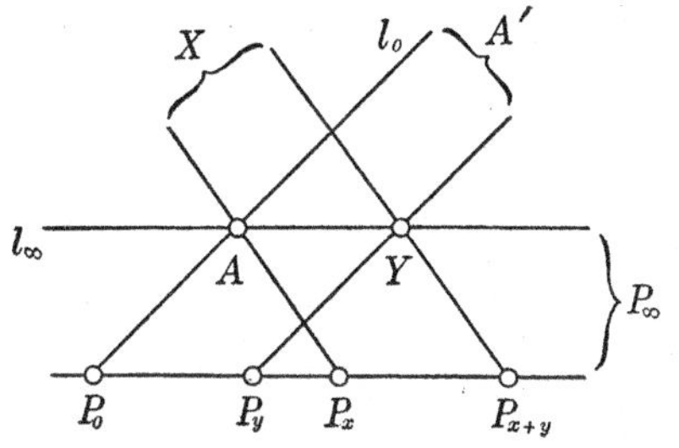


FIG. 72

* The historical origin of this construction will be evident on inspection of the attached figure. This is the figure which results, if we choose for $l'_\infty$ the "line at infinity" in the plane in the sense of ordinary Euclidean geometry (cf. p. 8). The construction is clearly equivalent to a translation of the vector $P_0P_y$ along the line $l$, which brings its initial point into coincidence with the terminal point of the vector $P_0P_x$, which is the ordinary construction for the sum of two vectors on a line.

Theorem 3, Chap. II, in case $P_x$ and $P_y$ are distinct from $P_0$ and $P_\infty$. If one of the points $P_x$, $P_y$ coincides with $P_0$ or $P_\infty$, it follows from Corollary 1.

COROLLARY 3. *The operation of addition is associative; i.e.*

$$P_x + (P_y + P_z) = (P_x + P_y) + P_z$$

*for any three points* $P_x$, $P_y$, $P_z$ *for which the above expressions are defined.* (A, E)

*Proof* (fig. 73). Let $P_x + P_y$ be determined as in the definition by means of three lines $l_\infty$, $l'_\infty$, $l_0$ and the line $XY$. Let the line $P_0Y$ be denoted by $l'_0$, and by means of $l_\infty$, $l'_\infty$, $l'_0$ construct the point $(P_x + P_y) + P_z$,

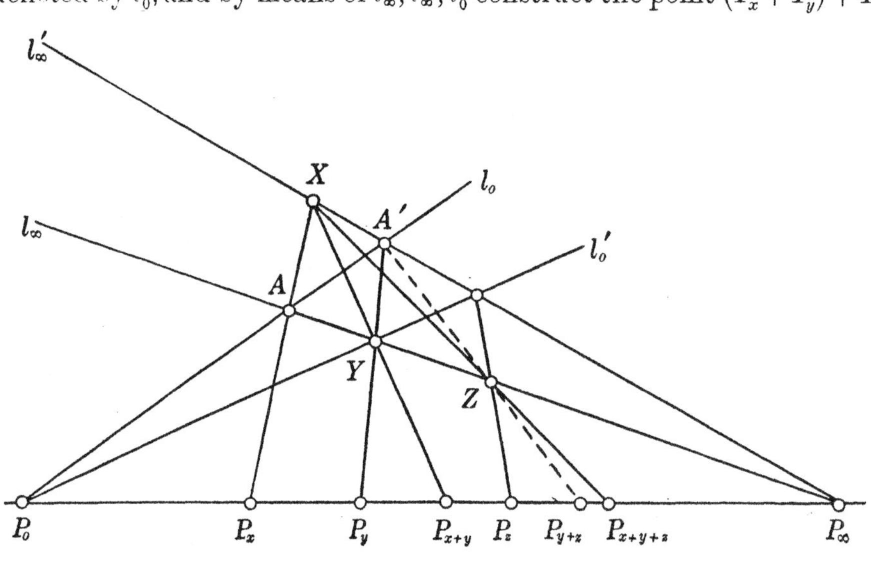


FIG. 73

which is determined by the line $XZ$, say. If now the point $P_y + P_z$ be constructed by means of the lines $l_\infty$, $l'_\infty$, $l'_0$, and then the point $P_x + (P_y + P_z)$ be constructed by means of the lines $l_\infty$, $l'_\infty$, $l_0$, it will be seen that the latter point is determined by the same line $XZ$.

COROLLARY 4. *The operation of addition is commutative; i.e.*

$$P_x + P_y = P_y + P_x$$

*for every pair of points* $P_x$, $P_y$ *for which the operation is defined.* (A, E)

*Proof.* By reference to the complete quadrangle $AXA'Y$ (fig. 71) there appears the quadrangular set $Q(P_\infty P_y P_0, P_\infty P_x P_{x+y})$, which by the theorem implies that $P_y + P_x = P_{x+y}$. But, by definition, $P_x + P_y = P_{x+y}$. Hence $P_y + P_x = P_x + P_y$.

THEOREM 2. *Any three points* $P_x$, $P_y$, $P_a (P_a \neq P_\infty)$ *satisfy the relation*

$$P_\infty P_0 P_x P_y \overline{\wedge} P_\infty P_a P_{x+a} P_{y+a};$$

*i.e. the correspondence established by making each point* $P_x$ *of* $l$ *correspond to* $P_x' = P_x + P_a$, *where* $P_a (\neq P_\infty)$ *is any fixed point of* $l$, *is projective.* (A, E)

*Proof.* The definition of addition (fig. 71) gives this projectivity as the result of two perspectivities:*

$$[P_x] \overset{A}{\overline{\overline{\wedge}}} [X] \overset{Y}{\overline{\overline{\wedge}}} [P_x'].$$

The set of all projectivities determined by all possible choices of $P_a$ in the formula $P_x' = P_x + P_a$ is the group described in Example 2, p. 70. The sum of two points $P_a$ and $P_b$ might indeed have been defined as the point into which $P_b$ is transformed when $P_0$ is transformed into $P_a$ by a projectivity of this group. The associative law for addition would thus appear as a special case of the associative law which holds for the composition of correspondences in general; and the commutative law for addition would be a consequence of the commutativity of this particular group.

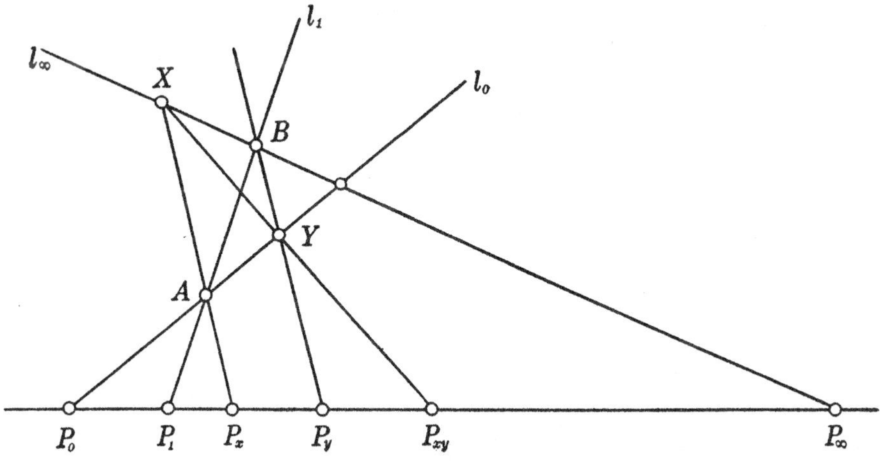


FIG. 74

**49. Multiplication of points.** DEFINITION. In any plane through $l$ let $l_0$, $l_1$, $l_\infty$ be any three lines through $P_0$, $P_1$, $P_\infty$ respectively, and let $l_1$ meet $l_0$ and $l_\infty$ in points $A$ and $B$ respectively (fig. 74). Let $P_x$, $P_y$ be any two points of $l$, and let the lines $P_xA$ and $P_yB$ meet $l_\infty$ and $l_0$ in the points $X$ and $Y$ respectively. The point $P_{xy}$ in which the line $XY$ meets $l$ is

* To make fig. 71 correspond to the notation of this theorem, $P_y$ must be identified with $P_a$.

called the *product of $P_x$ by $P_y$* (in symbols $P_x \cdot P_y = P_{xy}$) in the scale $P_0, P_1, P_\infty$ on $l$. The operation of obtaining the product of two points is called *multiplication*.* Each of the points $P_x$, $P_y$ is called a *factor* of the product $P_x \cdot P_y$.

THEOREM 3. *If $P_x$ and $P_y$ are any two points of $l$ distinct from $P_0, P_1, P_\infty$, $Q(P_0P_xP_1, P_\infty P_yP_{xy})$ is necessary and sufficient for the equality $P_x \cdot P_y = P_{xy}$.* (A, E)

This follows at once from the definition, $AXBY$ being the defining quadrangle.

COROLLARY 1. *For any point $P_x$ $(\neq P_\infty)$ on $l$ we have the relations $P_1 \cdot P_x = P_x \cdot P_1 = P_x$; $P_0 \cdot P_x = P_x \cdot P_0 = P_0$; $P_\infty P_x = P_x \cdot P_\infty = P_\infty$ $(P_x \neq P_0)$.*

This follows at once from the definition.

COROLLARY 2. *The operation of multiplication is one-valued for every pair of points $P_x$, $P_y$ of $l$, except $P_0 \cdot P_\infty$ and $P_\infty \cdot P_0$.* (A, E)

This follows from Corollary 1, if one of the points $P_x$, $P_y$ coincides with $P_0$, $P_1$, or $P_\infty$. Otherwise, it follows from the corollary, p. 50, in connection with the above theorem.

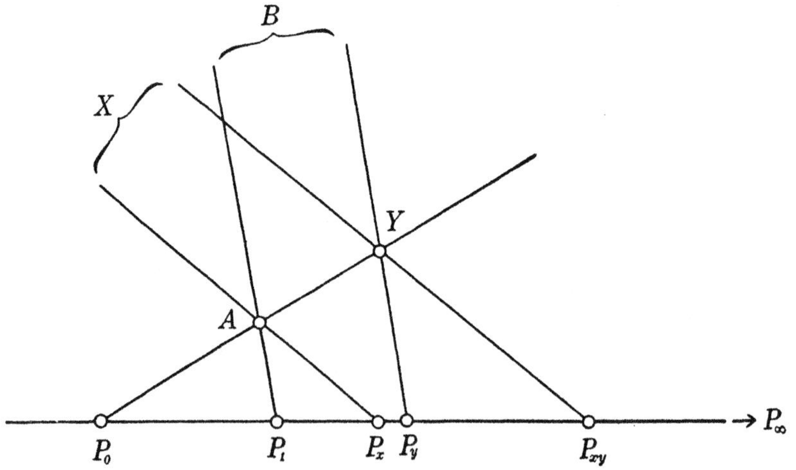


FIG. 75

* The origin of this construction may also be seen in a simple construction of metric Euclidean geometry, which results from the construction of the definition by letting the line $l_\infty$ be the "line at infinity" (cf. p. 8). In the attached figure which gives this metric construction we have readily, from similar triangles, the proportions:

$$\frac{P_0P_1}{P_0P_y} = \frac{P_0A}{P_0Y} = \frac{P_0P_x}{P_0P_{xy}},$$

which, on taking the segment $P_0P_1 = 1$, gives the desired result $P_0P_{xy} = P_0P_x \cdot P_0P_y$.

COROLLARY 3. *The operation of multiplication is associative; i.e. we have* $(P_x \cdot P_y) \cdot P_z = P_x \cdot (P_y \cdot P_z)$ *for every three points* $P_x, P_y, P_z$ *for which these products are defined.* (A, E)

*Proof* (fig. 76). The proof is entirely analogous to the proof for the associative law for addition. Let the point $P_x \cdot P_y$ be constructed

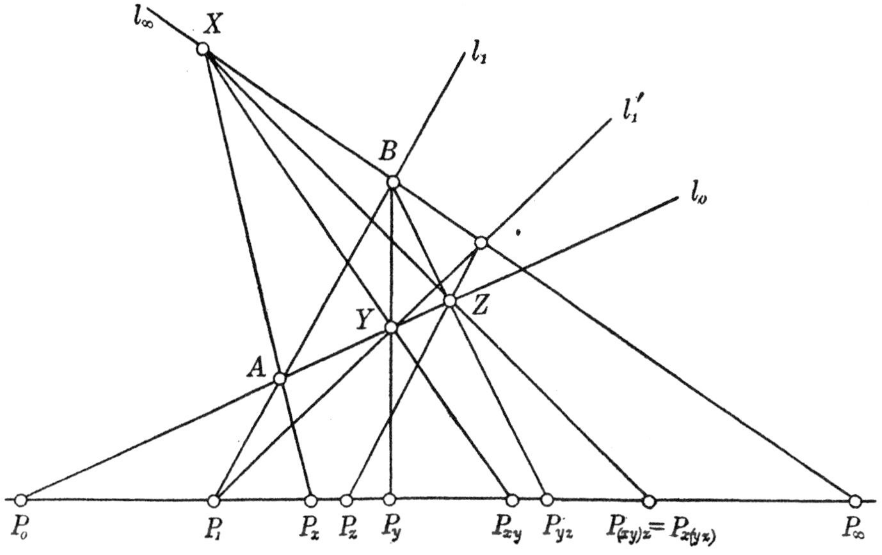


FIG. 76

as in the definition by means of three fundamental lines $l_0$, $l_1$, $l_\infty$, the point $P_{xy}$ being determined by the line $XY$. Denote the line $P_1Y$ by $l_1'$, and construct the point $P_{xy} \cdot P_z = (P_x \cdot P_y) \cdot P_z$, using the lines $l_0, l_1', l_\infty$ as fundamental. Further, let the point $P_y \cdot P_z = P_{yz}$ be constructed by means of the lines $l_0$, $l_1'$, $l_\infty$, and then let $P_x \cdot P_{yz} = P_x \cdot (P_y \cdot P_z)$ be constructed by means of $l_0$, $l_1$, $l_\infty$. It is then seen that the points $P_x \cdot P_{yz}$ and $P_{xy} \cdot P_z$ are determined by the same line.

By analogy with Theorem 1, Cor. 4, we should now prove that multiplication is also commutative. It will, however, appear presently that the commutativity of multiplication cannot be proved without the use of Assumption P (or its equivalent). It must indeed be clearly noted at this point that the definition of multiplication requires the first factor $P_x$ in a product to form with $P_0$ and $P_1$ a *point triple* of the quadrangular set on $l$ (cf. p. 49); the construction of the product is therefore not independent of the order of the factors. Moreover, the fact that in Theorem 3, Chap. II, the quadrangles giving the points of the set are similarly placed, was essential in the proof of that theorem. We cannot therefore use this theorem to prove the commutative law for multiplication as in the case of addition.

An important theorem analogous to Theorem 2 is, however, independent of Assumption P. It is as follows:

THEOREM 4. *If the relation* $P_x \cdot P_y = P_{xy}$ *holds between any three points* $P_x$, $P_y$, $P_{xy}$ *on* $l$ *distinct from* $P_0$, *we have* $P_\infty P_0 P_1 P_x \overline{\wedge} P_\infty P_0 P_y P_{xy}$ *and also* $P_\infty P_0 P_1 P_y \overline{\wedge} P_\infty P_0 P_x P_{xy}$; *i.e. the correspondence established by making each point* $P_x$ *of* $l$ *correspond to* $P_x' = P_x \cdot P_a$ (*or to* $P_x' = P_a \cdot P_x$), *where* $P_a$ *is any fixed point of* $l$ *distinct from* $P_0$, *is projective.* (A, E)

*Proof.* The definition of multiplication gives the first of the above projectivities as the result of two perspectivities (fig. 76):

$$[P_x] \overset{A}{\underset{\wedge}{=}} [X] \overset{Y}{\underset{\wedge}{=}} [P_{xy}].$$

The second one is obtained similarly. In fig. 76 we have

$$[P_y] \overset{B}{\underset{\wedge}{=}} [Y] \overset{X}{\underset{\wedge}{=}} [P_{xy}].$$

The set of all projectivities determined by all choices of $P_a$ in the formula $P_x' = P_x \cdot P_a$ is the group described in Example 1, p. 69. The properties of multiplication may be regarded as properties of that group in the same way that the properties of addition arise from the group described in Example 2, p. 70. In particular, this furnishes a second proof of the associative law for multiplication.

THEOREM 5. *Multiplication is distributive with respect to addition; i.e. if* $P_x$, $P_y$, $P_z$ *are any three points on* $l$ (*for which the operations below are defined*), *we have*

$$P_z \cdot (P_x + P_y) = P_z \cdot P_x + P_z \cdot P_y, \textit{ and } (P_x + P_y) \cdot P_z = P_x \cdot P_z + P_y \cdot P_z. \quad \text{(A, E)}$$

*Proof.* Place

$$P_x + P_y = P_{x+y},\ P_z \cdot P_x = P_{zx},\ P_z \cdot P_y = P_{zy}, \text{ and } P_z \cdot P_{x+y} = P_{z(x+y)}.$$

By Theorem 4 we then have

$$P_\infty P_0 P_1 P_x P_y P_{x+y} \overline{\wedge} P_\infty P_0 P_z P_{zx} P_{zy} P_{z(x+y)}.$$

But by Theorem 1 we also have $Q(P_\infty P_x P_0,\ P_\infty P_y P_{x+y})$. Hence, by Theorem 1, Cor., Chap. IV, we have $Q(P_\infty P_{zx} P_0,\ P_\infty P_{zy} P_{z(x+y)})$ which, by Theorem 1, implies $P_{zx} + P_{zy} = P_{z(x+y)}$. The relation

$$(P_x + P_y) \cdot P_z = P_x \cdot P_z + P_y \cdot P_z$$

is proved similarly.

**50. The commutative law for multiplication.** With the aid of Assumption P we will now derive finally the commutative law for multiplication:

THEOREM 6. *The operation of multiplication is commutative; i.e. we have* $P_x \cdot P_y = P_y \cdot P_x$ *for every pair of points* $P_x$, $P_y$ *of* $l$ *for which these two products are defined.* (A, E, P)

*Proof.* Let us place as before $P_x \cdot P_y = P_{xy}$, and $P_y \cdot P_x = P_{yx}$. Then, by the first relation of Theorem 4, and interchanging the points $P_x$, $P_y$, we have

$$P_\infty P_0 P_1 P_y \overline{\wedge} P_\infty P_0 P_x P_{yx};$$

and from the second relation of the same theorem we have

$$P_\infty P_0 P_1 P_y \overline{\wedge} P_\infty P_0 P_x P_{xy}.$$

By Theorem 17, Chap. IV, this requires $P_{yx} = P_{xy}$.

In view of the fact already noted, that the fundamental theorem of projective geometry (Theorem 17, Chap. IV) is equivalent to Assumption P, it follows (cf. § 3, Vol. II) that:

THEOREM 7. *Assumption P is necessary and sufficient for the commutative law for multiplication.** (A, E)

**51. The inverse operations.** DEFINITION. Given two points $P_a$, $P_b$ on $l$, the operation determining a point $P_x$ satisfying the relation $P_a + P_x = P_b$ is called *subtraction;* in symbols $P_b - P_a = P_x$. The point $P_x$ is called the *difference* of $P_b$ from $P_a$. Subtraction is the *inverse* of addition.

The construction for addition may readily be reversed to give a construction for subtraction. The preceding theorems on addition then give:

THEOREM 8. *Subtraction is a one-valued operation for every pair of points* $P_a$, $P_b$ *on* $l$, *except the pair* $P_\infty$, $P_\infty$. (A, E)

COROLLARY. *We have in particular* $P_a - P_a = P_0$ *for every point* $P_a (\neq P_\infty)$ *on* $l$. (A, E)

* The existence of algebras in which multiplication is not commutative is then sufficient to establish the fact that Assumption P is independent of the previous Assumptions A and E. For in order to construct a system (cf. p. 6) which satisfies Assumptions A and E without satisfying Assumption P, we need only construct an analytic geometry of three dimensions (as described in a later chapter) and use as a basis a noncommutative number system, e.g. the system of quaternions. That the fundamental theorem of projective geometry is equivalent to the commutative law for multiplication was first established by Hilbert, who, in his Foundations of Geometry, showed that the commutative law is equivalent to the theorem of Pappus (Theorem 21, Chap. IV). The latter is easily seen to be equivalent to the fundamental theorem.

DEFINITION. Given two points $P_a$, $P_b$ on $l$; the point $P_x$ determined by the relation $P_a \cdot P_x = P_b$ is called the *quotient* of $P_b$ by $P_a$ (also the *ratio* of $P_b$ to $P_a$); in symbols $P_b / P_a = P_x$, or $P_b : P_a = P_x$. The operation determining $P_b / P_a$ is called *division;* it is the inverse of multiplication.*

The construction for multiplication may also be reversed to give a construction for division. The preceding theorems on multiplication then give readily:

THEOREM 9. *Division is a one-valued operation for every pair of points $P_a$, $P_b$ on $l$ except the pairs $P_0$, $P_0$* and *$P_\infty$, $P_\infty$.* (A, E)

COROLLARY. *We have in particular $P_a / P_a = P_1$, $P_0 / P_a = P_0$, $P_a / P_0 = P_\infty$, etc., for every point $P_a$ on $l$ distinct from $P_0$ and $P_\infty$.* (A, E)

Addition, subtraction, multiplication, and division are known as the four *rational* operations.

**52. The abstract concept of a number system. Isomorphism.** The relation of the foregoing discussion of the algebra of points on a line to the foundations of analysis must now be briefly considered. With the aid of the notion of a group (cf. Chap. III, p. 66), the general concept of a number system is described simply as follows:

DEFINITION. A set N of elements is said to form a *number system,* provided two distinct operations, which we will denote by $\oplus$ and $\odot$ respectively, exist and operate on pairs of elements of N under the following conditions:

1. The set N forms a group with respect to $\oplus$.
2. The set N forms a group with respect to $\odot$, except that if $i_+$ is the identity element of N with respect to $\oplus$, no inverse with respect to $\odot$ exists for $i_+$.† If $a$ is any element of N, $a \odot i_+ = i_+ \odot a = i_+$.
3. Any three elements $a$, $b$, $c$ of N satisfy the relations $a \odot (b \oplus c) = (a \odot b) \oplus (a \odot c)$ and $(b \oplus c) \odot a = (b \odot a) \oplus (c \odot a)$.

The elements of a number system are called *numbers;* the two operations $\oplus$ and $\odot$ are called *addition* and *multiplication* respectively. If a number system forms commutative groups with respect to both addition and multiplication, the numbers are said to form a *field.*‡

* What we have defined is more precisely ***right-handed division.*** The left-handed quotient is defined similarly as the point $P_x$ determined by the relation $P_x \cdot P_a = P_b$. In a commutative algebra they are of course equivalent.

† The identity element $i_+$ in a number system is usually denoted by 0 (zero).

‡ The class of all ordinary rational numbers forms a field; also the class of real numbers; and the class of all integers reduced modulo $p$ ($p$ a prime), etc.

On the basis of this definition may be developed all the theory relating to the rational operations — i.e. addition, multiplication, subtraction, and division — in a number system. The ordinary algebra of the rational operations applying to the set of ordinary rational or ordinary real or complex numbers is a special case of such a theory. *The whole terminology of this algebra, in so far as it is definable in terms of the four rational operations, will in the future be assumed as defined.* We shall not, therefore, stop to define such terms as *reciprocal of a number, exponent, equation, satisfy, solution, root,* etc. The element of a number system represented by a letter as $a$ will be spoken of as the *value* of $a$. A letter which represents any one of a set of numbers is called a *variable;* variables will usually be denoted by the last letters of the alphabet.

Before applying the general definition above to our algebra of points on a line, it is desirable to introduce the notion of the abstract equivalence or isomorphism between two number systems.

DEFINITION. If two number systems are such that a one-to-one reciprocal correspondence exists between the numbers of the two systems, such that to the sum of any two numbers of one system there corresponds the sum of the two corresponding numbers of the other system; and that to the product of any two numbers of one there corresponds the product of the corresponding numbers of the other, the two systems are said to be *abstractly equivalent* or (simply) *isomorphic.**

When two number systems are isomorphic, if any series of operations is performed on numbers of one system and the same series of operations is performed on the corresponding numbers of the other, the resulting numbers will correspond.

**53. Nonhomogeneous coördinates.** By comparing the corollaries of Theorem 1 with the definition of group (p. 66), it is at once seen that the set of points of a line on which a scale has been established, forms a group with respect to addition, provided the point $P_\infty$ be excluded from the set. In this group $P_0$ is the identity element, and the existence of an inverse for every element follows from Theorem 8. In the same way it is seen that the set of points on a line on which a scale has been established, and from which the

* For the general idea of the isomorphism between groups, see Burnside's Theory of Groups, p. 22.

point $P_\infty$ has been excluded, forms a group with respect to multiplication, except that no inverse with respect to multiplication exists for $P_0$; $P_1$ is the identity element in this group, and Theorem 9 insures the existence of an inverse for every point except $P_0$. These considerations show that the first two conditions in the definition of a number system are satisfied by the points of a line, if the operations $\oplus$ and $\odot$ are identified with addition and multiplication as defined in §§ 48 and 49. The third condition in the definition of a number system is also satisfied in view of Theorem 5. Finally, in view of Theorem 1, Cor. 4, and Theorem 6, this number system of points on a line is commutative with respect to both addition and multiplication. This gives then:

THEOREM 10. *The set of all points on a line on which a scale has been established, and from which the point $P_\infty$ is excluded, forms a field with respect to the operations of addition and multiplication previously defined.* (A, E, P)

This provides a new way of regarding a point, viz., that of *regarding a point as a number of a number system.* This conception of a point will apply to any point of a line except the one chosen as $P_\infty$. It is desirable, however, both on account of the presence of such an exceptional point and also for other reasons, to keep the notion of point distinct from the notion of number, at least nominally. This we do by introducing *a field of numbers $a, b, c, \cdots, l, k, \cdots, x, y, z, \cdots$ which is isomorphic with the field of points on a line.* The numbers of the number field may, as we have seen, be the points of the line, or they may be mere symbols which combine according to the conditions specified in the definition of a number system; or they may be elements defined in some way in terms of points, lines, etc.*

In any number system the identity element with respect to addition is called *zero* and denoted by 0, and the identity element with respect to multiplication is called *one* or *unity,* and is denoted by 1. We shall, moreover, denote the numbers $1+1, 1+1+1, \cdots, 0-a, \cdots$ by the usual symbols $2, 3, \cdots, -a, \cdots$.† In the isomorphism of our system of numbers with the set of points on a line, the point $P_0$ must correspond to 0, the point $P_1$ to the number 1; and, in general, to every

* See, for example, § 55, on von Staudt's algebra of throws, where the numbers are thought of as sets of four points.

† Cf., however, in this connection § 57 below.

point will correspond a number (except to $P_\infty$), and to every number of the field will correspond a point. In this way every point of the line (except $P_\infty$) is *labeled* by a number. This number is called the (nonhomogeneous) *coördinate* of the point, to which it corresponds. This enables us to express relations between points by means of equations between their coördinates. The coördinates of points, or the points themselves when we think of them as numbers of a number system, we will denote by the small letters of the alphabet (or by numerals), and we shall frequently use the phrase "the point $x$" in place of the longer phrase "the point whose coördinate is $x$." It should be noted that this representation of the points of a line by numbers of a number system is not in any way dependent on the commutativity of multiplication; i.e. it holds in the general geometries for which Assumption P is not assumed.

Before leaving the present discussion it seems desirable to point out that the algebra of points on a line is merely representative, under the principle of duality, of the *algebra of the elements of any one-dimensional primitive form.* Thus three lines $l_0$, $l_1$, $l_\infty$ of a flat pencil determine a scale in the pencil of lines; and three planes $\alpha_0$, $\alpha_1$, $\alpha_\infty$ of an axial pencil determine a scale in this pencil of planes; to each corresponds the same algebra.

**54. The analytic expression for a projectivity in a one-dimensional primitive form.** Let a scale be established on a line $l$ by choosing three arbitrary points for $P_0$, $P_1$, $P_\infty$; and let the resulting field of points on a line be made isomorphic with a field of numbers $0, 1, a, \cdots$, so that $P_0$ corresponds to 0, $P_1$ to 1, and, in general, $P_a$ to $a$. For the exceptional point $P_\infty$, let us introduce a special symbol $\infty$ with exceptional properties, which will be assigned to it as the need arises. It should be noted here, however, that this new symbol $\infty$ does not represent a number of a field as defined on p. 149.

We may now derive the analytic relation between the coördinates of the points on $l$, which expresses a projective correspondence between these points. Let $x$ be the coördinate of any point of $l$. We have seen that if the point whose coördinate is $x$ is made to correspond to either of the points

(I) $$x' = x + a, \quad (a \neq \infty)$$

or (II) $$x' = ax, \quad (a \neq 0)$$

where $a$ is the coördinate of any given point on $l$, each of the resulting correspondences is projective (Theorem 2 and Theorem 4). It is readily seen, moreover, that if $x$ is made to correspond to

$$x' = \frac{1}{x}, \tag{III}$$

the resulting correspondence is likewise projective. For we clearly have the following construction for the point $1/x$ (fig. 77): With the same notation as before for the construction of the product of two

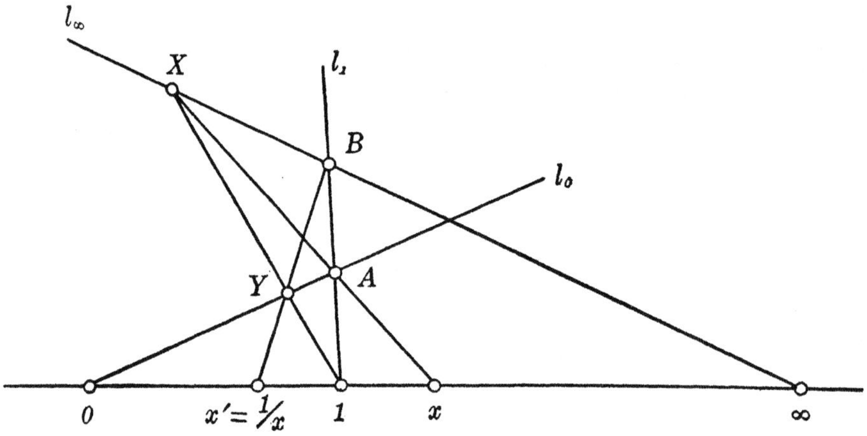


FIG. 77

numbers, let the line $xA$ meet $l_\infty$ in $X$. If $Y$ is determined as the intersection of $1X$ with $l_0$, the line $BY$ determines on $l$ a point $x'$, such that $xx' = 1$, by definition. We now have

$$[x] \overset{A}{\underset{\wedge}{=}} [X] \overset{1}{\underset{\wedge}{=}} [Y] \overset{B}{\underset{\wedge}{=}} [x'].$$

The three projectivities (I), (II), and (III) are of fundamental importance, as the next theorem will show. It is therefore desirable to consider their properties briefly; we will thus be led to define the behavior of the exceptional symbol $\infty$ with respect to the operations of addition, subtraction, multiplication, and division.

The projectivity $x' = x + a$, from its definition, leaves the point $P_\infty$, which we associated with $\infty$, invariant. We therefore place $\infty + a = \infty$ for all values of $a$ $(a \neq \infty)$. This projectivity, moreover, can have no other invariant point unless it leaves every point invariant; for the equation $x = x + a$ gives at once $a = 0$, if $x \neq \infty$. Further, by properly choosing $a$, any point $x$ can be made to correspond to any point $x'$;

but when one such pair of homologous points is assigned in addition to the double point $\infty$, the projectivity is completely determined. The resultant or product of any two projectivities $x' = x + a$ and $x' = x + b$ is clearly $x' = x + (a + b)$. Two such projectivities are therefore commutative.

The projectivity $x' = ax$, from its definition, leaves the points 0 and $\infty$ invariant, and by the fundamental theorem (Theorem 17, Chap. IV) cannot leave any other point invariant without reducing to the identical projectivity. As another property of the symbol $\infty$ we have therefore $\infty = a\infty\ (a \neq 0)$. Here, also, by properly choosing $a$, any point $x$ can be made to correspond to any point $x'$, but then the projectivity is completely determined. The fundamental theorem in this case shows, moreover, that any projectivity with the double points $0, \infty$ can be represented by this equation. The product of two projectivities $x' = ax$ and $x' = bx$ is clearly $x' = (ab)x$, so that any two projectivities of this type are also commutative (Theorem 6).

Finally, the projectivity $x' = 1/x$, by its definition, makes the point $\infty$ correspond to 0 and the point 0 to $\infty$. We are therefore led to assign to the symbol $\infty$ the following further properties: $1/\infty = 0$, and $1/0 = \infty$. This projectivity leaves 1 and $-1$ (defined as $0 - 1$) invariant. Moreover, it is an involution because the resultant of two applications of this projectivity is clearly the identity; i.e. if the projectivity is denoted by $\pi$, it satisfies the relation $\pi^2 = 1$.

THEOREM 11. *Any projectivity on a line is the product of projectivities of the three types (I), (II), and (III), and may be expressed in the form*

$$(1) \qquad x' = \frac{ax + b}{cx + d}.$$

*Conversely, every equation of this form represents a projectivity, if* $ad - bc \neq 0$. (A, E, P)

*Proof.* We will prove the latter part of the theorem first. If we suppose first that $c \neq 0$, we may write the equation of the given transformation in the form

$$(2) \qquad x' = \frac{a}{c} + \frac{b - \frac{ad}{c}}{cx + d}.$$

This shows first that the determinant $ad - bc$ must be different from 0; otherwise the second term on the right of (2) would vanish, which would make every $x$ correspond to the same point $a/c$, while a projectivity is a one-to-one correspondence. Equation (1), moreover, shows at once that the correspondence established by it is the resultant of the five:

$$x_1 = cx, \quad x_2 = x_1 + d, \quad x_3 = \frac{1}{x_2}, \quad x_4 = \left(b - \frac{ad}{c}\right) x_3, \quad x' = x_4 + \frac{a}{c}.$$

If $c = 0$, and $ad \neq 0$, this argument is readily modified to show that the transformation of the theorem is the resultant of projectivities of the types ($I$) and ($II$). Since the resultant of any series of projectivities is a projectivity, this proves the last part of the theorem.

It remains to show that every projectivity can indeed be represented by an equation $x' = \dfrac{ax + b}{cx + d}\cdot$ To do this simply, it is desirable to determine first what point is made to correspond to the point $\infty$ by this projectivity. If we follow the course of this point through the five projectivities into which we have just resolved this transformation, it is seen that the first two leave it invariant, the third transforms it into 0, the fourth leaves 0 invariant, and the fifth transforms it into $a/c$; the point $\infty$ is then transformed by (1) into the point $a/c$. *This leads us to attribute a further property to the symbol* $\infty$, viz.,

$$\frac{ax + b}{cx + d} = \frac{a}{c}, \quad \textit{when} \quad x = \infty.$$

According to the fundamental theorem (Theorem 17, Chap. IV), a projectivity is completely determined when any three pairs of homologous points are assigned. Suppose that in a given projectivity the points 0, 1, $\infty$ are transformed into the points $p$, $q$, $r$ respectively. Then the transformation

$$x' = \frac{r(q - p)x + p(r - q)}{(q - p)x + (r - q)}$$

clearly transforms 0 into $p$, 1 into $q$, and, by virtue of the relation just developed for $\infty$, it also transforms $\infty$ into $r$. It is, moreover, of the form of (1). The determinant $ad - bc$ is in this case $(q-p)(r-q)(r-p)$, which is clearly different from zero, if $p$, $q$, $r$ are all distinct. This transformation is therefore the given projectivity.

COROLLARY 1. *The projectivity* $x' = a/x \, (a \neq 0, \textit{or } \infty)$ *transforms* 0 *into* $\infty$ *and* $\infty$ *into* 0. (A, E, P)

For it is the resultant of the two projectivities, $x_1 = 1/x$ and $x' = ax_1$, of which the first interchanges 0 and $\infty$, while the second leaves them both invariant. *We are therefore led to define the symbols* $a/0$ *and* $a/\infty$ *as equal to* $\infty$ *and* 0 *respectively, when* $a$ *is neither* 0 *nor* $\infty$.

COROLLARY 2. *Any projectivity leaving the point* $\infty$ *invariant may be expressed in the form* $x' = ax + b$. (A, E, P)

COROLLARY 3. *Any projectivity may be expressed analytically by the bilinear equation* $cxx' + dx' - ax - b = 0$; *and conversely, any bilinear equation defines a projective correspondence between its two variables, provided* $ad - bc \neq 0$. (A, E, P)

COROLLARY 4. *If a projectivity leaves any points invariant, the coördinates of these double points satisfy the quadratic equation* $cx^2 + (d - a)x - b = 0$. (A, E, P)

DEFINITION. A system of $mn$ numbers arranged in a rectangular array of $m$ rows and $n$ columns is called a *matrix*. If $m = n$, it is called a *square matrix of order n*.*

The coefficients $\begin{pmatrix} a & b \\ c & d \end{pmatrix}$ of the projective transformation (1) form a square matrix of the second order, which may be conveniently used to denote the transformation. Two matrices $\begin{pmatrix} a & b \\ c & d \end{pmatrix}$ and $\begin{pmatrix} a' & b' \\ c' & d' \end{pmatrix}$ represent the same transformation, if and only if $a : a' = b : b' = c : c' = d : d'$.

The product of two projectivities

$$x' = \pi(x) = \frac{ax + b}{cx + d} \quad \text{and} \quad x'' = \pi_1(x') = \frac{a'x' + b'}{c'x' + d'}$$

is given by the equation

$$x'' = \pi_1\pi(x) = \frac{(aa' + cb')x + ba' + db'}{(ac' + cd')x + bc' + dd'}.$$

This leads at once to the rule for the multiplication of matrices, which is similar to that for determinants.

DEFINITION. The *product* of two matrices is defined by the equation

$$\begin{pmatrix} a' & b' \\ c' & d' \end{pmatrix}\begin{pmatrix} a & b \\ c & d \end{pmatrix} = \begin{pmatrix} aa' + cb' & ba' + db' \\ ac' + cd' & bc' + dd' \end{pmatrix}.$$

* For a development of the principal properties of matrices, cf. Bôcher, Introduction to Higher Algebra, pp. 20 ff.

This gives, in connection with the result just derived,

THEOREM 12. *The product of two projectivities*

$$\pi = \begin{pmatrix} a & b \\ c & d \end{pmatrix} \quad \textit{and} \quad \pi_1 = \begin{pmatrix} a' & b' \\ c' & d' \end{pmatrix}$$

*is represented by the product of their matrices; in symbols,*

$$\pi_1 \pi = \begin{pmatrix} a' & b' \\ c' & d' \end{pmatrix} \begin{pmatrix} a & b \\ c & d \end{pmatrix}. \quad \text{(A, E, P)}$$

COROLLARY 1. *The determinant of the product of two projectivities is equal to the product of their determinants.* (A, E, P)

COROLLARY 2. *The inverse of the projectivity* $\pi = \begin{pmatrix} a & b \\ c & d \end{pmatrix}$ *is given by* $\pi^{-1} = \begin{pmatrix} d & -b \\ -c & a \end{pmatrix} = \begin{pmatrix} A & C \\ B & D \end{pmatrix}$, *where* $A$, $B$, $C$, $D$ *are the cofactors of* $a$, $b$, $c$, $d$ *respectively in the determinant* $\begin{vmatrix} a & b \\ c & d \end{vmatrix}$. (A, E, P)

This follows at once from Corollary 3 of the last theorem by interchanging $x$, $x'$. We may also verify the relation by forming the product $\pi^{-1}\pi = \begin{pmatrix} ad-bc & 0 \\ 0 & ad-bc \end{pmatrix}$, which transformation is equivalent to $\begin{pmatrix} 1 & 0 \\ 0 & 1 \end{pmatrix}$. The latter is called the *identical matrix*.

COROLLARY 3. *Any involution is represented by* $\begin{pmatrix} a & b \\ c & -a \end{pmatrix}$, *that is by* $x' = \dfrac{ax+b}{cx-a}$, *with the condition that* $a^2 + bc \neq 0$. (A, E, P)

**55. Von Staudt's algebra of throws.** We will now consider the number system of points on a line from a slightly different point of view. On p. 60 we defined a throw as consisting of two ordered pairs of points on a line; and defined two throws as equal when they are projective. The class of all throws which are projective (i.e. equal) to a given throw constitutes a class which we shall call a *mark*. Every throw determines one and only one mark, but each mark determines a whole class of throws.

According to the fundamental theorem (Theorem 17, Chap. IV), if three elements $A$, $B$, $C$ of a throw and their places in the symbol $\mathsf{T}(AB, CD)$ are given, the throw is completely determined by the mark to which it belongs. A given mark can be denoted by the symbol of any one of the (projective) throws which define it. We shall also denote marks by the small letters of the alphabet. And so, since the equality sign (=) indicates that the two symbols between which it stands denote the same thing, we may write $\mathsf{T}(AB, CD) = a = b$, if $a$, $b$, $\mathsf{T}(AB, CD)$ are notations for the same mark. Thus $\mathsf{T}(AB, CD) = \mathsf{T}(BA, DC) = \mathsf{T}(CD, AB) = \mathsf{T}(DC, BA)$ are all symbols denoting the same mark (Theorem 2, Chap. III).

According to the original definition of a throw the four elements which compose it must be distinct. The term is now to be extended to include the following sets of two ordered pairs, where $A$, $B$, $C$ are distinct. The set of all throws of the type $\mathsf{T}(AB, CA)$ is called a mark and denoted by $\infty$; the set of all throws of the type $\mathsf{T}(AB, CB)$ is called a mark and is denoted by 0; the set of all throws of the type $\mathsf{T}(AB, CC)$ is a mark and is denoted by 1. It is readily seen that if $P_0$, $P_1$, $P_\infty$ are any three points of a line, there exists for every point $P$ of the line a unique throw $\mathsf{T}(P_\infty P_0, P_1 P)$ of the line; and conversely, for every mark there is a unique point $P$. The mark $\infty$, by what precedes, corresponds to the point $P_\infty$; the mark 0 to $P_0$; and the mark 1 to $P_1$.

DEFINITION. Let $\mathsf{T}(AB, CD_1)$ be a throw of the mark $a$, and let $\mathsf{T}(AB, CD_2)$ be a throw of the mark $b$; then, if $D_3$ is determined by $\mathsf{Q}(AD_1B, AD_2D_3)$, the mark $c$ of the throw $\mathsf{T}(AB, CD_3)$ is called the *sum* of the marks $a$ and $b$, and is denoted by $a + b$; in symbols, $a + b = c$. Also, the point $D_3'$ determined by $\mathsf{Q}(AD_1C, BD_2D_3')$ determines a mark with the symbol $\mathsf{T}(AB, CD_3') = c'$ (say), which is called the *product* of the marks $a$ and $b$; in symbols, $ab = c'$. As to the marks 0 and 1, to which these two definitions do not apply, we define further: $a + 0 = 0 + a = a$, $a \cdot 0 = 0 \cdot a = 0$, and $a \cdot 1 = 1 \cdot a = a$.

Since any three distinct points $A, B, C$ may be projected into a fixed triple $P_\infty$, $P_1$, $P_0$, it follows that the operation of adding or multiplying marks may be performed on their representative throws of the form $\mathsf{T}(P_\infty P_0, P_1 P)$. By reference to Theorems 1 and 3 it is then clear that the class of all marks on a line (except $\infty$) forms a number system, with respect to the operations of addition and multiplication just defined, which is isomorphic with the number system of points previously developed.

This is, in brief, the method used by von Staudt to introduce analytic methods into geometry on a purely geometric basis.* We have

* Cf. reference on p. 141. Von Staudt used the notion of an involution on a line in defining addition and multiplication; the definition in terms of quadrangular sets is, however, essentially the same as his by virtue of Theorem 27, Chap. IV.

given it here partly on account of its historical importance; partly because it gives a concrete example of a number system isomorphic with the points of a line*; and partly because it gives a natural introduction to the fundamental concept of the cross ratio of four points. This we proceed to derive in the next section.

**56. The cross ratio.** We have seen in the preceding section that it is possible to associate a number with every throw of four points on a line. By duality all the developments of this section apply also to the other one-dimensional primitive forms, i.e. the pencil of lines and the pencil of planes. With every throw of four elements of any one-dimensional primitive form there may be associated a definite number, which must be the same for every throw projective with the first, and is therefore an invariant under any projective transformation, i.e. a property of the throw that is not changed when the throw is replaced by any projective throw. This number is called the *cross ratio* of the throw. It is also called the *double ratio* or the *anharmonic ratio.* The reason for these names will appear presently.

In general, *four given points give rise to six different cross ratios.* For the 24 possible permutations of the letters in the symbol $\mathsf{T}(AB, CD)$ fall into sets of four which, by virtue of Theorem 2, Chap. III, have the same cross ratios. In the array below, the permutations in any line are projective with each other, two permutations of different lines being in general not projective:

| | | | |
|---|---|---|---|
| $AB, CD$ | $BA, DC$ | $DC, BA$ | $CD, AB$ |
| $AB, DC$ | $BA, CD$ | $CD, BA$ | $DC, AB$ |
| $AC, BD$ | $CA, DB$ | $DB, CA$ | $BD, AC$ |
| $AC, DB$ | $CA, BD$ | $BD, CA$ | $DB, AC$ |
| $AD, BC$ | $DA, CB$ | $CB, DA$ | $BC, AD$ |
| $AD, CB$ | $DA, BC$ | $BC, DA$ | $CB, AD$ |

If, however, the four points form a harmonic set $\mathsf{H}(AB, CD)$, the throws $\mathsf{T}(AB, CD)$ and $\mathsf{T}(AB, DC)$ are projective (Theorem 5, Cor. 2, Chap. IV). In this case the permutations in the first two rows of the array just given are all projective and hence have the same cross ratio. *The four elements of a harmonic set, therefore, give rise to only three cross ratios.* The values of these cross ratios are readily seen

* Cf. § 53. Here, with every point of a line on which a scale has been established, is associated a mark which is the coördinate of the point.

to be $-1$, $\frac{1}{2}$, 2 respectively, for the constructions of our number system give at once $\mathrm{H}(P_\infty P_0, P_1 P_{-1})$, $\mathrm{H}(P_\infty P_{\frac{1}{2}}, P_0 P_1)$, and $\mathrm{H}(P_\infty P_1, P_0 P_2)$.

We now proceed to develop an analytic expression for the cross ratio $\text{R}(x_1 x_2, x_3 x_4)$ of any four points on a line (or, in general, of any four elements of any one-dimensional primitive form) whose coördinates in a given scale are given. It seems desirable to precede this derivation by an explicit definition of this cross ratio, which is independent of von Staudt's algebra of throws.

DEFINITION. The *cross ratio* $\text{R}(x_1 x_2, x_3 x_4)$ of elements $x_1, x_2, x_3, x_4$ of any one-dimensional form is, if $x_1, x_2, x_3$ are distinct, the coördinate $\lambda$ of the element of the form into which $x_4$ is transformed by the projectivity which transforms $x_1, x_2, x_3$ into $\infty$, 0, 1 respectively; i.e. the number, $\lambda$, defined by the projectivity $x_1 x_2 x_3 x_4 \,\overline{\wedge}\, \infty 0 1 \lambda$. If two of the elements $x_1, x_2, x_3$ coincide and $x_4$ is distinct from all of them, we define $\text{R}(x_1 x_2, x_3 x_4)$ as that one of $\text{R}(x_2 x_1, x_4 x_3)$, $\text{R}(x_3 x_4, x_1 x_2)$, $\text{R}(x_4 x_3, x_2 x_1)$, for which the first three elements are distinct.

THEOREM 13. *The cross ratio* $\text{R}(x_1 x_2, x_3 x_4)$ *of the four elements whose coördinates are respectively* $x_1, x_2, x_3, x_4$ *is given by the relation*

$$\lambda = \text{R}(x_1 x_2, x_3 x_4) = \frac{(x_1 - x_3)}{(x_1 - x_4)} : \frac{(x_2 - x_3)}{(x_2 - x_4)}.$$

(A, E, P)

*Proof.* The transformation

$$x' = \frac{x_1 - x_3}{x_1 - x} : \frac{x_2 - x_3}{x_2 - x}$$

is evidently a projectivity, since it is reducible to the form of a linear fractional transformation, viz.,

$$x' = \frac{-(x_1 - x_3)x + x_2(x_1 - x_3)}{-(x_2 - x_3)x + x_1(x_2 - x_3)}$$

in which the determinant $(x_1 - x_3)(x_2 - x_3)(x_2 - x_1)$ is not zero, provided the points $x_1, x_2, x_3$ are distinct. This projectivity transforms $x_1, x_2, x_3$ into $\infty$, 0, 1 respectively. By definition, therefore, this projectivity transforms $x_4$ into the point whose coördinate is the cross ratio in question, i.e. into the expression given in the theorem. If $x_1, x_2, x_3$ are not all distinct, replace the symbol $\text{R}(x_1 x_2, x_3 x_4)$ by one of its equal cross ratios $\text{R}(x_2 x_1, x_4 x_3)$, etc.; one of these must have the first three elements of the symbol distinct, since in a cross ratio of four points at least three must be distinct (def.).

COROLLARY 1. *We have in particular*

$$R(x_1x_2, x_3x_1) = \infty, \quad R(x_1x_2, x_3x_2) = 0, \quad \text{and} \quad R(x_1x_2, x_3x_3) = 1,$$

*if* $x_1$, $x_2$, $x_3$ *are any three distinct elements of the form.* (A, E)

COROLLARY 2. *The cross ratio of a harmonic set* $H(x_1x_2, x_3x_4)$ *is* $R(x_1x_2, x_3x_4) = -1$, for we have $H(\infty\, 0, 1\, -1)$. (A, E, P)

COROLLARY 3. *If* $R(x_1x_2, x_3x_4) = \lambda$, *the other five cross ratios of the throws composed of the four elements* $x_1$, $x_2$, $x_3$, $x_4$ *are*

$$R(x_1x_2, x_4x_3) = \frac{1}{\lambda}, \qquad R(x_1x_4, x_2x_3) = \frac{\lambda - 1}{\lambda},$$

$$R(x_1x_3, x_2x_4) = 1 - \lambda, \qquad R(x_1x_4, x_3x_2) = \frac{\lambda}{\lambda - 1}.$$

$$R(x_1x_3, x_4x_2) = \frac{1}{1 - \lambda},$$

(A, E, P)

The proof is left as an exercise.

COROLLARY 4. *If* $x_1$, $x_2$, $x_3$, $x_4$ *form a harmonic set* $H(x_1x_2, x_3x_4)$, *we have*

$$\frac{2}{x_2 - x_1} = \frac{1}{x_3 - x_1} + \frac{1}{x_4 - x_1}.$$

(A, E, P)

The proof is left as an exercise.

COROLLARY 5. *If* $a$, $b$, $c$ *are any three distinct elements of a one-dimensional primitive form, and* $a'$, $b'$, $c'$ *are any three other distinct elements of the same form, then the correspondence established by the relation* $R(ab, cx) = R(a'b', c'x')$ *is projective.* (A, E, P)

*Proof.* Analytically this relation gives

$$\frac{a - c}{a - x} \cdot \frac{b - x}{b - c} = \frac{a' - c'}{a' - x'} \cdot \frac{b' - x'}{b' - c'},$$

which, when expanded, evidently leads to a bilinear equation in the variables $x$, $x'$, which defines a projective correspondence by Theorem 11, Cor. 3.

That the cross ratio

$$\frac{x_1 - x_3}{x_1 - x_4} : \frac{x_2 - x_3}{x_2 - x_4}$$

is invariant under any projective transformation may also be verified directly by observing that each of the three types (I), (II), (III) of projectivities on pp. 152, 153 leaves it invariant. That every projectivity leaves it invariant then follows from Theorem 11.

**57. Coördinates in a net of rationality on a line.** We now consider the numbers associated with the points of a net of rationality on a line. The connection between the developments of this chapter and the notion of a linear net of rationality is contained in the following theorem:

THEOREM 14. *The coördinates of the points of the net of rationality* $R(P_0P_1P_\infty)$ *form a number system, or field, which consists of all numbers each of which can be obtained by a finite number of rational algebraic operations on* 0 *and* 1, *and only these.* (A, E)

*Proof.* By Theorem 14, Chap. IV, the linear net is a line of the rational space constituted by the points of a three-dimensional net of rationality. By Theorem 20, Chap. IV, this three-dimensional net is a properly projective space. Hence, by Theorem 10 of the present chapter, the numbers associated with $R(01\infty)$ form a field.

All numbers obtainable from 0 and 1 by the operations of addition, subtraction, multiplication, and division are in $R(01\infty)$, because (Theorem 9, Chap. IV) whenever $x$ and $y$ are in $R(01\infty)$ the quadrangular sets determining $x+y$, $xy$, $x-y$, $x/y$ have five out of six elements in $R(01\infty)$. On the other hand, every number of $R(01\infty)$ can be obtained by a finite number of these operations. This follows from the fact that the harmonic conjugate of any point $a$ in $R(01\infty)$ with respect to two others, $b$, $c$, can be obtained by a finite number of rational operations on $a$, $b$, $c$. This fact is a consequence of Theorem 13, Cor. 2, which shows that $x$ is connected with $a$, $b$, $c$ by the relation

$$(x-b)(a-c)+(x-c)(a-b)=0.$$

Solving this equation for $x$, we have

$$x=-\frac{2\,bc-ab-ac}{2\,a-b-c},$$

a number * which is clearly the result of a finite number of rational operations on $a$, $b$, $c$. This completes the proof of the theorem. We have here the reason for the term *net of rationality.*

It is well to recall at this point that our assumptions are not yet sufficient to identify the numbers associated with a net of rationality with the system of all ordinary rational numbers. We need only recall the example of the miniature geometry described in the Introduction, § 2, which contained only

* The expression for $x$ cannot be indeterminate unless $b=c$.

three points on a line. If in that triple-system geometry we perform the construction for the number 1 + 1 on any line in which we have assigned the numbers 0, 1, ∞ to the three points of the line in any way, it will be found that this construction yields the point 0. This is due to the fact previously noted that in that geometry the diagonal points of a complete quadrangle are collinear. In every geometry to which Assumptions A, E, P apply we may construct the points 1 + 1, 1 + 1 + 1, ..., thus forming a sequence of points which, with the usual notation for these sums, we may denote by 0, 1, 2, 3, 4, .... Two possibilities then present themselves: either the points thus obtained are all distinct, in which case the net R(01∞) contains all the ordinary rational numbers; or some point of this sequence coincides with one of the preceding points of the sequence, in which case the number of points in a net of rationality is finite. We shall consider this situation in detail in a later chapter, and will then add further assumptions. Here it should be emphasized that our results hitherto, and all subsequent results depending only on Assumptions A, E, P, are valid not only in the ordinary real or complex geometries, but in a much more general class of spaces, which are characterized merely by the fact that the coördinates of the points on a line are the numbers of a field, finite or infinite.

**58. Homogeneous coördinates on a line.** The exceptional character of the point $P_\infty$, as the coördinate of which we introduced a symbol ∞ with exceptional properties, often proves troublesome, and is, moreover, contrary to the spirit of projective geometry in which the points of a line are all equivalent; indeed, the choice of the point $P_\infty$ was entirely arbitrary. It is exceptional only in its relation to the operations of addition, multiplication, etc., which we have defined in terms of it. In this section we will describe another method of denoting points on a line by numbers, whereby it is not necessary to use any exceptional symbol.

As before, let a scale be established on a line by choosing any three points to be the points $P_0$, $P_1$, $P_\infty$; and let each point of the line be denoted by its (nonhomogeneous) coördinate in a number system isomorphic with the points of the line. We will now associate with every point a *pair of numbers* $(x_1, x_2)$ of this system in a given order, such that if $x$ is the (nonhomogeneous) coördinate of any point distinct from $P_\infty$, the pair $(x_1, x_2)$ associated with the point $x$ satisfies the relation $x = x_1/x_2$. With the point $P_\infty$ we associate any pair of the form $(k, 0)$, where $k$ is any number $(k \neq 0)$ of the number system isomorphic with the line. *To every point of the line corresponds a pair of numbers, and to every pair of numbers in the field, except the pair*

$(0, 0)$, *corresponds a unique point of the line.* These two numbers are called *homogeneous coördinates* of the point with which they are associated, and the pair of numbers is said to represent the point. This representation of points on a line by pairs of numbers is not unique, since only the ratio of the two coördinates is determined; i.e. the pairs $(x_1, x_2)$ and $(mx_1, mx_2)$ represent the same point for all values of $m$ different from 0. The point $P_0$ is characterized by the fact that $x_1 = 0$; the point $P_\infty$ by the fact that $x_2 = 0$; and the point $P_1$ by the fact that $x_1 = x_2$.

THEOREM 15. *In homogeneous coördinates a projectivity on a line is represented by a linear homogeneous transformation in two variables,*

$$\begin{aligned} \rho x_1' &= ax_1 + bx_2, \\ \rho x_2' &= cx_1 + dx_2, \end{aligned} \qquad (ad - bc \neq 0) \tag{1}$$

*where $\rho$ is an arbitrary factor of proportionality.* (A, E, P)

*Proof.* By division, this clearly leads to the transformation

$$x' = \frac{ax + b}{cx + d}, \tag{2}$$

provided $x_2'$ and $x_2$ are both different from 0. If $x_2 = 0$, the transformation (1) gives the point $(x_1', x_2') = (a, c)$; i.e. the point $P_\infty = (1, 0)$ is transformed by (1) into the point whose nonhomogeneous coördinate is $a/c$. And if $x_2' = 0$, we have in (1) $(x_1, x_2) = (d, -c)$; i.e. (1) transforms the point whose nonhomogeneous coördinate is $-d/c$ into the point $P_\infty$. By reference to Theorem 11 the validity of the theorem is therefore established.

As before, the matrix $\begin{pmatrix} a & b \\ c & d \end{pmatrix}$ of the coefficients may conveniently be used to represent the projectivity. The double points of the projectivity, if existent, are obtained in homogeneous coördinates as follows: The coördinates of a double point $(x_1, x_2)$ must satisfy the equations

$$\begin{aligned} \rho x_1 &= ax_1 + bx_2, \\ \rho x_2 &= cx_1 + dx_2. \end{aligned}$$

These equations are compatible only if the determinant of the system

$$\begin{aligned} (a - \rho) x_1 + bx_2 &= 0, \\ cx_1 + (d - \rho) x_2 &= 0, \end{aligned} \tag{3}$$

vanishes. This leads to the equation

$$\begin{vmatrix} a - \rho & b \\ c & d - \rho \end{vmatrix} = 0$$

for the determination of the factor of proportionality $\rho$. This equation is called the *characteristic equation* of the matrix representing the projectivity. Every value of $\rho$ satisfying this equation then leads to a double point when substituted in one of the equations (3); viz., if $\rho_1$ be a solution of the characteristic equation, the point

$$(x_1, x_2) = (-b, a - \rho_1) = (d - \rho_1, -c)$$

is a double point.*

In homogeneous coördinates the cross ratio $R(AB, CD)$ of four points $A = (a_1, a_2)$, $B = (b_1, b_2)$, $C = (c_1, c_2)$, $D = (d_1, d_2)$ is given by

$$R(AB, CD) = \frac{(ac)}{(ad)} : \frac{(bc)}{(bd)},$$

where the expressions $(ac)$, etc., are used as abbreviations for $a_1c_2 - a_2c_1$, etc. This statement is readily verified by writing down the above ratio in terms of the nonhomogeneous coördinates of the four points.

We will close this section by giving to the two homogeneous coördinates of a point on a line an explicit geometrical significance. In view of the fact that the coördinates of such a point are not uniquely determined, a factor of proportionality being entirely arbitrary, there may be many such interpretations. On account of the existence of this arbitrary factor, we may impose a further condition on the coördinates $(x_1, x_2)$ of a point, in addition to the defining relation $x_1/x_2 = x$, where $x$ is the nonhomogeneous coördinate of the point in question. We choose the relation $x_1 + x_2 = 1$. If this relation is satisfied,

$$x_1 = \frac{\begin{vmatrix} 1 & -1 \\ 1 & 0 \end{vmatrix}}{\begin{vmatrix} 1 & -1 \\ x_1 & x_2 \end{vmatrix}} : \frac{\begin{vmatrix} 0 & 1 \\ 1 & 0 \end{vmatrix}}{\begin{vmatrix} 0 & 1 \\ x_1 & x_2 \end{vmatrix}} = R(-10, \infty x),$$

$$x_2 = \frac{\begin{vmatrix} 1 & -1 \\ 0 & 1 \end{vmatrix}}{\begin{vmatrix} 1 & -1 \\ x_1 & x_2 \end{vmatrix}} : \frac{\begin{vmatrix} 1 & 0 \\ 0 & 1 \end{vmatrix}}{\begin{vmatrix} 1 & 0 \\ x_1 & x_2 \end{vmatrix}} = R(-1\infty, 0x).$$

Thus homogeneous coördinates subject to the condition $x_1 + x_2 = 1$ can be defined by choosing three points $A$, $B$, $C$ arbitrarily, and letting $x_1 = R(AB, CX)$ and $x_2 = R(AC, BX)$. The ordinary homogeneous coördinates would then be defined as any two numbers proportional to these two cross ratios.

* This point is indeterminate only if $b = c = 0$ and $a = d$. The projectivity is then the identity.

**59. Projective correspondence between the points of two different lines.** Hitherto we have confined ourselves, in the development of analytic methods, to the points of a single line, or, under duality, to the elements of a single one-dimensional primitive form. Suppose now that we have two lines $l$ and $m$ with a scale on each, and let the nonhomogeneous coördinate of any point of $l$ be represented by $x$, and that of any point of $m$ by $y$. The question then arises as to how a projective correspondence between the point $x$ and the point $y$ may be expressed analytically. It is necessary, first of all, to give a meaning to the equation $y = x$. In other words: What is meant by saying that two points — $x$ on $l$, and $y$ on $m$ — have the same coördinate? The coördinate $x$ is a number of a field and corresponds to the point of which it is the coördinate in an isomorphism of this field with the field of points on the line $l$. We may think of this same field of numbers as isomorphic with the field of points on the line $m$. In bringing about this isomorphism nothing has been specified except that the fundamental points $P_0$, $P_1$, $P_\infty$ determining the scale on $m$ must correspond to the numbers 0, 1 and the symbol $\infty$ respectively. If the correspondence between the points of the line and the numbers of the field were entirely determined by the respective correspondences of the points $P_0$, $P_1$, $P_\infty$ just mentioned, then we should know precisely what points on the two lines $l$ and $m$ have the same coördinates. It is not true of all fields, however, that this correspondence is uniquely determined when the points corresponding to 0, 1, $\infty$ are assigned.* It is necessary, therefore, to specify more definitely how the isomorphism between the points of $m$ and the numbers of the field is brought about. One way to bring it about is to make use of the projectivity which carries the fundamental points 0, 1, $\infty$ of $l$ into the fundamental points 0, 1, $\infty$ of $m$, and to assign the coördinate $x$ of any point $A$ of $l$ to that point of $m$ into which $A$ is transformed by this projectivity. In this projectivity pairs of homologous points will then have the same coördinates. That the field of points and the field of numbers are indeed made *isomorphic* by this process follows directly from Theorems 1 and 3 in connection with Theorem 1, Cor., Chap. IV. We may now readily prove the following theorem:

* This is shown by the fact that the field of all ordinary complex numbers can be isomorphic with itself not only by making each number correspond to itself, but also by making each number $a + ib$ correspond to its conjugate $a - ib$.

THEOREM 16. *Any projective correspondence between the points* $[x]$ *and* $[y]$ *of two distinct lines may be represented analytically by the relation* $y = x$ *by properly choosing the coördinates on the two lines. If the coördinates on the two lines are so related that the relation* $y = x$ *represents a projective correspondence, then any projective correspondence between the points of the two lines is given by a relation*

$$y = \frac{ax + b}{cx + d}, \qquad (ad - bc \neq 0).$$

(A, E, P)

*Proof.* The first part of the theorem follows at once from the preceding discussion, since any projectivity is determined by three pairs of homologous points, and any three points of either line may be chosen for the fundamental points. In fact, we may represent any projectivity between the points of the two lines by the relation $y = x$, by choosing the fundamental points on $l$ arbitrarily; the fundamental points on $m$ are then uniquely determined. To prove the second part of the theorem, let $\pi$ be any given projective transformation of the points of the line $l$ into those of $m$, and let $\pi_0$ be the projectivity $y = x$, regarded as a transformation from $m$ to $l$. The resultant $\pi_0\pi = \pi_1$ is a projectivity on $l$, and may therefore be represented by $x' = (ax + b)/(cx + d)$. Since $\pi = \pi_0^{-1}\pi_1$, this gives readily the result that $\pi$ may be represented by the relation given in the theorem.

## EXERCISES

**1.** Give constructions for subtraction and division in the algebra of points on a line.

**2.** Give constructions for the sum and the product of two lines of a pencil of lines in which a scale has been established.

**3.** Develop the point algebra on a line by using the properties expressed in Theorems 2 and 4 as the definitions of addition and multiplication respectively. Is it necessary to use Assumption P from the beginning?

**4.** Using Cor. 3 of Theorem 9, Chap. III, show that addition and multiplication may be defined as follows: As before, choose three points $P_0$, $P_1$, $P_\infty$ on a line $l$ as fundamental points, and let any line through $P_\infty$ be labeled $l_\infty$. Then the sum of two numbers $P_x$ and $P_y$ is the point $P_{x+y}$ into which $P_y$ is transformed by the elation with axis $l_\infty$ and center $P_\infty$ which transforms $P_0$ into $P_x$; and the product $P_x \cdot P_y$ is the point $P_{xy}$ into which $P_y$ is transformed by the homology with axis $l_\infty$ and center $P_0$ which transforms $P_1$ into $P_x$. Develop the point algebra on this basis without using Assumption P, except in the proof of the commutativity of multiplication.

**5.** If the relation $ax = by$ holds between four points $a, b, x, y$ of a line, show that we have Q$(0ba, \infty yx)$. Is Assumption P necessary for this result?

**6.** Prove by direct computation that the expression $\frac{x_1 - x_3}{x_1 - x_4} : \frac{x_2 - x_3}{x_2 - x_4}$ is unchanged in value when the four points $x_1, x_2, x_3, x_4$ are subjected to any linear fractional transformation $x' = \frac{ax + b}{cx + d}$.

**7.** Prove that the transformations

$$\lambda' = \lambda,\ \lambda' = \frac{1}{\lambda},\ \lambda' = 1 - \lambda,\ \lambda' = \frac{1}{1 - \lambda},\ \lambda' = \frac{\lambda}{\lambda - 1},\ \lambda' = \frac{\lambda - 1}{\lambda}$$

form a group. What are the periods of the various transformations of this group? (Cf. Theorem 13, Cor. 3.)

**8.** If $A, B, C, P_1, P_2, \cdots, P_n$ are any $n + 3$ points of a line, show that every cross ratio of any four of these points can be expressed rationally in terms of the $n$ cross ratios $\lambda_i = \text{R}(AB, CP_i)$, $i = 1, 2, \cdots, n$. When $n = 1$ this reduces to Theorem 13, Cor. 3. Discuss in detail the case $n = 2$.

**9.** If $\text{R}(x_1x_2, x_3x_4) = \lambda$, show that

$$\frac{1 - \lambda}{x_3 - x_4} = \frac{1}{x_3 - x_2} - \frac{\lambda}{x_3 - x_1}.$$

The relation of Cor. 3 of Theorem 13 is a special case of this relation.

**10.** Show that if $\text{R}(AB, CD) = \text{R}(AB, DC)$, the points form a harmonic set H$(AB, CD)$.

**11.** If the cross ratio $\text{R}(AB, CD) = \lambda$ satisfies the equation $\lambda^2 - \lambda + 1 = 0$,

then $\text{R}(AB, CD) = \text{R}(AC, DB) = \text{R}(AD, BC) = \lambda$,

and $\text{R}(AB, DC) = \text{R}(AC, BD) = \text{R}(AD, CB) = -\lambda^2$.

**12.** If $A, B, X, Y, Z$ are any five distinct points on a line, show that

$$\text{R}(AB, XY) \cdot \text{R}(AB, YZ) \cdot \text{R}(AB, ZX) = 1.$$

**13.** State the corollaries of Theorem 11 in homogeneous coördinates.

**14.** By direct computation show that the two methods of determining the double points of a projectivity described in §§ 54 and 58 are equivalent.

**15.** If Q$(ABC, XYZ)$, then

$$\text{R}(AX, YC) + \text{R}(BY, ZA) + \text{R}(CZ, XB) = 1.$$

**16.** If $M_1, M_2, M_3$ are any three points in the plane of a line $l$ but not on $l$, the cross ratios of the lines $l, PM_1, PM_2, PM_3$ are different for any two points $P$ on $l$.

**17.** If $A, B$ are any two fixed points on a line $l$, and $X, Y$ are two variable points such that $\text{R}(AB, XY)$ is constant, the set $[X]$ is projective with the set $[Y]$.

# CHAPTER VII

## COÖRDINATE SYSTEMS IN TWO- AND THREE-DIMENSIONAL* FORMS

**60. Nonhomogeneous coördinates in a plane.** In order to represent the points and lines of a plane analytically we proceed as follows: Choose any two distinct lines of the plane, which we will call the *axes of coördinates*, and determine on each a scale (§ 48) arbitrarily, except that the point of intersection $O$ of the lines shall be the 0-point on each scale (fig. 78). This point we call the *origin*. Denote the fundamental points on one of the lines, which we call the *x-axis*, by $0_x$, $1_x$, $\infty_x$; and on the other line, which we will call the *y-axis*, by $0_y$, $1_y$, $\infty_y$. Let the line $\infty_x\infty_y$ be denoted by $l_\infty$.

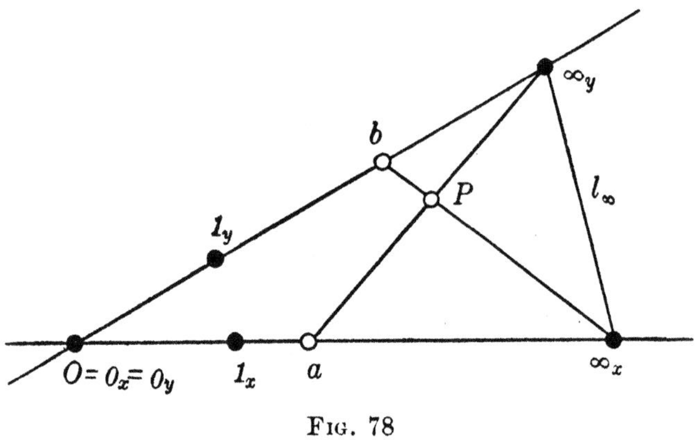


FIG. 78

Now let $P$ be any point in the plane not on $l_\infty$. Let the lines $P\infty_y$ and $P\infty_x$ meet the $x$-axis and the $y$-axis in points whose nonhomogeneous coördinates are $a$ and $b$ respectively, in the scales just established. The two numbers $a$, $b$ uniquely determine and are uniquely determined by the point $P$. Thus every point in the plane not on $l_\infty$ is represented by a pair of numbers; and, conversely, every pair of numbers of which one belongs to the scale on the $x$-axis and the other to the scale on the $y$-axis determines a point in the plane (the pair of symbols $\infty_x$, $\infty_y$ being excluded). The exceptional character of the points on $l_\infty$ will be removed presently (§ 63) by considerations similar to those used to remove the exceptional character of

* All the developments of this chapter are on the basis of Assumptions A, E, P.

the point ∞ in the case of the analytic treatment of the points of a line (§ 58). The two numbers just described, determining the point $P$, are called the *nonhomogeneous coördinates* of $P$ with reference to

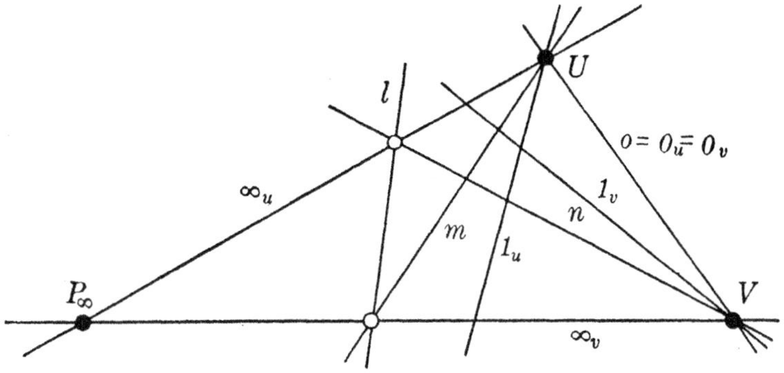


FIG. 79

the two scales on the $x$- and the $y$-axes. The point $P$ is then represented analytically by the symbol $(a, b)$. The number $a$ is called the *x-coördinate* or the *abscissa* of the point, and is always written first in the symbol representing the point; the number $b$ is called the *y-coördinate* or the *ordinate* of the point, and is always written last in this symbol.

The plane dual of the process just described leads to the corresponding analytic representation of a line in the plane. For this purpose, choose any two distinct points in the plane, which we will call the *centers of coördinates*; and in each of the pencils of lines with these centers determine a scale arbitrarily, except that the line $o$ joining the two points shall be the 0-line in each scale. This line we call the *origin*. Denote the fundamental lines on one of the points, which we will call the *u-center*, by $0_u$, $1_u$, $\infty_u$; and on the other point, which we will call the *v-center*, by $0_v$, $1_v$, $\infty_v$. Let the point of intersection of the lines $\infty_u$, $\infty_v$ be denoted by $P_\infty$ (fig. 79).

Now let $l$ be any line in the plane not on $P_\infty$. Let the points $l\infty_v$ and $l\infty_u$ be on the lines of the $u$-center and the $v$-center, whose nonhomogeneous coördinates are $m$ and $n$ respectively in the scales just established. The two numbers $m$, $n$ uniquely determine and are uniquely determined by the line $l$. Thus every line in the plane not on $P_\infty$ is represented by a pair of numbers; and, conversely, every pair of numbers of which one belongs to the scale on the $u$-center and the other to the scale on the $v$-center determines a line in the plane (the pair of symbols $\infty_u$, $\infty_v$ being excluded). The exceptional character of the lines on $P_\infty$ will also be removed presently. The two numbers just described, determining the line $l$, are called the *nonhomogeneous coördinates* of $l$ with reference to the two scales on the $u$- and $v$-centers. The line $l$ is then represented analytically by the symbol $[m, n]$. The number $m$ is called the *u-coördinate* of the line, and is always written first in the symbol just given; the number $n$ is called the *v-coördinate* of the line, and is always written second in this symbol. A variable point of the plane will frequently be represented by the symbol $(x, y)$; a variable line by the symbol $[u, v]$. The coördinates of a point referred to two axes are called *point coördinates;* the coördinates of a line referred to two centers are called *line coördinates.* The line $l_\infty$ and the point $P_\infty$ are called the *singular line* and the *singular point* respectively.

**61. Simultaneous point and line coördinates.** In developing further our analytic methods we must agree upon a convenient relation between the axes and centers of the point and line coördinates respectively. Let us consider any triangle in the plane, say with vertices

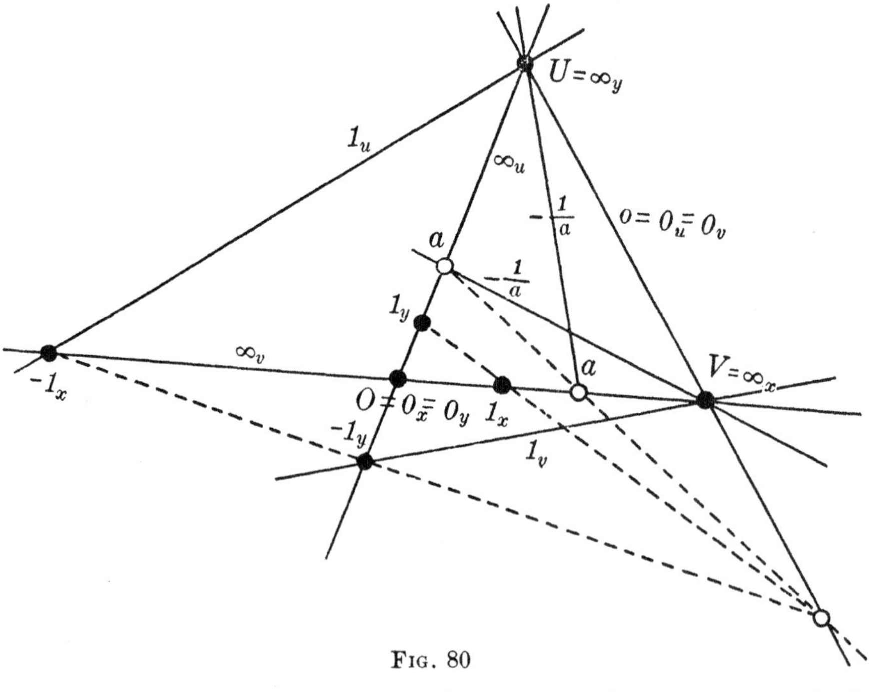


FIG. 80

$O, U, V$. Let the lines $OU$ and $OV$ be the $y$- and $x$-axes respectively, and in establishing the scales on these axes let the points $U, V$ be the points $\infty_y$, $\infty_x$ respectively (fig. 80). Further, let the points $U, V$ be the $u$-center and the $v$-center respectively, and in establishing the scales on these centers let the lines $UO$, $VO$ be the lines $\infty_u$, $\infty_v$ respectively. The scales are now established except for the choice of the 1 points or lines in each scale. Let us choose arbitrarily a point $1_x$ on the $x$-axis and a point $1_y$ on the $y$-axis (distinct, of course, from the points $O$, $U$, $V$). The scales on the axes now being determined, we determine the scales on the centers as follows: Let the line on $U$ and the point $-1_x$ on the $x$-axis be the line $1_u$; and let the line on $V$ and the point $-1_y$ on the $y$-axis be the line $1_v$. All the scales are now fixed. Let $\pi$ be the projectivity (§ 59, Chap. VI) between the points of the $x$-axis and the lines of the $u$-center in which points and lines correspond when their $x$- and $u$-coördinates respectively are the same. If $\pi'$ is the perspectivity in which every line on the $u$-center corresponds to the point in which it meets the $x$-axis, the product $\pi'\pi$ transforms the $x$-axis into itself and interchanges $O$ and $\infty_x$, and $1_x$ and $-1_x$. Hence $\pi'\pi$ is the involution $x' = -1/x$. Hence it follows that *the line on $U$ whose coördinate is $u$ is on the point of the $x$-axis whose coördinate is $-1/u$;* and *the point on the $x$-axis whose coördinate is $x$ is on the line of the $u$-center whose coördinate is $-1/x$.* This is the relation between the scales on the $x$-axis and the $u$-center.

Similar considerations with reference to the $y$-axis and the $v$-center lead to the corresponding result in this case: *The line on $V$ whose coördinate is $v$ is on that point of the $y$-axis whose coördinate is $-1/v$;* and *the point of the $y$-axis whose coördinate is $y$ is on that line of the $v$-center whose coördinate is $-1/y$.*

**62. Condition that a point be on a line.** Suppose that, referred to a system of point-and-line coördinates described above, a point $P$ has coördinates $(a, b)$ and a line $l$ has coördinates $[m, n]$. The condition that $P$ be on $l$ is now readily obtainable. Let us suppose, first, that none of the coördinates $a, b, m, n$ are zero. We may proceed in either one of two dual ways. Adopting one of these, we know from the results of the preceding section that the line $[m, n]$ meets the $x$-axis in a point whose $x$-coördinate is $-1/m$, and meets the $y$-axis in a point whose $y$-coördinate is $-1/n$ (fig. 81). Also, by definition, the line joining $P = (a, b)$ to $U$ meets the $x$-axis in a point whose $x$-coördinate is $a$; and the line joining $P$ to $V$ meets the $y$-axis in a point whose $y$-coördinate is $b$. If $P$ is on $l$, we clearly have the following perspectivity:

(1) $$-\frac{1}{m} O a \infty_x \overset{P}{\underset{\wedge}{=}} -\frac{1}{n} O \infty_y b.$$

Hence we have

(2) $$\text{R}\left(-\frac{1}{m} O,\ a\infty_x\right) = \text{R}\left(-\frac{1}{n} O,\ \infty_y b\right),$$

which, when expanded (Theorem 13, Chap. VI), gives for the desired condition

(3) $$ma + nb + 1 = 0.$$

This condition has been shown to be necessary. It is also sufficient, for, if it is satisfied, relation (2) must hold, and hence would follow (Theorem 13, Cor. 5, Chap. VI)

$$-\frac{1}{m} O a \infty_x \underset{\wedge}{\overline{\ \ }} -\frac{1}{n} O \infty_y b.$$

But since this projectivity has the self-corresponding element $O$, it is a perspectivity which leads to relation (1). But this implies that $P$ is on $l$.

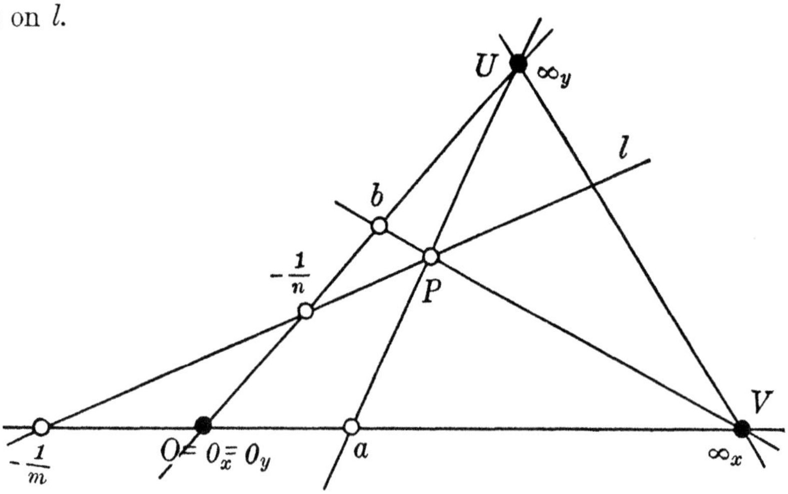


FIG. 81

If now $a = 0\ (b \neq 0)$, we have at once $b = -1/n$; and if $b = 0\ (a \neq 0)$, we have likewise $a = -1/m$ for the condition that $P$ be on $l$. But each of these relations is equivalent to (3) when $a = 0$ and $b = 0$ respectively. The combination $a = 0$, $b = 0$ gives the origin 0 which is never on a line $[m, n]$ where $m \neq \infty \neq n$. It follows in the same way directly from the definition that relation (3) gives the desired condition, if we have either $m = 0$ or $n = 0$. The condition (3) is then valid for all cases, and we have

THEOREM 1. *The necessary and sufficient condition that a point* $P=(a, b)$ *be on a line* $l=[m, n]$ *is that the relation* $ma+nb+1=0$ *be satisfied.*

DEFINITION. The equation which is satisfied by the coördinates of all the points on a given line and no others is called the *point equation of the line.*

DEFINITION. The equation which is satisfied by the coördinates of all the lines on a given point and no others is called the *line equation of the point.*

COROLLARY 1. *The point equation of the line* $[m, n]$ *is*

$$mx+ny+1=0.$$

COROLLARY 1′. *The line equation of the point* $(a, b)$ *is*

$$au+bv+1=0.$$

**EXERCISE**

Derive the condition of Theorem 1 by dualizing the proof given.

**63. Homogeneous coördinates in the plane.** In the analytic representation of points and lines developed in the preceding sections the points on the line $UV=o$ and the lines on the point $O$ were left unconsidered. To remove the exceptional character of these points and lines, we may recall that in the case of a similar problem in the analytic representation of the elements of a one-dimensional form we found it convenient to replace the nonhomogeneous coördinate $x$ of a point on a line by a pair of numbers $x_1$, $x_2$ whose ratio $x_1/x_2$ was equal to $x\,(x \neq \infty)$, and such that $x_2=0$ when $x=\infty$.

A similar system of *homogeneous coördinates* can be established for the plane. Denote the vertices $O$, $U$, $V$ of any triangle, which we will call the *triangle of reference,* by the "coördinates" (0, 0, 1), (0, 1, 0), (1, 0, 0) respectively, and an arbitrary point $T$, not on a side of the triangle of reference, by (1, 1, 1). The complete quadrangle $OUVT$ is called the *frame of reference* * of the system of coördinates to be established. The three lines $UT$, $VT$, $OT$ meet the other sides of the triangle of reference in points which we denote by $1_x=(1, 0, 1)$, $1_y=(0, 1, 1)$, $1_z=(1, 1, 0)$ respectively (fig. 82).

We will now show how it is possible to denote every point in the plane by a set of coördinates $(x_1, x_2, x_3)$. Observe first that we have thus far determined three points on each of the sides of the triangle

* *Frame of reference* is a general term that may be applied to the fundamental elements of any coördinate system.

of reference, viz.: (0, 0, 1), (0, 1, 1), (0, 1, 0) on $OU$; (0, 0, 1), (1, 0, 1), (1, 0, 0) on $OV$; and (0, 1, 0), (1, 1, 0), (1, 0, 0) on $UV$. The coördinates which we have assigned to these points are all of the form $(x_1, x_2, x_3)$. The three points on $OU$ are characterized by the fact that $x_1 = 0$. Fixing attention on the remaining coördinates, we choose the points (0, 0, 1), (0, 1, 1), (0, 1, 0) as the fundamental points (0, 1), (1, 1), (1, 0) of a system of homogeneous coördinates on the line $OU$. If in this system a point has coördinates $(l, m)$, we denote it in our planar system by $(0, l, m)$. In like manner, to the points of the other two sides of the triangle of reference may be assigned coördinates of the form $(k, 0, m)$ and $(k, l, 0)$ respectively. We have thus assigned coördinates of the form $(x_1, x_2, x_3)$ to all the points of the sides of the triangle of reference. Moreover, the coördinates of every point on these sides satisfy one of the three relations $x_1 = 0$, $x_2 = 0$, $x_3 = 0$.

Now let $P$ be any point in the plane not on a side of the triangle of reference. $P$ is uniquely determined if the coördinates of its projections from any two of the vertices of the triangle of reference on the opposite sides are known. Let its projections from $U$ and $V$ on the sides $OV$ and $OU$ be $(k, 0, n)$ and $(0, l', n')$ respectively. Since under the hypothesis none of the numbers $k$, $n$, $l'$, $n'$ is zero, it is clearly possible to choose three numbers $(x_1, x_2, x_3)$ such that $x_1 : x_3 = k : n$, and $x_2 : x_3 = l' : n'$. We may then denote $P$ by the coördinates $(x_1, x_2, x_3)$. To make this system of coördinates effective, however, we must show that the same set of three numbers $(x_1, x_2, x_3)$ can be obtained by projecting $P$ on any other pair of sides of the triangle of reference. In other words, we must show that the projection of $P = (x_1, x_2, x_3)$ from $O$ on the line $UV$ is the point $(x_1, x_2, 0)$. Since this is clearly true of the point $T = (1, 1, 1)$, we assume $P$ distinct from $T$. Since the numbers $x_1$, $x_2$, $x_3$ are all different from 0, let us place $x_1 : x_3 = x$, and $x_2 : x_3 = y$, so that $x$ and $y$ are the nonhomogeneous coördinates of $(x_1, 0, x_3)$ and $(0, x_2, x_3)$ respectively in the scales on $OV$ and $OU$ defined by $O = 0_x$, $1_x$, $V = \infty_x$ and $O = 0_y$, $1_y$, $U = \infty_y$. Finally, let $OP$ meet $UV$ in the point whose nonhomogeneous coördinate in the scale defined by $U = 0_z$, $1_z$, $V = \infty_z$ is $z$; and let $OP$ meet the line $1_x U$ in $A$. We now have

$$\infty_z 0_z 1_z z \overset{O}{\underset{\wedge}{=}} 1_x 0_z T A \overset{V}{\underset{\wedge}{=}} 0_y \infty_y 1_y C,$$

where $C$ is the point in which $VA$ meets $OU$. This projectivity between the lines $UV$ and $OU$ transforms $0_z$ into $\infty_y$, $\infty_z$ into $0_y$, and $1_z$ into $1_y$. It follows that $C$ has the coördinate $1/z$ in the scale on $OU$. We have also

$$\infty_x 0_x 1_x x \overset{U}{\underset{\wedge}{=}} z 0_x A P \overset{V}{\underset{\wedge}{=}} \infty_y 0_y \frac{1}{z} y,$$

which gives

$$x = \text{R}(\infty_x 0_x,\ 1_x x) = \text{R}\left(\infty_y 0_y,\ \frac{1}{z} y\right) = zy.$$

Substituting $x = x_1 : x_3$, and $y = x_2 : x_3$, this gives the desired relation $z = x_1 : x_2$. The results of this discussion may be summarized as follows:

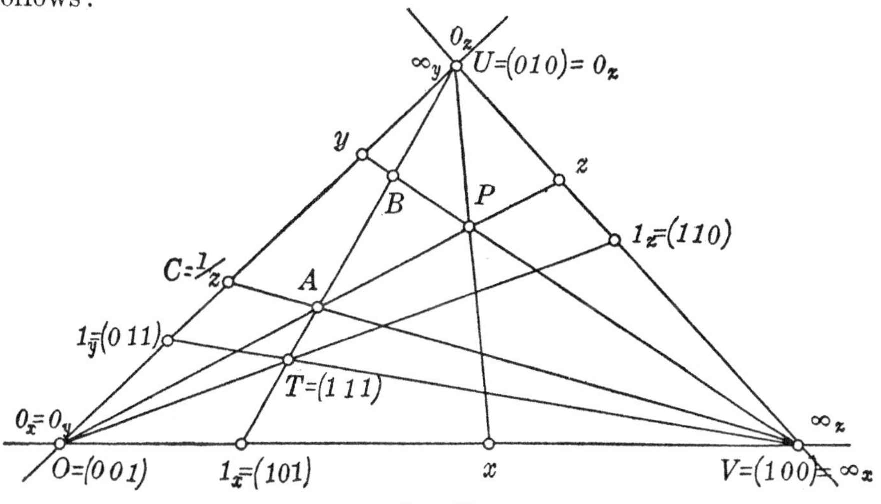


FIG. 82

THEOREM 2. DEFINITION. *If $P$ is any point not on a side of the triangle of reference $OUV$, there exist three numbers $x_1, x_2, x_3$ (all different from 0) such that the projections of $P$ from the vertices $O$, $U$, $V$ on the opposite sides have coördinates $(x_1, x_2, 0)$, $(x_1, 0, x_3)$, $(0, x_2, x_3)$ respectively. These three numbers are called the homogeneous coördinates of $P$, and $P$ is denoted by $(x_1, x_2, x_3)$. Any set of three numbers (not all equal to 0) determine uniquely a point whose (homogeneous) coördinates they are.*

The truth of the last sentence in the above theorem follows from the fact that, if one of the coördinates is 0, they determine uniquely a point on one of the sides of the triangle of reference; whereas, if none is equal to 0, the lines joining $U$ to $(x_1, 0, x_3)$ and $V$ to $(0, x_2, x_3)$ meet in a point whose coördinates by the reasoning above are $(x_1, x_2, x_3)$.

COROLLARY. *The coördinates* $(x_1, x_2, x_3)$ *and* $(kx_1, kx_2, kx_3)$ *determine the same point, if k is not* 0.

Homogeneous line coördinates arise by dualizing the above discussion in the plane. Thus we choose any quadrilateral in the plane as frame of reference, denoting the sides by [1, 0, 0], [0, 1, 0], [0, 0, 1], [1, 1, 1] respectively. The points of intersection with [1, 1, 1] of the lines [1, 0, 0], [0, 1, 0], [0, 0, 1] are joined to the vertices of the triangle of reference opposite to [1, 0, 0], [0, 1, 0], [0, 0, 1] respectively by lines that are denoted by [0, 1, 1], [1, 0, 1], [1, 1, 0]. The three lines [1, 0, 0], [1, 1, 0], [0, 1, 0] are then taken as the fundamental lines [1, 0], [1, 1], [0, 1] of a homogeneous system of coördinates in a flat pencil. If in this system a line is denoted by $[u_1, u_2]$, it is denoted in the planar system by $[u_1, u_2, 0]$. In like manner, to the lines on the other vertices are assigned coördinates of the forms $[0, u_2, u_3]$ and $[u_1, 0, u_3]$ respectively. As the plane dual of the theorem and definition above we then have at once

THEOREM 2′. DEFINITION. *If l is any line not on a vertex of the triangle of reference, there exist three numbers* $u_1, u_2, u_3$ *all different from zero, such that the traces of l on the three sides of the triangle of reference are projected from the respective opposite vertices by the lines* $[u_1, u_2, 0]$, $[u_1, 0, u_3]$, $[0, u_2, u_3]$. *These three numbers are called the homogeneous coördinates of l, and l is denoted by* $[u_1, u_2, u_3]$. *Any set of three numbers (not all zero) determine uniquely a line whose coördinates they are.*

Homogeneous point and line coördinates may be put into such a relation that the condition that a point $(x_1, x_2, x_3)$ be on a line $[u_1, u_2, u_3]$ is that the relation $u_1x_1 + u_2x_2 + u_3x_3 = 0$ be satisfied. We have seen that if $(x_1, x_2, x_3)$ is a point not on a side of the triangle of reference, and we place $x = x_1/x_3$, and $y = x_2/x_3$, the numbers $(x, y)$ are the nonhomogeneous coördinates of the point $(x_1, x_2, x_3)$ referred to $OV$ as the $x$-axis and to $OU$ as the $y$-axis of a system of nonhomogeneous coördinates in which the point $T = (1, 1, 1)$ is the point (1, 1) ($O$, $U$, $V$ being used in the same significance as in the proof of Theorem 2). By duality, if $[u_1, u_2, u_3]$ is any line not on any vertex of the triangle of reference, and we place $u = u_1/u_3$ and $v = u_2/u_3$, the numbers $[u, v]$ are the nonhomogeneous coördinates of the line $[u_1, u_2, u_3]$ referred to two of the vertices of the triangle of reference as $U$-center and $V$-center respectively, and in which the line $[1, 1, 1]$ is the line $[1, 1]$. If, now, we superpose these two systems of nonhomogeneous coördinates in the way described in the preceding section, the condition that the point $(x, y)$ be on the line $[u, v]$ is that the relation $ux + vy + 1 = 0$ be satisfied (Theorem 1). It is now easy to recognize the resulting relation between the systems of homogeneous coördinates with which we started. Clearly the point $(0, 1, 0) = U$ is the $U$-center, $(1, 0, 0) = V$ is the $V$-center, and $(0, 0, 1) = O$ is the third

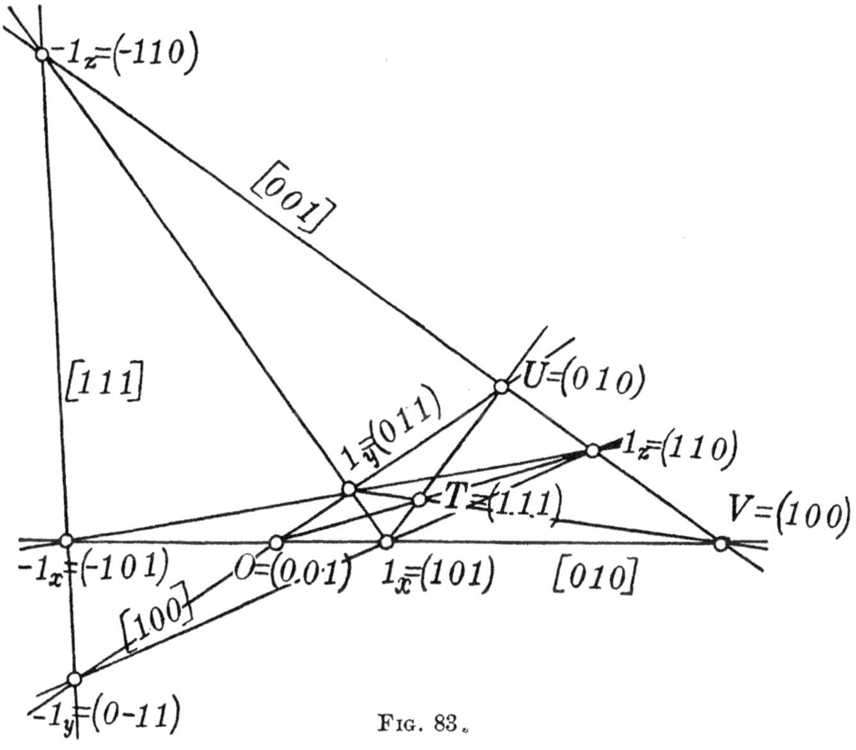


FIG. 83.

vertex of the triangle of reference in the homogeneous system of line coördinates. Also the line whose points satisfy the relation $x_1 = 0$ is the line $[1, 0, 0]$, the line for which $x_2 = 0$ is the line $[0, 1, 0]$, and the line for which $x_3 = 0$ is the line $[0, 0, 1]$. Finally, the line $[1, 1] = [1, 1, 1]$, whose equation in nonhomogeneous coördinates is $x + y + 1 = 0$, meets the line $x_1 = 0$ in the point $(0, -1, 1)$, and the line $x_2 = 0$ in the point $(-1, 0, 1)$. The two coördinate systems are then completely determined (fig. 83).

It now follows at once from the result of the preceding section that the condition that $(x_1, x_2, x_3)$ be on the line $[u_1, u_2, u_3]$ is $u_1x_1 + u_2x_2 + u_3x_3 = 0$, if none of the coördinates $x_1, x_2, x_3, u_1, u_2, u_3$

is zero. To see that the same condition holds also when one (or more) of the coördinates is zero, we note first that the points $(0, -1, 1)$, $(-1, 0, 1)$, and $(-1, 1, 0)$ are collinear. They are, in fact (fig. 83), on the axis of perspectivity of the two perspective triangles $OUV$ and $1_x 1_y 1_z$, the center of perspectivity being $T$. It is now clear that

the line $[1, 0, 0]$ passes through the point $(0, 1, 0)$,
the line $[0, 1, 0]$ passes through the point $(1, 0, 0)$,
the line $[1, 1, 0]$ passes through the point $(-1, 1, 0)$.

There is thus an involution between the points $(x_1, x_2, 0)$ of the line $x_3 = 0$ and the traces $(x_1', x_2', 0)$ of the lines with the same coördinates, and this involution is given by the equations

$$x_1' = x_2,$$
$$x_2' = -x_1.$$

In other words, the line $[u_1, u_2, 0]$ passes through the point $(-u_2, u_1, 0)$. Any other point of this line (except $(0, 0, 1)$) has, by definition, the coördinates $(-u_2, u_1, x_3)$. Hence all points $(x_1, x_2, x_3)$ of the line $[u_1, u_2, 0]$ satisfy the relation $u_1x_1 + u_2x_2 + u_3x_3 = 0$. The same argument applied when any one of the other coördinates is zero establishes this condition for all cases. A system of point and a system of line coördinates, when placed in the relation described above, will be said to form *a system of homogeneous point and line coördinates* in the plane. The result obtained may then be stated as follows:

THEOREM 3. *In a system of homogeneous point and line coördinates in a plane the necessary and sufficient condition that a point* $(x_1, x_2, x_3)$ *be on a line* $[u_1, u_2, u_3]$ *is that the relation* $u_1x_1 + u_2x_2 + u_3x_3 = 0$ *be satisfied.*

COROLLARY. *The equation of a line through the origin of a system of nonhomogeneous coördinates is of the form* $mx + ny = 0$.

### EXERCISES

**1.** The line $[1, 1, 1]$ is the polar of the point $(1, 1, 1)$ with regard to the triangle of reference (cf. p. 46).

**2.** The same point is represented by $(a_1, a_2, a_3)$ and $(b_1, b_2, b_3)$ if and only if the two-rowed determinants of the matrix $\begin{pmatrix} a_1 & a_2 & a_3 \\ b_1 & b_2 & b_3 \end{pmatrix}$ are all zero.

**3.** Describe nonhomogeneous and homogeneous systems of line and plane coördinates in a bundle by dualizing in space the preceding discussion. In such a bundle what is the condition that a line be on a plane?

**64. The line on two points. The point on two lines.** Given two points, $A = (a_1, a_2, a_3)$ and $B = (b_1, b_2, b_3)$, the question now arises as to what are the coördinates of the line joining them; and the dual of this problem, namely, given two lines, $m = [m_1, m_2, m_3]$ and $n = [n_1, n_2, n_3]$, to find the coördinates of the point of intersection of the two lines.

THEOREM 4. *The equation of the line joining the points* $(a_1, a_2, a_3)$ *and* $(b_1, b_2, b_3)$ *is*

$$\begin{vmatrix} x_1 & x_2 & x_3 \\ a_1 & a_2 & a_3 \\ b_1 & b_2 & b_3 \end{vmatrix} = 0.$$

THEOREM 4'. *The equation of the point of intersection of the lines* $[m_1, m_2, m_3]$ *and* $[n_1, n_2, n_3]$ *is*

$$\begin{vmatrix} u_1 & u_2 & u_3 \\ m_1 & m_2 & m_3 \\ n_1 & n_2 & n_3 \end{vmatrix} = 0.$$

*Proof.* When these determinants are expanded, we get

$$\begin{vmatrix} a_2 & a_3 \\ b_2 & b_3 \end{vmatrix} x_1 + \begin{vmatrix} a_3 & a_1 \\ b_3 & b_1 \end{vmatrix} x_2 + \begin{vmatrix} a_1 & a_2 \\ b_1 & b_2 \end{vmatrix} x_3 = 0,$$

$$\begin{vmatrix} m_2 & m_3 \\ n_2 & n_3 \end{vmatrix} u_1 + \begin{vmatrix} m_3 & m_1 \\ n_3 & n_1 \end{vmatrix} u_2 + \begin{vmatrix} m_1 & m_2 \\ n_1 & n_2 \end{vmatrix} u_3 = 0,$$

respectively. The one above is the equation of a line, the one below the equation of a point. Moreover, the determinants above both evidently vanish when the variable coördinates are replaced by the coördinates of the given elements. The expanded form just given leads at once to the following:

COROLLARY 1. *The coördinates of the line joining the points* $(a_1, a_2, a_3)$, $(b_1, b_2, b_3)$ *are*

$$u_1 : u_2 : u_3 = \begin{vmatrix} a_2 & a_3 \\ b_2 & b_3 \end{vmatrix} : \begin{vmatrix} a_3 & a_1 \\ b_3 & b_1 \end{vmatrix} : \begin{vmatrix} a_1 & a_2 \\ b_1 & b_2 \end{vmatrix}.$$

COROLLARY 1'. *The coördinates of the point of intersection of the lines* $[m_1, m_2, m_3]$, $[n_1, n_2, n_3]$ *are*

$$x_1 : x_2 : x_3 = \begin{vmatrix} m_2 & m_3 \\ n_2 & n_3 \end{vmatrix} : \begin{vmatrix} m_3 & m_1 \\ n_3 & n_1 \end{vmatrix} : \begin{vmatrix} m_1 & m_2 \\ n_1 & n_2 \end{vmatrix}.$$

There also follows immediately from this theorem:

COROLLARY 2. *The condition that three points A, B, C be collinear is*

$$\begin{vmatrix} a_1 & a_2 & a_3 \\ b_1 & b_2 & b_3 \\ c_1 & c_2 & c_3 \end{vmatrix} = 0.$$

COROLLARY 2'. *The condition that three lines m, n, p be concurrent is*

$$\begin{vmatrix} m_1 & m_2 & m_3 \\ n_1 & n_2 & n_3 \\ p_1 & p_2 & p_3 \end{vmatrix} = 0.$$

EXAMPLE. Let us verify the theorem of Desargues (Theorem 1, Chap. II) analytically. Choose one of the two perspective triangles as triangle of reference, say $A' = (0, 0, 1)$, $B' = (0, 1, 0)$, $C' = (1, 0, 0)$, and let the center of perspectivity be $P = (1, 1, 1)$. If the other triangle is $ABC$, we may place

$A = (1, 1, a)$, $B = (1, b, 1)$, $C = (c, 1, 1)$; for the equation of the line $PA'$ is $x_1 - x_2 = 0$; and since $A$ is, by hypothesis, on this line, its first two coördinates must be equal, and may therefore be assumed equal to 1; the third coördinate is arbitrary. Similarly for the other points. Now, from the above theorems and their corollaries we readily obtain in succession the following:

The coördinates of the line $A'B'$ are $[1, 0, 0]$.

The coördinates of the line $AB$ are $[1 - ab, a - 1, b - 1]$.

Hence the coördinates of their intersection $C''$ are

$$C'' = (0, 1 - b, a - 1).$$

Similarly, we find the coördinates of the intersection $A''$ of the lines $B'C'$, $BC$ to be

$$A'' = (1 - c, b - 1, 0);$$

and, finally, the coördinates of the intersection $B''$ of the lines $C'A'$, $CA$ to be

$$B'' = (c - 1, 0, 1 - a).$$

The points $A''$, $B''$, $C''$ are readily seen to satisfy the condition for collinearity.

## EXERCISES

**1.** Work through the dual of the example just given, choosing the sides of one of the triangles and the axis of perspectivity as the fundamental lines of the system of coördinates. Show that the work may be made identical, step for step, with that above, except for the interpretation of the symbols.

**2.** Show that the system of coördinates may be so chosen that a quadrangle-quadrilateral configuration is represented by all the sets of coördinates that can be formed from the numbers 0 and 1. Dualize.

**3.** Derive the equation of the polar line of any point with regard to the triangle of reference. Dualize.

**65. Pencils of points and lines. Projectivity.** A convenient analytic representation of the points of a pencil of points or the lines of a pencil of lines is given by the following dual theorems:

THEOREM 5. *Any point of a pencil of points may be represented by*

$$P = (\lambda_2 a_1 + \lambda_1 b_1, \lambda_2 a_2 + \lambda_1 b_2, \lambda_2 a_3 + \lambda_1 b_3),$$

*where* $A = (a_1\ a_2, a_3)$ *and* $B = (b_1, b_2, b_3)$ *are any two distinct points of the pencil.*

THEOREM 5′. *Any line of a pencil of lines may be represented by*

$$p = [\mu_2 m_1 + \mu_1 n_1, \mu_2 m_2 + \mu_1 n_2, \mu_2 m_3 + \mu_1 n_3],$$

*where* $m = [m_1, m_2, m_3]$ *and* $n = [n_1, n_2, n_3]$ *are any two distinct lines of the pencil.*

*Proof.* We may confine ourselves to the proof of the theorem on the left. By Theorem 4, Cor. 2, any point $(x_1, x_2, x_3)$ of the pencil of points on the line $AB$ satisfies the relation

(1)
$$\begin{vmatrix} x_1 & x_2 & x_3 \\ a_1 & a_2 & a_3 \\ b_1 & b_2 & b_3 \end{vmatrix} = 0.$$

We may then determine three numbers $\rho$, $\lambda_2'$, $\lambda_1'$, such that we have

(2)
$$\rho x_i = \lambda_2' a_i + \lambda_1' b_i. \qquad (i = 1, 2, 3)$$

The number $\rho$ cannot be 0 under the hypothesis, for then we should have from (2) the proportion $a_1 : a_2 : a_3 = b_1 : b_2 : b_3$, which would imply that the points $A$ and $B$ coincide. We may therefore divide by $\rho$. Denoting the ratios $\lambda_2'/\rho$ and $\lambda_1'/\rho$ by $\lambda_2$ and $\lambda_1$, we see that every point of the pencil may be represented in the manner specified. Conversely, every point whose coördinates are of the form specified clearly satisfies relation (1) and is therefore a point of the pencil.

The points $A$ and $B$ in the above representation are called the *base points* of this so-called *parametric representation* of the elements of a pencil of points. Evidently any two distinct points may be chosen as base points in such a representation. The ratio $\lambda_1/\lambda_2$ is called the *parameter* of the point it determines. It is here written in *homogeneous form,* which gives the point $A$ for the value $\lambda_1 = 0$ and the point $B$ for the value $\lambda_2 = 0$. In many cases, however, it is more convenient to write this parameter in nonhomogeneous form,

$$P = (a_1 + \lambda b_1,\ a_2 + \lambda b_2,\ a_3 + \lambda b_3),$$

which is obtained from the preceding by dividing by $\lambda_2$ and replacing $\lambda_1/\lambda_2$ by $\lambda$. In this representation the point $B$ corresponds to the value $\lambda = \infty$. We may also speak of any point of the pencil under this representation as *the point* $\lambda_1 : \lambda_2$ or *the point* $\lambda$ when it corresponds to the value $\lambda_1/\lambda_2 = \lambda$ of the parameter. Similar remarks and the corresponding terminology apply, of course, to the parametric representation of the lines of a flat pencil. It is sometimes convenient, moreover, to adopt the notation $A + \lambda B$ to denote any point of the pencil whose base points are $A$, $B$ or to denote the pencil itself; also, to use the notation $m + \mu n$ to denote the pencil of lines or any line of this pencil whose base lines are $m$, $n$.

In order to derive an analytic representation of a projectivity between two one-dimensional primitive forms in the plane, we seek first the condition that the point $\lambda$ of a pencil of points $A + \lambda B$ be on the line $\mu$ of a pencil of lines $m + \mu n$. By Theorem 3 the condition that the point $\lambda$ be on the line $\mu$ is the relation

$$\sum_{i=1}^{i=3}(m_i+\mu n_i)(a_i+\lambda b_i)=0.$$

When expanded this relation gives

$$\mu\lambda\sum_{i=1}^{i=3}n_ib_i+\mu\sum_{i=1}^{i=3}n_ia_i+\lambda\sum_{i=1}^{i=3}m_ib_i+\sum_{i=1}^{i=3}m_ia_i=0.$$

This is a bilinear equation whose coefficients depend only on the coördinates of the base points and base lines of the two pencils and not on the individual points for which the condition is sought. Placing

$$\sum n_ib_i=C,\ \sum n_ia_i=D,\ \sum m_ib_i=-A,\ \sum m_ia_i=-B,$$

this equation becomes $C\mu\lambda+D\mu-A\lambda-B=0,$

which may also be written *

(1) $$\mu=\frac{A\lambda+B}{C\lambda+D}.$$

The result may be stated as follows: *Any perspective relation between two one-dimensional primitive forms of different kinds is obtained by establishing a projective correspondence between the parameters of the two forms.* Since any projective correspondence between two one-dimensional primitive forms is obtained as the resultant of a sequence of such perspectivities, and since the resultant of any two linear fractional transformations of type (1) is a transformation of the same type, we have the following theorem:

THEOREM 6. *Any projective correspondence between two one-dimensional primitive forms in the plane is obtained by establishing a projective relation*

$$\mu=\frac{\alpha\lambda+\beta}{\gamma\lambda+\delta}\qquad(\alpha\delta-\beta\gamma\neq 0)$$

*between the parameters $\mu$, $\lambda$ of the two forms.*

In particular we have

COROLLARY 1. *Any projectivity in a one-dimensional primitive form in the plane is given by a relation of the form*

$$\lambda'=\frac{\alpha\lambda+\beta}{\gamma\lambda+\delta},\qquad(\alpha\delta-\beta\gamma\neq 0)$$

*where $\lambda$ is the parameter of the form.*

* The determinant $\begin{vmatrix}A & B\\ C & D\end{vmatrix}$ does not vanish because the correspondence between $\lambda$ and $\mu$ is (1, 1).

COROLLARY 2. *If* $\lambda_1, \lambda_2, \lambda_3, \lambda_4$ *are the parameters of four elements* $A_1, A_2, A_3, A_4$ *of a one-dimensional primitive form, the cross ratio* $R(A_1A_2, A_3A_4)$ *is given by*

$$R(A_1A_2, A_3A_4) = R(\lambda_1\lambda_2, \lambda_3\lambda_4) = \frac{\lambda_1 - \lambda_3}{\lambda_1 - \lambda_4} : \frac{\lambda_2 - \lambda_3}{\lambda_2 - \lambda_4}.$$

A projectivity between two different one-dimensional forms may be represented in a particularly simple form by a judicious choice of the base elements of the parametric representation. To fix ideas, let us take the case of two projective pencils of points. Choose any two distinct points $A$, $B$ of the first pencil to be the base points, and let the homologous points of the second pencil be base points of the latter. Then to the values $\lambda = 0$ and $\lambda = \infty$ of the first pencil must correspond the values $\mu = 0$ and $\mu = \infty$ respectively of the second. In this case the relation of Theorem 6, however, assumes the form $\mu = k\lambda$. Hence, since the same argument applies to any distinct forms, we have

COROLLARY 3. *If two distinct projective one-dimensional primitive forms in the plane are represented parametrically so that the base elements form two homologous pairs, the projectivity is represented by a relation of the form* $\mu = k\lambda$ *between the parameters* $\mu$, $\lambda$ *of the two forms.*

This relation may be still further simplified. Taking again the case discussed above of two projective pencils of points, we have seen that, in general, to the point $(a_1 + b_1, a_2 + b_2, a_3 + b_3)$, i.e. to $\lambda = 1$, corresponds the point $(a_1' + kb_1', a_2' + kb_2', a_3' + kb_3')$, i.e. the point $\mu = k$. Since the point $B' = (b_1', b_2', b_3')$ is also represented by the set of coördinates $(kb_1', kb_2', kb_3')$, it follows that if we choose the latter values for the coördinates of the base point $B'$, to the value $\lambda = 1$ will correspond the value $\mu = 1$, and hence we have always $\mu = \lambda$. In other words, we have

COROLLARY 4. *If two distinct one-dimensional forms are projective, the base elements may be so chosen that the parameters of any two homologous elements are equal.*

Before closing this section it seems desirable to call attention explicitly to the forms of the equation of any line of a pencil and of the equation of any point of a pencil which is implied by Theorem 5′ and Theorem 5 respectively. If we place $m = m_1x_1 + m_2x_2 + m_3x_3$ and

$n = n_1x_1 + n_2x_2 + n_3x_3$, it follows from the first theorem mentioned that the equation of any line of the pencil whose center is the intersection of the lines $m = 0$, $n = 0$ is given by an equation of the form $m + \mu n = 0$. Similarly, the equation of any point of the line joining $A = a_1u_1 + a_2u_2 + a_3u_3 = 0$ and $B = b_1u_1 + b_2u_2 + b_3u_3 = 0$ is of the form $A + \lambda B = 0$.

**66. The equation of a conic.** The results of § 65 lead readily to the equation of a conic. By this is meant an equation in point (line) coördinates which is satisfied by all the points (lines) of a conic, and by no others. To derive this equation, let $A$, $B$ be two distinct points on a conic, and let

$$\begin{aligned} m &= m_1x_1 + m_2x_2 + m_3x_3 = 0, \\ n &= n_1x_1 + n_2x_2 + n_3x_3 = 0, \\ p &= p_1x_1 + p_2x_2 + p_3x_3 = 0 \end{aligned} \tag{1}$$

be the equations of the tangent at $A$, the tangent at $B$, and the line $AB$ respectively. The conic is then generated as a point locus by two projective pencils of lines at $A$ and $B$, in which $m$, $p$ at $A$ are homologous with $p$, $n$ at $B$ respectively. This projectivity between the pencils

$$\begin{aligned} m + \lambda p &= 0, \\ p + \mu n &= 0 \end{aligned} \tag{2}$$

is given (Theorem 6, Cor. 3) by a relation

$$\mu = k\lambda \tag{3}$$

between the parameters $\mu$, $\lambda$ of the two pencils. To obtain the equation which is satisfied by all the points of intersection of pairs of homologous lines of these pencils, and by no others, we need simply eliminate $\mu$, $\lambda$ between the last three relations. The result of this elimination is

$$p^2 - kmn = 0, \tag{4}$$

which is the equation required. By multiplying the coördinates of one of the lines by a constant we may make $k = 1$.

Conversely, it is obvious that the points which satisfy any equation of type (4) are the points of intersection of homologous lines in the pencils (2), provided that $\mu = k\lambda$. If $m$, $n$, $p$ are fixed, the condition that the conic (4) shall pass through a point $(a_1, a_2, a_3)$ is a linear equation in $k$. Hence we have

THEOREM 7. *If $m = 0$, $n = 0$, $p = 0$ are the equations of two distinct tangents of a conic and the line joining their points of contact respectively, the point equation of the conic is of the form*

$$p^2 - kmn = 0.$$

*The coefficient $k$ is determined by any third point on the conic. Conversely, the points which satisfy an equation of the above form constitute a conic of which $m = 0$ and $n = 0$ are tangents at points on $p = 0$.*

COROLLARY. *By properly choosing the triangle of reference, the point equation of any conic may be put in the form*

$$x_2^2 - kx_1x_3 = 0,$$

*where $x_1 = 0$, $x_3 = 0$ are two tangents, and $x_2 = 0$ is the line joining their points of contact.*

THEOREM 7.′ *If $A = 0$, $B = 0$, $C = 0$ are the equations of two distinct points of a conic and the intersection of the tangents at these points respectively, the line equation of the conic is of the form*

$$C^2 - kAB = 0.$$

*The coefficient $k$ is determined by any third line of the conic. Conversely, the lines which satisfy an equation of the above form constitute a conic of which $A = 0$ and $B = 0$ are points of contact of the tangents through $C = 0$.*

COROLLARY. *By properly choosing the triangle of reference, the line equation of any conic may be put in the form*

$$u_2^2 - ku_1u_3 = 0,$$

*where $u_1 = 0$, $u_3 = 0$ are two points, and $u_2 = 0$ is the intersection of the tangents at these points.*

It is clear that if we choose the point $(1, 1, 1)$ on the conic, we have $k = 1$. Supposing the choice to have been thus made, we inquire regarding the condition that a line $[u_1, u_2, u_3]$ be tangent to the conic

$$x_2^2 - x_1x_3 = 0.$$

This condition is equivalent to the condition that the line whose equation is

$$u_1x_1 + u_2x_2 + u_3x_3 = 0$$

shall have one and only one point in common with the conic. Eliminating $x_3$ between this equation and that of the conic, the points common to the line and the conic are determined by the equation

$$u_1x_1^2 + u_2x_1x_2 + u_3x_2^2 = 0.$$

The roots of this equation are equal, if and only if we have

$$u_2^2 - 4\,u_1u_3 = 0.$$

Since this is the line equation of all tangents to the conic, and since it is of the form given in Theorem 7′, Cor., above, we have here a new proof of the fact that *the tangents to a point conic form a line conic* (cf. Theorem 11, Chap. V).

When the linear expressions for $m, n, p$ are substituted in the equation $p^2 - kmn = 0$ of any conic, there results, when multiplied out, a homogeneous equation of the second degree in $x_1, x_2, x_3$, which may be written in the form

$$(1) \quad a_{11}x_1^2 + a_{22}x_2^2 + a_{33}x_3^2 + 2\,a_{12}x_1x_2 + 2\,a_{13}x_1x_3 + 2\,a_{23}x_2x_3 = 0.$$

We have seen that the equation of every conic is of this form. We have not shown that every equation of this form represents a conic (see § 85, Chap. IX).

**EXERCISE**

Show that the conic

$$a_{11}x_1^2 + a_{22}x_2^2 + a_{33}x_3^2 + 2\,a_{12}x_1x_2 + 2\,a_{13}x_1x_3 + 2\,a_{23}x_2x_3 = 0$$

degenerates into (distinct or coincident) straight lines, if and only if we have

$$\begin{vmatrix} a_{11} & a_{12} & a_{13} \\ a_{12} & a_{22} & a_{23} \\ a_{13} & a_{23} & a_{33} \end{vmatrix} = 0.$$

Dualize. (A, E, P, $H_0$)

**67. Linear transformations in a plane.** We inquire now concerning the geometric properties of a linear transformation

$$(1) \quad \begin{aligned} \rho x_1' &= a_{11}x_1 + a_{12}x_2 + a_{13}x_3, \\ \rho x_2' &= a_{21}x_1 + a_{22}x_2 + a_{23}x_3, \\ \rho x_3' &= a_{31}x_1 + a_{32}x_2 + a_{33}x_3. \end{aligned}$$

Such a transformation transforms any point $(x_1, x_2, x_3)$ of the plane into a unique point $(x_1', x_2', x_3')$ of the plane. Reciprocally, to every point $x'$ will correspond a unique point $x$, provided the *determinant of the transformation*

$$A = \begin{vmatrix} a_{11} & a_{12} & a_{13} \\ a_{21} & a_{22} & a_{23} \\ a_{31} & a_{32} & a_{33} \end{vmatrix}$$

is not 0. For we may then solve equations (1) for the ratios $x_1 : x_2 : x_3$ in terms of $x_1' : x_2' : x_3'$ as follows:

$$(2) \quad \begin{aligned} \rho' x_1 &= A_{11}x_1' + A_{21}x_2' + A_{31}x_3', \\ \rho' x_2 &= A_{12}x_1' + A_{22}x_2' + A_{32}x_3', \\ \rho' x_3 &= A_{13}x_1' + A_{23}x_2' + A_{33}x_3'; \end{aligned}$$

here the coefficients $A_{ij}$ are the cofactors of the elements $a_{ij}$ respectively in the determinant $A$.

Further, equations (1) transform every line in the plane into a unique line. In fact, the points $x$ satisfying the equation

$$u_1x_1 + u_2x_2 + u_3x_3 = 0$$

are, by reference to equations (2), transformed into points $x'$ satisfying the equation

$$(A_{11}u_1 + A_{12}u_2 + A_{13}u_3)\,x_1' + (A_{21}u_1 + A_{22}u_2 + A_{23}u_3)\,x_2' + (A_{31}u_1 + A_{32}u_2 + A_{33}u_3)\,x_3' = 0,$$

which is the equation of a line. If the coördinates of this new line be denoted by $[u_1', u_2', u_3']$, we clearly have the following relations between the coördinates $[u_1, u_2, u_3]$ of any line and the coördinates $[u_1', u_2', u_3']$ of the line into which it is transformed by (1):

$$(3)\qquad \begin{aligned} \sigma u_1' &= A_{11}u_1 + A_{12}u_2 + A_{13}u_3,\\ \sigma u_2' &= A_{21}u_1 + A_{22}u_2 + A_{23}u_3,\\ \sigma u_3' &= A_{31}u_1 + A_{32}u_2 + A_{33}u_3. \end{aligned}$$

We have seen thus far that (1) represents a collineation in the plane in point coördinates. The equations (3) represent the same collineation in line coördinates.

It is readily seen, finally, that this collineation is projective. For this purpose it is only necessary to show that it transforms any pencil of lines into a projective pencil of lines. But it is clear that if $m = 0$ and $n = 0$ are the equations of any two lines, and if (1) transforms them respectively into the lines whose equations are $m' = 0$ and $n' = 0$, any line $m + \lambda n = 0$ is transformed into $m' + \lambda n' = 0$, and the correspondence thus established between the lines of the pencils has been shown to be projective (Theorem 6).

Having shown that every transformation (1) represents a projective collineation, we will now show conversely that every projective collineation in a plane may be represented by equations of the form (1). To this end we recall that every such collineation is completely determined as soon as the homologous elements of any complete quadrangle are assigned (Theorem 18, Chap. IV). If we can show that likewise there is one and only one transformation of the form (1) changing a given quadrangle into a given quadrangle, it will follow that, since the linear transformation is a projective collineation, it is the given projective collineation.

Given any projective collineation in a plane, let the fundamental points (0, 0, 1), (0, 1, 0), (1, 0, 0), and (1, 1, 1) of the plane (which form a quadrangle) be transformed respectively into the points $A=(a_1, a_2, a_3)$, $B=(b_1, b_2, b_3)$, $C=(c_1, c_2, c_3)$, and $D=(d_1, d_2, d_3)$, forming a quadrangle. Suppose, now, we seek to determine the coefficients of a transformation (1) so as to effect the correspondences just indicated. Clearly, if (0, 0, 1) is to be transformed into $(a_1, a_2, a_3)$, we must have

$$a_{13}=\lambda a_1,\quad a_{23}=\lambda a_2,\quad a_{33}=\lambda a_3,$$

$\lambda$ being an arbitrary factor of proportionality, the value ($\neq 0$) of which we may choose at pleasure. Similarly, we obtain

$$a_{12}=\mu b_1,\quad a_{22}=\mu b_2,\quad a_{32}=\mu b_3,$$
$$a_{11}=\nu c_1,\quad a_{21}=\nu c_2,\quad a_{31}=\nu c_3.$$

Since, by hypothesis, the three points $A$, $B$, $C$ are not collinear, it follows from these equations and the condition of Theorem 4, Cor. 2, that the determinant $A$ of a transformation determined in this way is not 0. Substituting the values thus obtained in (1), it is seen that if the point (1, 1, 1) is to be transformed into $(d_1, d_2, d_3)$, the following relations must hold:

$$\rho d_1 = c_1\nu + b_1\mu + a_1\lambda,$$
$$\rho d_2 = c_2\nu + b_2\mu + a_2\lambda,$$
$$\rho d_3 = c_3\nu + b_3\mu + a_3\lambda.$$

Placing $\rho=1$ and solving this system of equations for $\nu$, $\mu$, $\lambda$, we obtain the coefficients $a_{ij}$ of the transformation. This solution is unique, since the determinant of the system is not zero. Moreover, none of the values $\lambda$, $\mu$, $\nu$ will be 0; for the supposition that $\nu=0$, for example, would imply the vanishing of the determinant

$$\begin{vmatrix} d_1 & b_1 & a_1 \\ d_2 & b_2 & a_2 \\ d_3 & b_3 & a_3 \end{vmatrix},$$

which in turn would imply that the three points $D$, $B$, $A$ are collinear, contrary to the hypothesis that the four points $A$, $B$, $C$, $D$ form a complete quadrangle.

Collecting the results of this section, we have

THEOREM 8. *Any projective collineation in the plane may be represented in point coördinates by equations of form* (1) *or in line coördinates by equations of form* (3), *and in each case the determinant of the transformation is different from* 0; *conversely, any transformation of one of these forms in which the determinant is different from* 0 *represents a projective collineation in the plane.*

COROLLARY 1. *In nonhomogeneous point coördinates the equations of a projective collineation are*

$$x' = \frac{a_{11}x + a_{12}y + a_{13}}{a_{31}x + a_{32}y + a_{33}}, \qquad \begin{vmatrix} a_{11} & a_{12} & a_{13} \\ a_{21} & a_{22} & a_{23} \\ a_{31} & a_{32} & a_{33} \end{vmatrix} \neq 0.$$
$$y' = \frac{a_{21}x + a_{22}y + a_{23}}{a_{31}x + a_{32}y + a_{33}},$$

COROLLARY 2. *If the singular line of the system of nonhomogeneous point coördinates is transformed into itself, these equations can be written*

$$x' = a_1x + b_1y + c_1, \qquad \begin{vmatrix} a_1 & b_1 \\ a_2 & b_2 \end{vmatrix} \neq 0.$$
$$y' = a_2x + b_2y + c_2,$$

**68. Collineations between two different planes.** The analytic form of a collineation between two different planes is now readily derived. Let the two planes be $\alpha$ and $\beta$, and let a system of coördinates be established in each, the point coördinates in $\alpha$ being $(x_1, x_2, x_3)$ and the point coördinates in $\beta$ being $(y_1, y_2, y_3)$. Further, let the isomorphism between the number systems in the two planes be established in such a way that the correspondence established by the equations

$$y_1 = x_1, \quad y_2 = x_2, \quad y_3 = x_3,$$

is projective. It then follows, by an argument (cf. § 59, p. 166), which need not be repeated here, that any collineation between the two planes may be obtained as the resultant of a projectivity in the plane $\alpha$, which transforms a point $X$, say, into a point $X'$, and the projectivity $Y = X'$ between the two planes. *The analytic form of any projective collineation between the two planes is therefore:*

$$y_1 = a_{11}x_1 + a_{12}x_2 + a_{13}x_3,$$
$$y_2 = a_{21}x_1 + a_{22}x_2 + a_{23}x_3,$$
$$y_3 = a_{31}x_1 + a_{32}x_2 + a_{33}x_3,$$

*with the determinant* $\Delta$ *of the coefficients different from* 0. *And, conversely, every such transformation in which* $\Delta \neq 0$ *represents a projective collineation between the two planes.*

**69. Nonhomogeneous coördinates in space.** Point coördinates in space are introduced in a way entirely analogous to that used for the introduction of point coördinates in the plane. Choose a *tetrahedron of reference* $OUVW$ and label the vertices $O = 0_x = 0_y = 0_z$, $U = \infty_x$,

$V = \infty_y$, $W = \infty_z$ (fig. 84); and on the lines $0_x\infty_x$, $0_y\infty_y$, $0_z\infty_z$, called respectively the *x-axis*, the *y-axis*, the *z-axis*, establish three scales by choosing the points $1_x$, $1_y$, $1_z$. The planes $O\infty_x\infty_y$, $O\infty_x\infty_z$, $O\infty_y\infty_z$ are called the *xy-plane, xz-plane, yz-plane* respectively. The point $O$ is called the *origin*. If $P$ is any point not on the plane $\infty_x\infty_y\infty_z$, which is called the *singular plane* of the coördinate system, the plane $P\infty_y\infty_z$ meets the $x$-axis in a point whose nonhomogeneous coördinate in the scale $(0_x, 1_x, \infty_x)$ we call $a$. Similarly, let the plane $P\infty_x\infty_z$ meet the $y$-axis in a point whose nonhomogeneous coördinate in the scale $(0_y, 1_y, \infty_y)$ is $b$; and let the plane $P\infty_x\infty_y$ meet the $z$-axis in a point whose nonhomogeneous coördinate in the scale $(0_z, 1_z, \infty_z)$ is $c$. The numbers $a, b, c$ are then the nonhomogeneous *x-, y-, and z-coördinates* of the point $P$. Conversely, any three numbers $a, b, c$ determine three points $A$, $B$, $C$ on the $x$-, $y$-, and $z$-axes respectively, and the three planes $A\infty_y\infty_z$, $B\infty_x\infty_z$, $C\infty_x\infty_y$ meet in a point $P$ whose coördinates are $a$, $b$, $c$. Thus every point not on the singular plane of the coördinate system determines and is determined by three coördinates. The point $P$ is then represented by the symbol $(a, b, c)$.

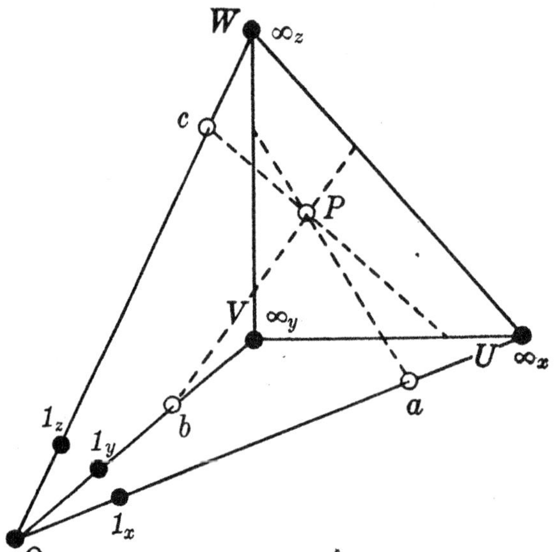


FIG. 84

The dual process gives rise to the coördinates of a plane. Point and plane coördinates may then be put into a convenient relation, as was done in the case of point and line coördinates in the plane, thus giving rise to a system of *simultaneous point and plane coördinates* in space. We will describe the system of plane coördinates with reference to this relation. Given the system of nonhomogeneous point coördinates described above, establish in each of the pencils of planes on the lines $VW$, $UW$, $UV$ a scale by choosing the plane $UVW$ as the zero plane $0_u = 0_v = 0_w$ in each of the scales, and letting the planes $OVW$, $OUW$, $OUV$ be the planes $\infty_u$, $\infty_v$, $\infty_w$ respectively. In the $u$-scale let that plane through $VW$ be the plane $1_u$, which meets the $x$-axis in the point $-1_x$. Similarly, let the plane $1_v$ meet the $y$-axis in the point $-1_y$; and let the plane $1_w$ meet the $z$-axis in the point $-1_z$. The $u$-scale, $v$-scale, and $w$-scale being now completely determined, any plane $\pi$ not on the point $O$ (which is called the *singular point* of this system of plane coördinates) meets the $x$-, $y$-, and $z$-axes in three points $L, M, N$ which determine in the $u$-, $v$-, and $w$-scales planes whose coördinates, let us say, are $l, m, n$. These three numbers are called the nonhomogeneous plane coördinates of $\pi$. They completely determine and are completely determined by the plane $\pi$. The plane $\pi$ is then denoted by the symbol $[l, m, n]$.

In this system of coördinates it is now readily seen that *the condition that the point $(a, b, c)$ be on the plane $[l, m, n]$ is that the relation $la + mb + nc + 1 = 0$ be satisfied.* It follows readily, as in the planar case, that the plane $[l, m, n]$ meets the $x$-, $y$-, and $z$-axes in points whose coördinates on these axes are $-1/l, -1/m$, and $-1/n$ respectively.* In deriving the above condition we will suppose that the plane $\pi = [l, m, n]$ does not contain two of the points $U, V, W$, leaving the other case as an exercise for the reader. Suppose, then, that $U = \infty_x$ and $V = \infty_y$ are not on $\pi$. By projecting the $yz$-plane with $U$ as center upon the plane $\pi$, and then projecting $\pi$ with $V$ as center on the $xz$-plane, we obtain the following perspectivities:

$$[(0, y, z)] \overset{U}{\underset{\wedge}{=}} [(x, y, z)] \overset{V}{\underset{\wedge}{=}} [(x, 0, z)],$$

where $(x, y, z)$ represents any point on $\pi$. The product of these two perspectivities is a projectivity between the $yz$-plane and the $xz$-plane, by which the singular line of the former is transformed into the singular line of the latter. Denoting the $z$-coördinate of points in the $yz$-plane by $z'$, this projectivity is represented (according to Theorem 8, Cor. 2, and § 68) by relations of the form

$$(1) \qquad \begin{aligned} y &= a_1x + b_1z + c_1, \\ z' &= z. \end{aligned}$$

We proceed to determine the coefficients $a_1, b_1, c_1$. The point of intersection of $\pi$ with the $y$-axis is $(0, -1/m, 0)$, and is clearly

* This statement remains valid even if one or two of the numbers $l, m, n$ are zero (they cannot all be zero unless the plane in question is the singular plane which we exclude from consideration), provided the negative reciprocal of 0 be denoted by the symbol $\infty$.

transformed by the projectivity in question into the point (0, 0, 0). Hence (1) gives

$$c_1 = -\frac{1}{m}.$$

The point of intersection of $\pi$ with the $z$-axis is, if $n \neq 0$, $(0, 0, -1/n)$ and is transformed into itself. Hence (1) gives

$$-\frac{b_1}{n} - \frac{1}{m} = 0,$$

or

$$b_1 = -\frac{n}{m}.$$

If $n = 0$, we have at once $b_1 = 0$.

Finally, the point of intersection of $\pi$ with the $x$-axis is $(-1/l, 0, 0)$, and the transform of the point (0, 0, 0). Hence we have

$$-\frac{a_1}{l} - \frac{1}{m} = 0,$$

or

$$a_1 = -\frac{l}{m}.$$

Hence (1) becomes

$$y = -\frac{l}{m}x - \frac{n}{m}z - \frac{1}{m},$$

a relation which must be satisfied by the coördinates $(x, y, z)$ of any point on $\pi$. This relation is equivalent to

$$lx + my + nz + 1 = 0.$$

Hence $(a, b, c)$ is on $[l, m, n]$, if

$$(2) \qquad la + mb + nc + 1 = 0.$$

Conversely, if (2) is satisfied by a point $(a, b, c)$, the point $(0, b, c) = P$ is transformed by the projectivity above into $(a, 0, c) = Q$, and hence the lines $PU$ and $QV$ which meet in $(a, b, c)$ meet on $\pi$.

DEFINITION. An equation which is satisfied by all the points $(x, y, z)$ of a plane and by no other points is called the *point equation* of the plane.

DEFINITION. An equation which is satisfied by all the planes $[u, v, w]$ on a point and by no other planes is called the *plane equation* of the point.

The result of the preceding discussion may then be stated as follows:

THEOREM 9. *The point equation of the plane* $[l, m, n]$ *is*

$$lx + my + nz + 1 = 0.$$

THEOREM 9′. *The plane equation of the point* $(a, b, c)$ *is*

$$au + bv + cw + 1 = 0.$$

**70. Homogeneous coördinates in space.** Assign to the vertices $O, U, V, W$ of any tetrahedron of reference the symbols $(0, 0, 0, 1)$, $(1, 0, 0, 0)$, $(0, 1, 0, 0)$, $(0, 0, 1, 0)$ respectively, and assign to any fifth point $T$ not on a face of this tetrahedron the symbol $(1, 1, 1, 1)$. The five points $O, U, V, W, T$ are called the *frame of reference* of the system of homogeneous coördinates now to be described. The four lines joining $T$ to the points $O, U, V, W$ meet the opposite faces in four points, which we denote respectively by $(1, 1, 1, 0)$, $(0, 1, 1, 1)$, $(1, 0, 1, 1)$, $(1, 1, 0, 1)$. The planar four-point $(0, 0, 0, 1)$, $(0, 0, 1, 0)$, $(0, 1, 0, 0)$, $(0, 1, 1, 1)$ we regard as the frame of reference $(0, 0, 1)$, $(0, 1, 0)$, $(1, 0, 0)$, $(1, 1, 1)$ of a system of homogeneous coördinates in the plane. To any point in this plane we assign the coördinates $(0, x_2, x_3, x_4)$, if its coördinates in the planar system just indicated are $(x_2, x_3, x_4)$. In like manner, to the points of the other three faces of the tetrahedron of reference we assign coördinates of the forms $(x_1, 0, x_3, x_4)$, $(x_1, x_2, 0, x_4)$, and $(x_1, x_2, x_3, 0)$. The coördinates of the points in the faces opposite the vertices $(1, 0, 0, 0)$, $(0, 1, 0, 0)$, $(0, 0, 1, 0)$, $(0, 0, 0, 1)$ satisfy respectively the equations $x_1 = 0$, $x_2 = 0$, $x_3 = 0$, $x_4 = 0$.

To the points of each edge of the tetrahedron of reference a notation has been assigned corresponding to each of the two faces which meet in the edge. Consider, for example, the line of intersection of the planes $x_1 = 0$ and $x_2 = 0$. Regarding this edge as a line of $x_1 = 0$, the coördinate system on the edge has as its fundamental points $(0, 0, 1, 0)$, $(0, 0, 0, 1)$, $(0, 0, 1, 1)$. The first two of these are vertices of the tetrahedron of reference, and the third is the trace of the line joining $(0, 1, 0, 0)$ to $(0, 1, 1, 1)$. On the other hand, regarding this edge as a line of $x_2 = 0$, the coördinate system has the vertices $(0, 0, 1, 0)$ and $(0, 0, 0, 1)$ as two fundamental points, and has as $(0, 0, 1, 1)$ the trace of the line joining $(1, 0, 0, 0)$ to $(1, 0, 1, 1)$. But by construction the plane $(0, 1, 0, 0)(1, 0, 0, 0)(1, 1, 1, 1)$ contains both $(0, 1, 1, 1)$ and $(1, 0, 1, 1)$, so that the two determinations of $(0, 0, 1, 1)$ are identical. Hence the symbols denoting points in the two planes $x_1 = 0$ and $x_2 = 0$ are identical along their line of intersection. A similar result holds for the other edges of the tetrahedron of reference.

THEOREM 10. DEFINITION. *If $P$ is any point not on a face of the tetrahedron of reference, there exist four numbers $x_1$, $x_2$, $x_3$, $x_4$, all different from zero, such that the projections of $P$ from the four vertices* $(1, 0, 0, 0)$, $(0, 1, 0, 0)$, $(0, 0, 1, 0)$, $(0, 0, 0, 1)$ *respectively upon their* *opposite faces are* $(0, x_2, x_3, x_4)$, $(x_1, 0, x_3, x_4)$, $(x_1, x_2, 0, x_4)$, $(x_1, x_2, x_3, 0)$. *These four numbers are called the homogeneous coördinates of* $P$ *and* $P$ *is denoted by* $(x_1, x_2, x_3, x_4)$. *Any ordered set of four numbers, not all zero, determine uniquely a point in space whose coördinates they are.*

*Proof.* The line joining $P$ to $(1, 0, 0, 0)$ meets the opposite face in a point $(0, x_2, x_3, x_4)$, which is not an edge of the tetrahedron of reference, and such therefore that none of the numbers $x_2, x_3, x_4$ is zero. Likewise the line joining $P$ to $(0, 1, 0, 0)$ meets the opposite face in a point $(x_1', 0, x_3', x_4')$, such that none of the numbers $x_1', x_3', x_4'$ is zero. But the plane $P(1, 0, 0, 0)(0, 1, 0, 0)$ meets $x_1 = 0$ in the line joining $(0, 1, 0, 0)$ to $(0, x_2, x_3, x_4)$, and meets $x_2 = 0$ in the line joining $(1, 0, 0, 0)$ to $(x_1', 0, x_3', x_4')$. By the analytic methods already developed for the plane, the first of these lines meets the edge common to $x_1 = 0$ and $x_2 = 0$ in the point $(0, 0, x_3, x_4)$, and the second meets it in the point $(0, 0, x_3', x_4')$. But the points $(0, 0, x_3, x_4)$ and $(0, 0, x_3', x_4')$ are identical, and hence, by the preceding paragraph, we have $x_3/x_4 = x_3'/x_4'$. Hence, if we place $x_1 = x_1'x_4/x_4'$, the point $(x_1', 0, x_3', x_4')$ is identical with $(x_1, 0, x_3, x_4)$. The line joining $P$ to $(0, 0, 1, 0)$ meets the face $x_3 = 0$ in a point $(x_1'', x_2'', 0, x_4'')$. By the same reasoning as that above it follows that we have $x_1''/x_4'' = x_1/x_4$ and $x_2''/x_4'' = x_2/x_4$, so that the point $(x_1'', x_2'', 0, x_4'')$ is identical with $(x_1, x_2, 0, x_4)$. Finally, the line joining $P$ to $(0, 0, 0, 1)$ meets the face $x_4 = 0$ in a point which a like argument shows to be $(x_1, x_2, x_3, 0)$.

Conversely, if the coördinates $(x_1, x_2, x_3, x_4)$ are given, and one of them is zero, they determine a point on a face of the tetrahedron of reference. If none of them is zero, the lines joining $(1, 0, 0, 0)$ to $(0, x_2, x_3, x_4)$ and $(0, 1, 0, 0)$ to $(x_1, 0, x_3, x_4)$ are in the plane $(1, 0, 0, 0)(0, 1, 0, 0)(0, 0, x_3, x_4)$, and hence meet in a point which, by the reasoning above, has the coördinates $(x_1, x_2, x_3, x_4)$.

COROLLARY. *The notations* $(x_1, x_2, x_3, x_4)$ *and* $(kx_1, kx_2, kx_3, kx_4)$ *denote the same point for any value of* $k$ *not equal to zero.*

Homogeneous plane coördinates in space arise by the dual of the above process. The four faces of a *tetrahedron of reference* are denoted respectively by [1, 0, 0, 0], [0, 1, 0, 0], [0, 0, 1, 0], and [0, 0, 0, 1]. These, together with any plane [1, 1, 1, 1] not on a vertex of the tetrahedron, form the *frame of reference.* The four lines of intersection of the plane [1, 1, 1, 1] with the other four planes in the order above are projected from the opposite vertices by planes which are denoted by [0, 1, 1, 1], [1, 0, 1, 1], [1, 1, 0, 1], [1, 1, 1, 0] respectively. The four planes [0, 1, 0, 0], [0, 0, 1, 0], [0, 0, 0, 1], and [0, 1, 1, 1] form, if the first 0 in each of these symbols is suppressed, the frame of reference of a system of homogeneous coördinates in a bundle (the space dual of such a system in a plane). The center of this bundle is the vertex of the tetrahedron of reference opposite to [1, 0, 0, 0]. To any plane on this point is assigned the notation $[0, u_2, u_3, u_4]$, if its coördinates in the bundle are $[u_2, u_3, u_4]$. In like manner, to the planes on the other vertices are assigned coördinates of the forms $[u_1, 0, u_3, u_4]$, $[u_1, u_2, 0, u_4]$, $[u_1, u_2, u_3, 0]$. The space dual of the last theorem then gives:

THEOREM 10′. DEFINITION. *If $\pi$ is any plane not on a vertex of the tetrahedron of reference, there exist four numbers $u_1, u_2, u_3, u_4$, all different from zero, such that the traces of $\pi$ on the four faces* [1, 0, 0, 0], [0, 1, 0, 0], [0, 0, 1, 0], [0, 0, 0, 1] *respectively are projected from the opposite vertices by the planes* $[0, u_2, u_3, u_4]$, $[u_1, 0, u_3, u_4]$, $[u_1, u_2, 0, u_4]$, $[u_1, u_2, u_3, 0]$. *These four numbers are called the homogeneous coördinates of $\pi$, and $\pi$ is denoted by $[u_1, u_2, u_3, u_4]$. Any ordered set of four numbers, not all zero, determine uniquely a plane whose coördinates they are.*

By placing these systems of point and plane coördinates in a proper relation we may now readily derive the necessary and sufficient condition that a point $(x_1, x_2, x_3, x_4)$ be on a plane $[u_1, u_2, u_3, u_4]$. This condition will turn out to be

$$u_1x_1 + u_2x_2 + u_3x_3 + u_4x_4 = 0.$$

We note first that in a system of point coördinates as described above the six points $(-1, 1, 0, 0)$, $(-1, 0, 1, 0)$, $(-1, 0, 0, 1)$, $(0, -1, 1, 0)$, $(0, 0, -1, 1)$, $(0, -1, 0, 1)$ are coplanar, each being the harmonic conjugate, with respect to two vertices of the tetrahedron of reference, of the point into which (1, 1, 1, 1) is projected by the line joining the other two vertices. The plane containing these is, in fact, the polar of (1, 1, 1, 1) with respect to the tetrahedron of reference (cf. Ex. 3, p. 47). Now choose

as the plane [1, 0, 0, 0] the plane $x_1 = 0$,
as the plane [0, 1, 0, 0] the plane $x_2 = 0$,
as the plane [0, 0, 1, 0] the plane $x_3 = 0$,
as the plane [0, 0, 0, 1] the plane $x_4 = 0$,

as the plane [1, 1, 1, 1] the plane containing the points (− 1, 1, 0, 0), (− 1, 0, 1, 0), (− 1, 0, 0, 1).

With this choice of coördinates the planes [1, 0, 0, 0], [0, 1, 0, 0], [0, 0, 1, 0], and [1, 1, 1, 0] through the vertex $V_4$, say, whose point coördinates are (0, 0, 0, 1), meet the opposite face $x_4 = 0$ in lines whose equations in that plane are

$$x_1 = 0, \quad x_2 = 0, \quad x_3 = 0, \quad x_1 + x_2 + x_3 = 0.$$

Hence the first three coördinates of any plane $[u_1, u_2, u_3, 0]$ on $V_4$ are the line coördinates of its trace on $x_4 = 0$, in a system so chosen that the point $(x_1, x_2, x_3)$ is on the line $[u_1, u_2, u_3]$ if and only if the relation $u_1x_1 + u_2x_2 + u_3x_3 = 0$ is satisfied. Hence a point $(x_1, x_2, x_3, 0)$ lies on a plane $[u_1, u_2, u_3, 0]$ if and only if we have $u_1x_1 + u_2x_2 + u_3x_3 = 0$. But any point $(x_1, x_2, x_3, x_4)$ on the plane $[u_1, u_2, u_3, 0]$ has, by definition, its first three coördinates identical with the first three coördinates of some point on the trace of this plane with the plane $x_4 = 0$. Hence any point $(x_1, x_2, x_3, x_4)$ on $[u_1, u_2, u_3, 0]$ satisfies the condition $u_1x_1 + u_2x_2 + u_3x_3 + u_4x_4 = 0$. Applying this reasoning to each of the four vertices of the tetrahedron of reference and dualizing, we find that *if one coördinate of $[u_1, u_2, u_3, u_4]$ is zero, the necessary and sufficient condition that this plane contain a point $(x_1, x_2, x_3, x_4)$ is that the relation*

$$u_1x_1 + u_2x_2 + u_3x_3 + u_4x_4 = 0$$

*be satisfied; and if one coördinate of $(x_1, x_2, x_3, x_4)$ is zero, the necessary and sufficient condition that this point be on the plane $[u_1, u_2, u_3, u_4]$ is likewise that the relation just given be satisfied.*

Confining our attention now to points and planes no coördinate of which is zero, let $x_1/x_4 = x$, $x_2/x_4 = y$, $x_3/x_4 = z$, and let $u_1/u_4 = u$, $u_2/u_4 = v$, $u_3/u_4 = w$. Since $x$, $y$, $z$ are the ratios of homogeneous coördinates on the lines $x_2 = x_3 = 0$, $x_1 = x_3 = 0$, and $x_1 = x_2 = 0$ respectively, they satisfy the definition of nonhomogeneous coördinates given in § 69. And since the homogeneous coördinates have been so chosen that the plane $(u_1, u_2, u_3, u_4)$ meets the line $x_2 = x_3 = 0$ in the point $(-u_4, 0, 0, u_1) = (-1/u, 0, 0, 1)$, it follows that $u$, $v$, $w$ are nonhomogeneous plane coördinates so chosen that a point $(x, y, z)$, none of whose coördinates is zero, is on a plane $[u, v, w]$ none of whose coördinates is zero, if and only if we have (Theorem 9)

$$ux + vy + wz + 1 = 0;$$

that is, if and only if we have

$$u_1x_1 + u_2x_2 + u_3x_3 + u_4x_4 = 0.$$

This completes for all cases the proof of

THEOREM 11. *The necessary and sufficient condition that a point* $(x_1, x_2, x_3, x_4)$ *be on a plane* $[u_1, u_2, u_3, u_4]$ *is that the relation*

$$u_1x_1 + u_2x_2 + u_3x_3 + u_4x_4 = 0$$

*be satisfied.*

By methods analogous to those employed in §§ 64 and 65 we may now derive the results of Exs. 1–8 below.

### EXERCISES

**1.** The equation of the plane through the three points $A = (a_1, a_2, a_3, a_4)$, $B = (b_1, b_2, b_3, b_4)$, $C = (c_1, c_2, c_3, c_4)$ is

$$\begin{vmatrix} x_1 & x_2 & x_3 & x_4 \\ a_1 & a_2 & a_3 & a_4 \\ b_1 & b_2 & b_3 & b_4 \\ c_1 & c_2 & c_3 & c_4 \end{vmatrix} = 0.$$

Dualize.

**2.** The necessary and sufficient condition that four points $A$, $B$, $C$, $D$ be coplanar is the vanishing of the determinant

$$\begin{vmatrix} a_1 & a_2 & a_3 & a_4 \\ b_1 & b_2 & b_3 & b_4 \\ c_1 & c_2 & c_3 & c_4 \\ d_1 & d_2 & d_3 & d_4 \end{vmatrix}.$$

**3.** The necessary and sufficient condition that three points $A$, $B$, $C$ be collinear is the vanishing of the three-rowed determinants of the matrix

$$\begin{pmatrix} a_1 & a_2 & a_3 & a_4 \\ b_1 & b_2 & b_3 & b_4 \\ c_1 & c_2 & c_3 & c_4 \end{pmatrix}.$$

**4.** Any point of a pencil of points containing $A$ and $B$ may be represented by

$$P = (\lambda_2 a_1 + \lambda_1 b_1, \lambda_2 a_2 + \lambda_1 b_2, \lambda_2 a_3 + \lambda_1 b_3, \lambda_2 a_4 + \lambda_1 b_4).$$

**5.** Any plane of a pencil of planes containing $m = [m_1, m_2, m_3, m_4]$ and $n = [n_1, n_2, n_3, n_4]$ may be represented by

$$\pi = [\lambda_2 m_1 + \lambda_1 n_1, \lambda_2 m_2 + \lambda_1 n_2, \lambda_2 m_3 + \lambda_1 n_3, \lambda_2 m_4 + \lambda_1 n_4].$$

**6.** Any projectivity between two one-dimensional primitive forms (of points or planes) in space is expressed by a relation between their parameters $\lambda$, $\mu$ of the form

$$\mu = \frac{\alpha\lambda + \beta}{\gamma\lambda + \delta}.$$

If the base elements of the pencil are homologous, this relation reduces to $\mu = \rho\lambda$.

7. If $\lambda_1, \lambda_2, \lambda_3, \lambda_4$ are the parameters of four points or planes of a pencil, their cross ratio is

$$R\,(\lambda_1\lambda_2,\ \lambda_3\lambda_4) = \frac{\lambda_1 - \lambda_3}{\lambda_1 - \lambda_4} : \frac{\lambda_2 - \lambda_3}{\lambda_2 - \lambda_4}.$$

8. Any point (plane) of a plane of points (bundle of planes) containing the noncollinear points $A$, $B$, $C$ (planes $\alpha$, $\beta$, $\gamma$) may be represented by

$$P = (\lambda_1 a_1 + \lambda_2 b_1 + \lambda_3 c_1,\ \lambda_1 a_2 + \lambda_2 b_2 + \lambda_3 c_2,\ \lambda_1 a_3 + \lambda_2 b_3 + \lambda_3 c_3,\ \lambda_1 a_4 + \lambda_2 b_4 + \lambda_3 c_4).$$

9. Derive the equation of the polar plane of any point with regard to the tetrahedron of reference.

10. Derive the equation of a cone.

*11. Derive nonhomogeneous and homogeneous systems of coördinates in a space of four dimensions.

**71. Linear transformations in space.** The properties of a linear transformation in space

$$(1)\qquad \begin{aligned} \rho x_1' &= a_{11}x_1 + a_{12}x_2 + a_{13}x_3 + a_{14}x_4,\\ \rho x_2' &= a_{21}x_1 + a_{22}x_2 + a_{23}x_3 + a_{24}x_4,\\ \rho x_3' &= a_{31}x_1 + a_{32}x_2 + a_{33}x_3 + a_{34}x_4,\\ \rho x_4' &= a_{41}x_1 + a_{42}x_2 + a_{43}x_3 + a_{44}x_4 \end{aligned}$$

are similar to those found in § 68 for the linear transformations in a plane. If the determinant of the transformation

$$A = \begin{vmatrix} a_{11} & a_{12} & a_{13} & a_{14}\\ a_{21} & a_{22} & a_{23} & a_{24}\\ a_{31} & a_{32} & a_{33} & a_{34}\\ a_{41} & a_{42} & a_{43} & a_{44} \end{vmatrix}$$

is different from zero, the transformation (1) will have a unique inverse, viz.:

$$(2)\qquad \begin{aligned} \rho' x_1 &= A_{11}x_1' + A_{21}x_2' + A_{31}x_3' + A_{41}x_4',\\ \rho' x_2 &= A_{12}x_1' + A_{22}x_2' + A_{32}x_3' + A_{42}x_4',\\ \rho' x_3 &= A_{13}x_1' + A_{23}x_2' + A_{33}x_3' + A_{43}x_4',\\ \rho' x_4 &= A_{14}x_1' + A_{24}x_2' + A_{34}x_3' + A_{44}x_4', \end{aligned}$$

where the coefficients $A_{ij}$ are the cofactors of the elements $a_{ij}$ respectively in the determinant $A$.

The transformation is evidently a collineation, as it transforms the plane

$$u_1x_1 + u_2x_2 + u_3x_3 + u_4x_4 = 0$$

into the plane

$$\begin{aligned} &(A_{11}u_1 + A_{12}u_2 + A_{13}u_3 + A_{14}u_4)\,x_1'\\ +&(A_{21}u_1 + A_{22}u_2 + A_{23}u_3 + A_{24}u_4)\,x_2'\\ +&(A_{31}u_1 + A_{32}u_2 + A_{33}u_3 + A_{34}u_4)\,x_3'\\ +&(A_{41}u_1 + A_{42}u_2 + A_{43}u_3 + A_{44}u_4)\,x_4' = 0. \end{aligned}$$

Hence the collineation (1) produces on the planes of space the transformation

$$(3)\qquad \begin{aligned} \sigma u_1' &= A_{11}u_1 + A_{12}u_2 + A_{13}u_3 + A_{14}u_4, \\ \sigma u_2' &= A_{21}u_1 + A_{22}u_2 + A_{23}u_3 + A_{24}u_4, \\ \sigma u_3' &= A_{31}u_1 + A_{32}u_2 + A_{33}u_3 + A_{34}u_4, \\ \sigma u_4' &= A_{41}u_1 + A_{42}u_2 + A_{43}u_3 + A_{44}u_4. \end{aligned}$$

To show that the transformation (1) is projective consider any pencil of planes

$$(a_1x_1 + a_2x_2 + a_3x_3 + a_4x_4) + \lambda(b_1x_1 + b_2x_2 + b_3x_3 + b_4x_4) = 0.$$

In accordance with (2) this pencil is transformed into a pencil of the form

$$(a_1'x_1 + a_2'x_2 + a_3'x_3 + a_4'x_4) + \lambda(b_1'x_1 + b_2'x_2 + b_3'x_3 + b_4'x_4) = 0,$$

and these two pencils of planes are projective (Ex. 6, p. 198).

Finally, as in § 67, we see that there is one and only one transformation (1) changing the points (0, 0, 0, 1), (0, 0, 1, 0), (0, 1, 0, 0), (1, 0, 0, 0), and (1, 1, 1, 1) into the vertices of an arbitrary complete five-point in space. Since this transformation is a projective collineation, and since there is only one projective collineation transforming one five-point into another (Theorem 19, Chap. IV), it follows that every projective collineation in space may be represented by a linear transformation of the form (1). This gives

THEOREM 12. *Any projective collineation of space may be represented in point coördinates by equations of the form* (1), *or in plane coördinates by equations of the form* (3). *In each case the determinant of the transformation is different from zero. Conversely, any transformation of this form in which the determinant is different from zero represents a projective collineation of space.*

COROLLARY 1. *In nonhomogeneous point coördinates a projective collineation is represented by the linear fractional equations*

$$x' = \frac{a_{11}x + a_{12}y + a_{13}z + a_{14}}{a_{41}x + a_{42}y + a_{43}z + a_{44}},$$

$$y' = \frac{a_{21}x + a_{22}y + a_{23}z + a_{24}}{a_{41}x + a_{42}y + a_{43}z + a_{44}},$$

$$z' = \frac{a_{31}x + a_{32}y + a_{33}z + a_{34}}{a_{41}x + a_{42}y + a_{43}z + a_{44}},$$

*in which the determinant $A$ is different from zero.*

COROLLARY 2. *If the singular plane of the nonhomogeneous system is transformed into itself, these equations reduce to*

$$x' = a_1x + a_2y + a_3z + a_4,\qquad \begin{vmatrix} a_1 & a_2 & a_3 \\ b_1 & b_2 & b_3 \\ c_1 & c_2 & c_3 \end{vmatrix} \neq 0.$$
$$y' = b_1x + b_2y + b_3z + b_4,$$
$$z' = c_1x + c_2y + c_3z + c_4,$$

**72. Finite spaces.** It will be of interest at this point to emphasize again the generality of the theory which we are developing. Since all the developments of this chapter are on the basis of Assumptions A, E, and P only, and since these assumptions imply nothing regarding the number system of points on a line, except that it be commutative, it follows that we may assume the points of a line, or, indeed, the elements of any one-dimensional form, to be in one-to-one reciprocal correspondence with the elements of *any* commutative number system. We may, moreover, study our geometry entirely by analytic methods. From this point of view, any point in a plane is simply a set of three numbers $(x_1, x_2, x_3)$, it being understood that the sets $(x_1, x_2, x_3)$ and $(kx_1, kx_2, kx_3)$ are equivalent for all values of $k$ in the number system, provided $k$ is different from 0. Any line in the plane is the set of all these points which satisfy any equation of the form $u_1x_1 + u_2x_2 + u_3x_3 = 0$, the set of all lines being obtained by giving the coefficients (coördinates) $[u_1, u_2, u_3]$ all possible values in the number system (except $[0, 0, 0]$), with the obvious agreement that $[u_1, u_2, u_3]$ and $[ku_1, ku_2, ku_3]$ represent the same line $(k \neq 0)$. By letting the number system consist of all ordinary rational numbers, or all ordinary real numbers, or all ordinary complex numbers, we obtain respectively the analytic form of ordinary rational, or real, or complex projective geometry in the plane. All of our theory thus far applies equally to each of these geometries as well as to the geometry obtained by choosing as our number system any field whatever (any ordinary algebraic field, for example).

In particular, we may also choose a *finite* field, i.e. one which contains only a finite number of elements. The simplest of these are the *modular fields*, the modulus being any prime number $p$.* If we

* A modular field with modulus $p$ is obtained as follows: Two integers $n$, $n'$ (positive, negative, or zero) are said to be *congruent modulo* $p$, written $n \equiv n'$, mod. $p$, if the difference $n - n'$ is divisible by $p$. Every integer is then congruent to one and only one of the numbers $0, 1, 2, \cdots, p-1$. *These numbers are taken as the elements of our field, and any number obtained from these by addition, subtraction,*

consider, for example, the case $p = 2$, our number system contains only the elements 0 and 1. There are then seven points, which we will label $A, B, C, D, E, F, G$, as follows: $A=(0, 0, 1)$, $B=(0, 1, 0)$, $C=(1, 0, 0)$, $D=(0, 1, 1)$, $E=(1, 1, 0)$, $F=(1, 1, 1)$, $G=(1, 0, 1)$. The reader will readily verify that these seven points are arranged in lines according to the table

$$\begin{array}{ccccccc} A & B & C & D & E & F & G \\ B & C & D & E & F & G & A \\ D & E & F & G & A & B & C, \end{array}$$

each column constituting a line. For example, the line $x_1 = 0$ clearly consists of the points $(0, 0, 1) = A$, $(0, 1, 0) = B$, and $(0, 1, 1) = D$, these being the only points whose first coördinate is 0. We have labeled the points of this *finite plane* in such a way as to exhibit clearly its abstract identity with the system of triples used for illustrative purposes in the Introduction, § 2.*

## EXERCISES

**1.** Verify analytically that two sides of a complete quadrangle containing a diagonal point are harmonic with the other two diagonal points.

**2.** Show analytically that if two projective pencils of lines in a plane have a self-corresponding line, they are perspective. (This is equivalent to Assumption P.)

**3.** Show that the lines whose equations are $x_1 + \lambda x_2 = 0$, $x_2 + \mu x_3 = 0$, and $x_3 + \nu x_1 = 0$ are concurrent if $\lambda\mu\nu = -1$; and that they meet the opposite sides of the triangle of reference respectively in collinear points, if $\lambda\mu\nu = 1$.

**4.** Find the equations of the lines joining $(c_1, c_2, c_3)$ to the four points $(1, \pm 1, \pm 1)$, and determine the cross ratios of the pencil.

*and multiplication, if not equal to one of these elements, is replaced by the element to which it is congruent.* The modular field with modulus 5, for example, consists of the elements 0, 1, 2, 3, 4, and we have as examples of addition, subtraction, and multiplication $1 + 3 = 4$, $2 + 3 = 0$ (since $5 \equiv 0$, mod. 5), $1 - 4 = 2$, $2 \cdot 3 = 1$, etc. Furthermore, if $a, b$ are any two elements of this field ($a \neq 0$), there is a unique element $x$ determined by the congruence $ax \equiv b$, mod. $p$; this element is defined as the *quotient* $b/a$. (For the proof of this proposition the reader may refer to any standard text on the theory of numbers.) In the example discussed we have, for example, $4/3 = 3$.

* For references and a further discussion of finite projective geometries see a paper by O. Veblen and W. H. Bussey, Finite Projective Geometries, Transactions of the American Mathematical Society, Vol. VII (1906), pp. 241–259. Also a subsequent paper by O. Veblen, Collineations in a Finite Projective Geometry, Transactions of the American Mathematical Society, Vol. VIII (1907), pp. 266–268.

**5.** Show that the throw of lines determined on $(c_1, c_2, c_3)$ by the four points $(1, \pm 1, \pm 1)$ is projective with (equal to) the throw of lines determined on $(b_1, b_2, b_3)$ by the points $(a_1, \pm a_2, \pm a_3)$, if the following relations hold:

$$a_1 + a_2 + a_3 = 0,$$
$$a_1c_1^2 + a_2c_2^2 + a_3c_3^2 = 0,$$
$$a_2a_3b_1^2 + a_1a_3b_2^2 + a_1a_2b_3^2 = 0,$$

and that the six cross ratios are $-a_2/a_3$, $-a_3/a_1$, $-a_1/a_2$, $-a_3/a_2$, $-a_1/a_3$, $-a_2/a_1$ (C. A. Scott, Mod. Anal. Geom., p. 50).

**6.** Write the equations of transformation for the five types of planar collineations described in § 40, Chap. IV, choosing points of the triangle of reference as fixed points.

**7.** Generalize Ex. 6 to space.

**8.** Show that the set of values of the parameter $\lambda$ of the pencil of lines $m + \lambda n = 0$ is isomorphic with the scale determined in this pencil by the lines for which the fundamental lines are respectively the lines $\lambda = 0, 1, \infty$.

**9.** Show directly from the discussion of § 61 that the points whose nonhomogeneous coördinates $x$, $y$ satisfy the equation $y = x$ are on the line joining the origin to the point (1, 1).

**10.** There is then established on this line a scale whose fundamental points are respectively the origin, the point (1, 1), and the point in which the line meets the line $l_\infty$. The lines joining any point $P$ in the plane to the points $\infty_y$, $\infty_x$ meet the line $y = x$ in two points whose coördinates in the scale just determined are the nonhomogeneous coördinates of $P$, so that any point in the plane (not on $l_\infty$) is represented by a pair of points on the line $y = x$. Hence, show that in general the points $(x, y)$ of any line in the plane determine on the line $y = x$ a projectivity with a double point on $l_\infty$; and hence that the equation of any such line is of the form $y = ax + b$. What lines are exceptions to this proposition?

**11.** Discuss the modular plane geometry in which the modulus is $p = 3$; and by properly labeling the points show that it is abstractly identical with the system of quadruples exhibited as System (2) on p. 6.

**12.** Show in general that the modular projective plane with modulus $p$ contains $p^2 + p + 1$ points and the same number of lines; and that there are $p + 1$ points (lines) on every line (point).

**13.** The diagonal points of a complete quadrangle in a modular plane projective geometry are collinear if and only if $p = 2$.

**14.** Show that the points and lines of a modular plane all belong to the same net of rationality. Such a plane is then properly projective without the use of Assumption P.

**15.** Show how to construct a modular three-space. If the modulus is 2, show that its points may be labeled 0, 1, . . . , 14 in such a way that the planes are the sets of seven obtained by cyclic permutation from the set 0 1 4 6 11 12 13 (i.e. 1 2 5 7 12 13 14, etc.), and that the lines are obtained from the lines 0 1 4, 0 2 8, 0 5 10 by cyclic permutations. (For a study of this space, see G. M. Conwell, Annals of Mathematics, Vol. 11 (1910), p. 60.)

**16.** Show that the ten diagonal points of a complete five-point in space (0, 0, 0, 1), (0, 0, 1, 0), (0, 1, 0, 0), (1, 0, 0, 0), (1, 1, 1, 1) are given by the remaining sets of coördinates in which occur only the digits 0 and 1.

**17.** Show that the ten diagonal points in Ex. 16 determine in all 45 planes, of which each of a set of 25 contains four diagonal points, while each of the remaining 20 contains only three diagonal points. Through any diagonal point pass 16 of these planes. The diagonal lines, i.e. lines joining two diagonal points, are of two kinds: through each of the diagonal lines of the first kind pass five diagonal planes; through each line of the second kind pass four diagonal planes.

**18.** Show how the results of Ex. 17 are modified in a modular space with modulus 2; with modulus 3. Show that in the modular space with modulus 5 the results of Ex. 17 hold without modification.

*** 19.** Derive homogeneous and nonhomogeneous coördinate systems for a space of $n$ dimensions, and establish the formulas for an $n$-dimensional projective collineation.

# CHAPTER VIII

## PROJECTIVITIES IN ONE-DIMENSIONAL FORMS *

### 73. Characteristic throw and cross ratio.

THEOREM 1. *If $M$, $N$ are double points of a projectivity on a line, and $AA'$, $BB'$ are any two pairs of homologous points (i.e. if $MNAB \;\overline{\wedge}\; MNA'B'$), then $MNAA' \;\overline{\wedge}\; MNBB'$.*

*Proof.* Let $S$, $S'$ be any two distinct points on a line through $M$ (fig. 85), and let the lines $SA$ and $S'A'$ meet in $A''$, and $SB$ and

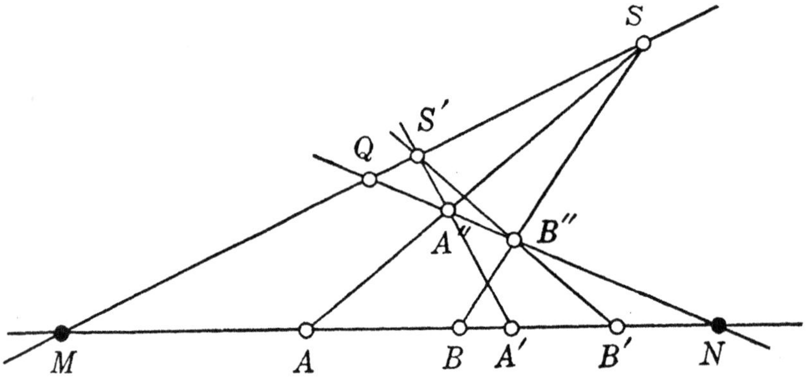


FIG. 85

$S'B'$ meet in $B''$. The points $A''$, $B''$, $N$ are then collinear (Theorem 23, Chap. IV). If the line $A''B''$ meets $SS'$ in a point $Q$, we have

$$MNAA' \overset{A''}{\underset{\wedge}{=}} MQSS' \overset{B''}{\underset{\wedge}{=}} MNBB'.$$

This proves the theorem, which may also be stated as follows:

*The throws consisting of the pair of double points in a given order and any pair of homologous points are all equal.*

DEFINITION. The throw $\mathsf{T}(MN, AA')$, consisting of the double points and a pair of homologous points of a projectivity, is called the *characteristic throw* of the projectivity; and the cross ratio of this throw is called the *characteristic cross ratio* of the projectivity.†

* All the developments of this chapter are on the basis of Assumptions A, E, P, $H_0$.

† Since the double points enter symmetrically, the throws $\mathsf{T}(MN, AA')$ and $\mathsf{T}(NM, AA')$ may be used equally well for the characteristic throw. The corresponding cross ratios ℞$(MN, AA')$ and ℞$(NM, AA')$ are reciprocals of each other (cf. Theorem 13, Cor. 3, Chap. VI).

COROLLARY 1. *A projectivity on a line with two given distinct double points is uniquely determined by its characteristic throw or cross ratio.*

COROLLARY 2. *The characteristic cross ratio of any involution with double points is* $-1$.

This follows directly from Theorem 27, Cor. 1, Chap. IV, and Theorem 13, Cor. 2, Chap. VI.

If $m, n$ are nonhomogeneous coördinates of the double points, and $k$ is the characteristic cross ratio of a projectivity on a line, we have

$$\frac{x'-m}{x'-n} : \frac{x-m}{x-n} = k$$

for every pair of homologous points $x, x'$. This is the analytic expression of the above theorem, and leads at once to the following analytic expression for a projectivity on a line with two distinct double points $m, n$:

COROLLARY 3. *Any projectivity on a line with two distinct double points $m, n$ may be represented by the equation*

$$\frac{x'-m}{x'-n} = k\frac{x-m}{x-n},$$

*$x', x$ being any pair of homologous points.*

For when cleared of fractions this is a bilinear equation in $x', x$ which obviously has $m, n$ as roots. Moreover, since any projectivity with two given distinct double points is uniquely determined by one additional pair of homologous elements, it follows that any projectivity of the kind described can be so represented, in view of the fact that one such pair of homologous points will always determine the multiplier $k$. These considerations offer an analytic proof of Theorem 1, for the case when the double points $M, N$ are distinct.

It is to be noted, however, that the proof of Theorem 1 applies equally well when the points $M, N$ coincide, and leads to the following theorem:

THEOREM 2. *If in a parabolic projectivity with double point $M$ the points $AA'$ and $BB'$ are two pairs of homologous points, the parabolic projectivity with double point $M$ which puts $A$ into $B$ also puts $A'$ into $B'$.*

COROLLARY. *The characteristic cross ratio of any parabolic projectivity is unity.*

The characteristic cross ratio together with the double point is therefore not sufficient to characterize a parabolic projectivity completely. Also, the analytic form for a projectivity with double points $m$, $n$, obtained above, breaks down when $m = n$. We may, however, readily derive a characteristic property of parabolic projectivities, from which will follow an analytic form for these projectivities.

THEOREM 3. *If a parabolic projectivity with double point $M$ transforms a point $A$ into $A'$ and $A'$ into $A''$, the pair of points $A$, $A''$ is harmonic with the pair $A'M$; i.e. we have* $\text{H}(MA', AA'')$.

*Proof.* By Theorem 23, Chap. IV, Cor., we have $\text{Q}(MAA', MA''A')$.

Analytically, if the coördinates of $M$, $A$, $A'$, $A''$ are $m$, $x$, $x'$, $x''$ respectively, we have, by Theorem 13, Cor. 4, Chap. VI,

$$\frac{2}{x'-m} = \frac{1}{x-m} + \frac{1}{x''-m}.$$

This gives

$$\frac{1}{x'-m} - \frac{1}{x-m} = \frac{1}{x''-m} - \frac{1}{x'-m},$$

which shows that if each member of this equation be placed equal to $t$, the relation

$$(1) \qquad \frac{1}{x'-m} = \frac{1}{x-m} + t$$

is satisfied by every pair of homologous points of the sequence obtained by applying the projectivity successively to the points $A$, $A'$, $A''$, $\cdots$. It is, however, readily seen that this relation is satisfied by every pair of homologous points on the line. For relation (1), when cleared of fractions, clearly gives a bilinear form in $x'$ and $x$, and is therefore a projectivity; and this projectivity clearly has only the one double point $m$. It therefore represents a parabolic projectivity with the double point $m$, and must represent the projectivity in question, since the relation is satisfied by the coördinates of the pair of homologous points $A$, $A'$, which are sufficient with the double point to determine the projectivity.

We have then:

COROLLARY 1. *Any parabolic projectivity with a double point, $M$, may be represented by the relation* (1).

DEFINITION. The number $t$ is called the *characteristic constant* of the projectivity (1).

COROLLARY 2. *Conversely, if a projectivity with a double point* $M$ *transforms a point* $A$ *into* $A'$, *and* $A'$ *into* $A''$, *such that we have* $H(MA', AA'')$, *the projectivity is parabolic.*

*Proof.* The double point $M$ and the two pairs of homologous points $AA'$, $A'A''$ are sufficient to determine the projectivity uniquely; and there is a parabolic projectivity satisfying the given conditions.

**74. Projective projectivities.** Let $\pi$ be a projectivity on a line $l$, and let $\pi_1$ be a projectivity transforming the points of $l$ into the points of another or the same line $l'$. The projectivity $\pi_1\pi\pi_1^{-1}$ is then a projectivity on $l'$. For $\pi_1^{-1}$ transforms any point of $l'$ into a point of $l$, $\pi$ transforms this point into another point of $l$, which in turn is transformed into a point of $l'$ by $\pi_1$. Thus, to every point of $l'$ is made to correspond a unique point of $l'$, and this correspondence is projective, since it is the product of projective correspondences. Clearly, also, the projectivity $\pi_1$ transforms any pair of homologous points of $\pi$ into a pair of homologous points of $\pi_1\pi\pi_1^{-1}$.

DEFINITION. The projectivity $\pi_1\pi\pi_1^{-1}$ is called the *transform of* $\pi$ *by* $\pi_1$; two projectivities are said to be *projective* or *conjugate* if one is a transform of the other by a projectivity.

The question now arises as to the conditions under which two projectivities are projective or conjugate. A necessary condition is evident. If one of two conjugate projectivities has two distinct double points, the other must likewise have two distinct double points; if one has no double points, the other likewise can have no double points; and if one is parabolic, the other must be parabolic. The further conditions are readily derivable in the case of two projectivities with distinct double points and in the case of two parabolic projectivities. They are stated in the two following theorems:

THEOREM 4. *Two projectivities each of which has two distinct double points are conjugate if and only if their characteristic throws are equal.*

*Proof.* The condition is necessary. For if $\pi$, $\pi'$ are two conjugate projectivities, any projectivity $\pi_1$ transforming $\pi$ into $\pi'$ transforms the double points $M$, $N$ of $\pi$ into the double points $M'$, $N'$ of $\pi'$, and also transforms any pair of homologous points $A$, $A_1$ of $\pi$ into a pair of homologous points $A'$, $A_1'$ of $\pi'$; i.e.

$$\pi_1(MNAA_1) = M'N'A'A_1'.$$

But this states that their characteristic throws are equal.

The condition is also sufficient; for if it is satisfied, the projectivity $\pi_1$ defined by

$$\pi_1(MNA) = M'N'A'$$

clearly transforms $\pi$ into $\pi'$.

COROLLARY. *Any two involutions with double points are conjugate.*

THEOREM 5. *Any two parabolic projectivities are conjugate.*

*Proof.* Let the two parabolic projectivities be defined by

$$\pi(MMA) = MMA_1, \text{ and } \pi'(M'M'A') = M'M'A_1'.$$

Then the projectivity $\pi_1$ defined by

$$\pi_1(MAA_1) = M'A'A_1'$$

clearly transforms $\pi$ into $\pi'$.

Since the characteristic cross ratio of any parabolic projectivity is unity, the condition of Theorem 4 may also be regarded as holding for parabolic projectivities.

**75. Groups of projectivities on a line.** DEFINITION. Two groups G and G′ of projectivities on a line are said to be *conjugate* if there exists a projectivity $\pi_1$ which transforms every projectivity of G into a projectivity of G′, and conversely. We may then write $\pi_1 G \pi_1^{-1} = G'$; and G′ is said to be the *transform of* G *by* $\pi_1$.

We have already seen (Theorem 8, Chap. III) that the set of all projectivities on a line form a group, which is called the *general projective group on the line.* The following are important subgroups:

1. *The set of all projectivities leaving a given point of the line invariant.*

*Any two groups of this type are conjugate.* For any projectivity transforming the invariant point of one group into the invariant point of the other clearly transforms every projectivity of the one into some projectivity of the other. Analytically, if we choose $x = \infty$ as the invariant point of the group, the group consists of all projectivities of the form

$$x' = ax + b.$$

2. *The set of all projectivities leaving two given distinct points invariant.*

*Any two groups of this type are conjugate.* For any projectivity transforming the two invariant points of the one into the invariant points of the other clearly transforms every projectivity of the one into a projectivity of the other. Analytically, if $x_1$, $x_2$ are the two invariant points, the group consists of all projectivities of the form

$$\frac{x' - x_1}{x' - x_2} = k\frac{x - x_1}{x - x_2}.$$

The product of two such projectivities with multipliers $k$ and $k'$ is clearly given by

$$\frac{x' - x_1}{x' - x_2} = kk' \cdot \frac{x - x_1}{x - x_2}.$$

This shows that any two projectivities of this group are commutative. This result gives

THEOREM 6. *Any two projectivities which have two double points in common are commutative.*

This theorem is equivalent to the commutative law for multiplication. If the double points are the points 0 and $\infty$, the group consists of all projectivities of the form $x' = ax$.

3. *The set of all parabolic projectivities with a common double point.* In order to show that this set of projectivities is a group, it is only necessary to show that the product of two parabolic projectivities with the same double point is parabolic. This follows readily from the analytic representation. The set of projectivities above described consists of all transformations of the form

$$\frac{1}{x' - x_1} = \frac{1}{x - x_1} + t,$$

where $x_1$ is the common double point (Theorem 3, Cor. 1). If

$$\frac{1}{x' - x_1} = \frac{1}{x - x_1} + t_1, \text{ and } \frac{1}{x' - x_1} = \frac{1}{x - x_1} + t_2$$

are two projectivities of this set, the product of the first by the second is given by

$$\frac{1}{x' - x_1} = \frac{1}{x - x_1} + t_1 + t_2,$$

which is clearly a projectivity of the set. It shows, moreover, that any two projectivities of this group are commutative. Whence

THEOREM 7. *Any two parabolic projectivities on a line with the same double point are commutative.*

This theorem is independent of Assumption P, although this assumption is implied in the proof we have given. The theorem has already been proved without this assumption in Example 2, p. 70.

*Any two groups of this type are conjugate.* For every projectivity transforming the double point of one group into the double point of the other transforms the one group into the other, since the projective transform of a parabolic projectivity is parabolic.

DEFINITION. Two subgroups of a group G are said to be *conjugate under* G if there exists a transformation of G which transforms one of the subgroups into the other. A subgroup of G is said to be *self-conjugate* or *invariant under* G if it is transformed into itself by every transformation of G; i.e. if every transformation in G transforms any transformation of the subgroup into another (or the same) transformation of the subgroup.

We have seen that any two groups of any one of the three types are conjugate subgroups of the general projective group on the line. We may now give an example of a self-conjugate subgroup.

*The set of all parabolic projectivities in a group of Type* 1 *above is a self-conjugate subgroup of this group.* It is clearly a subgroup, since it is a group of Type 3. And it is self-conjugate, since any conjugate of a parabolic projectivity is parabolic, and since every projectivity of the group leaves the common double point invariant.

### EXERCISES

**1.** Write the equations of all the projective transformations which permute among themselves (*a*) the points (0, 1), (1, 0), (1, 1); (*b*) the points (0, 1), (1, 0), (1, 1), (*a*, *b*); (*c*) the points (0, 1), (1, 0), (1, 1), (– 1, 1). What are the equations of the self-conjugate subgroup of the group of transformations (*a*)?

**2.** If a projectivity $x' = (ax + b)/(cx + d)$ having two distinct double elements be written in the form of Cor. 3, Theorem 1, show that

$$k = \frac{a - cx_1}{a - cx_2} = \frac{x_2}{x_1} \cdot \frac{b - dx_1}{b - dx_2}; \text{ and that } \frac{(1 + k)^2}{k} = \frac{(a + d)^2}{ad - bc}.$$

**3.** If a parabolic projectivity $x' = (ax + b)/(cx + d)$ be written in the form of Theorem 3, Cor. 1, show that $m = (a - d)/2c$, and $t = 2c/(a + d)$.

**4.** Show that a projectivity with distinct double points $x_1$, $x_2$ and characteristic cross ratio $k$ can be written in the form

$$x' = \frac{\begin{vmatrix} x & 0 & 1 \\ x_1 & x_1 & 1 \\ x_2 & kx_2 & 1 \end{vmatrix}}{\begin{vmatrix} x & 0 & 1 \\ x_1 & 1 & 1 \\ x_2 & k & 1 \end{vmatrix}}.$$

5. Show that the parabolic projectivity of Theorem 3, Cor. 1, may be written in the form

$$x' = \frac{\begin{vmatrix} x & 0 & 1 \\ x_1 & x_1 & 1 \\ 1 & tx_1 + 1 & 0 \end{vmatrix}}{\begin{vmatrix} x & 0 & 1 \\ x_1 & 1 & 1 \\ 1 & t & 0 \end{vmatrix}}.$$

6. If by means of a suitably chosen transformation of a group any of the elements transformed may be transformed into any other element, the group is said to be *transitive*. If by a suitably chosen transformation of a group any set of $n$ distinct elements may be transformed into any other set of $n$ distinct elements, and if this is not true for all sets of $n+1$ distinct elements, the group is said to be *n-ply transitive*. Show that the general projective group on a line is triply transitive, and that of the subgroups listed in § 75 the first is doubly transitive and the other two are simply transitive.

7. If two projectivities on a line, each having two distinct double points, have one double point in common, the characteristic cross ratio of their product is equal to the product of their characteristic cross ratios.

**76. Projective transformations between conics.** We have considered hitherto projectivities between one-dimensional forms of the first degree only. We shall now see how projectivities exist also between one-dimensional forms of the second degree, and also between a one-dimensional form of the first and one of the second degree. Many familiar theorems will hereby appear in a new light.

As typical for the one-dimensional forms of the second degree we choose the conic. The corresponding theorems for the cone then follow by the principle of duality.

Let $\pi_1$ be a projective collineation between two planes $\alpha$, $\alpha_1$, and let $C^2$ be any conic in $\alpha$. Any two projective pencils of lines in $\alpha$ are then transformed by $\pi_1$ into two projective pencils of lines in $\alpha_1$, such that any two homologous lines of the pencils in $\alpha$ are transformed into a pair of homologous lines in $\alpha_1$; for if $\pi$ be the projectivity between the pencils in $\alpha$, $\pi_1\pi\pi_1^{-1}$ will be a projectivity between the pencils in $\alpha_1$ (cf. § 74). Two projective pencils of lines generating the conic $C^2$ thus correspond to two pencils of lines in $\alpha_1$ generating a conic $C_1^2$. The transformation $\pi_1$ then transforms every point of $C^2$ into a unique point of $C_1^2$. Similarly, it is seen that $\pi_1$ transforms every tangent of $C^2$ into a unique tangent of $C_1^2$.

DEFINITION. Two conics are said to be *projective* if to every point of one corresponds a point of the other, and to every tangent of one corresponds a tangent of the other, in such a way that this correspondence may be brought about by a projective collineation between the planes of the conics. The projective collineation is then said to *generate* the projectivity between the conics.

Two conics in different planes are projective, for example, if one is the projection of the other from a point on neither of the two planes. If the second of these is projected back on the plane of the first from a new center, we obtain two conics in the same plane that are projective. We will see presently that two projective conics may also coincide, in which case we obtain a projectivity on a conic.

THEOREM 8. *Two conics that are projective with a third are projective.*

*Proof.* This is an immediate consequence of the definition and the fact that the resultant of two collineations is a collineation.

We proceed now to prove the fundamental theorem for projectivities between two conics.

THEOREM 9. *A projectivity between two conics is uniquely determined if three distinct points (or tangents) of one are made to correspond to three distinct points (or tangents) of the other.*

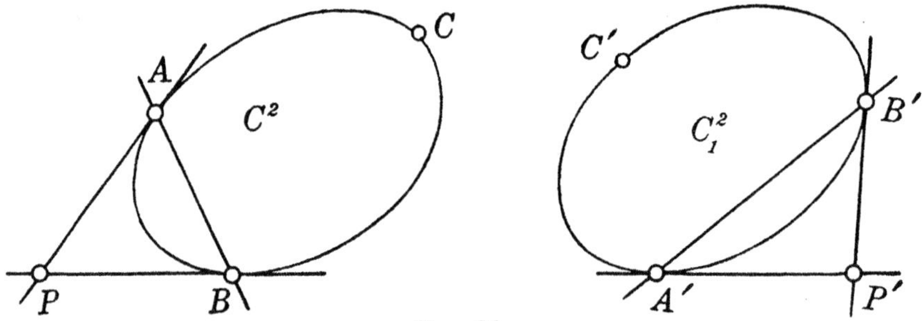


FIG. 86

*Proof.* Let $C^2$, $C_1^2$ be the two conics (fig. 86), and let $A$, $B$, $C$ be three points of $C^2$, and $A'$, $B'$, $C'$ the corresponding points of $C_1^2$. Let $P$ and $P'$ be the poles of $AB$ and $A'B'$ with respect to $C^2$ and $C_1^2$ respectively. If now the collineation $\pi$ is defined by the relation $\pi(ABCP) = A'B'C'P'$ (Theorem 18, Chap. IV), it is clear that the conic $C^2$ is transformed by $\pi$ into a conic through the points $A'$, $B'$, $C'$, with tangents $A'P'$ and $B'P'$. This conic is uniquely determined by these specifications, however, and is therefore identical with $C_1^2$. The collineation $\pi$ then transforms $C^2$ into $C_1^2$ in such a way that the points $A$, $B$, $C$ are transformed into $A'$, $B'$, $C'$ respectively. Moreover, suppose $\pi'$ were a second collineation transforming $C^2$ into $C_1^2$ in the way specified. Then $\pi'^{-1}\pi$ would be a collineation leaving $A, B, C, P$ invariant; i.e. $\pi = \pi'$.

The argument applies equally well if $A'B'C'$ are on the conic $C^2$, i.e. when the two conics $C^2$, $C_1^2$ coincide. In this case the projectivity is on the conic $C$. This gives

COROLLARY 1. *A projectivity on a conic is uniquely determined when three pairs of homologous elements (points or tangents) are given.*

Also from the proof of the theorem follows

COROLLARY 2. *A collineation in a plane which transforms three distinct points of a conic into three distinct points of the same conic and which transforms the pole of the line joining two of the first three points into the pole of the line joining the two corresponding points transforms the conic into itself.*

The two following theorems establish the connection between projectivities between two conics and projectivities between one-dimensional forms of the first degree.

THEOREM 10. *If $A$ and $B'$ are any two points of two projective conics $C^2$ and $C_1^2$ respectively, the pencils of lines with centers at $A$ and $B'$ are projective if every pair of homologous lines of these pencils pass through a pair of homologous points on the two conics respectively.*

THEOREM 10′. *If $a$ and $b'$ are any two tangents of two projective conics $C^2$ and $C_1^2$ respectively, the pencils of points on $a$ and $b'$ are projective if every pair of homologous points on these lines is on a pair of homologous tangents of the conics respectively.*

*Proof.* It will suffice to prove the theorem on the left. Let $A'$ be the point of $C_1^2$ homologous with $A$. The collineation which generates the projectivity between the conics then makes the pencils of lines at $A$ and $A'$ projective, in such a way that every pair of homologous lines contains a pair of homologous points of the two conics. The pencil of lines at $B'$ is projective with that at $A'$ if they correspond in such a way that pairs of homologous lines intersect on $C_1^2$ (Theorem 2, Chap. V). This establishes a projective correspondence between the pencils at $A$ and $B'$ in which any two homologous lines pass through two homologous points of the conics and proves the theorem.

It should be noted that in this projectivity the tangent to $C^2$ at $A$ corresponds to the line of the pencil at $B'$ passing through $A'$.

COROLLARY. *Conversely, if two conics correspond in such a way that every pair of homologous points is on a pair of homologous lines of two projective pencils of lines whose centers are on the conics, they are projective.*

COROLLARY. *Conversely, if two conics correspond in such a way that every pair of homologous tangents is on a pair of homologous points of two projective pencils of points whose axes are tangents of the conics, they are projective.*

*Proof.* This follows from the fact that the projectivity between the pencils of lines is uniquely determined by three pairs of homologous lines. A projectivity between the conics is also determined by the three pairs of points (Theorem 9), in which three pairs of homologous lines of the pencils meet the conics. But by what precedes and the theorem above, this projectivity is the same as that described in the corollary on the left. The corollary on the right may be proved similarly. If the two conics are in the same plane, it is simply the plane dual of the one on the left.

By means of these two theorems the construction of a projectivity between two conics is reduced to the construction of a projectivity between two primitive one-dimensional forms.

It is now in the spirit of our previous definitions to adopt the following:

DEFINITION. A point conic and a pencil of lines whose center is a point of the conic are said to be *perspective* if they correspond in such a way that every point of the conic is on the homologous line of the pencil. A point conic and a pencil of points are said to be *perspective* if every two homologous points are on the same line of a pencil of lines whose center is a point of the conic.

DEFINITION. A line conic and a pencil of points whose axis is a line of the conic are said to be *perspective* if they correspond in such a way that every line of the conic passes through the homologous point of the pencil of points. A line conic and a pencil of lines are said to be *perspective* if every two homologous lines meet in a point of a pencil of points whose axis is a line of the conic.

The reader will now readily verify that with this extended use of the term *perspective*, any sequence of perspectivities leads to a projectivity. For example, two pencils of lines perspective with the same point conic are projective by Theorem 2, Chap. V; two point conics perspective with the same pencil of lines or with the same pencil of points are projective by Theorem 10, Cor., etc.

Another illustration of this extension of the notion of perspectivity leads readily to the following important theorem:

THEOREM 11. *Two conics which are not in the same plane and have a common tangent at a point A are sections of one and the same cone.*

*Proof.* If the two conics $C^2$, $C_1^2$ (fig. 87) are made to correspond in such a way that every tangent $x$ of one is associated with that tangent $x'$ of the other which meets $x$ in a point of the common tangent $a$ of the conics, they are projective. For the tangents of the conics are then perspective with the same pencil of points

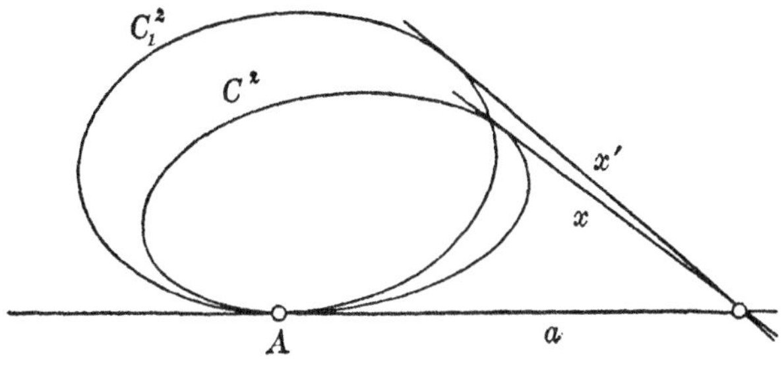

FIG. 87

(cf. Theorem 10′, Cor.). Every pair of homologous tangents of the two conics determines a plane. If we consider the point $O$ of intersection of three of these planes, say, those determined by the pairs of tangents $bb'$, $cc'$, $dd'$, and project the conic $C_1^2$ on the plane of $C^2$ from $O$, there results a conic in the plane of $C^2$. This conic has the lines $b$, $c$, $d$ for tangents and is tangent to $a$ at $A$; it therefore coincides with $C^2$ (Theorem 6′, Chap. V). The two conics $C^2$, $C_1^2$ then have the same projection from $O$, which proves the theorem.*

## EXERCISES

**1.** State the theorems concerning cones dual to the theorems of the preceding sections.

**2.** By dualizing the definitions of the last article, define what is meant by the perspectivity between cones and the primitive one-dimensional forms.

**3.** If two projective conics have three self-corresponding points, they are perspective with a common pencil of lines.

**4.** If two projective conics have four self-corresponding elements, they coincide.

**5.** State the space duals of the last two propositions.

* It will be seen later that this theorem leads to the proposition that any conic may be obtained as the projection of a circle tangent to it in a different plane.

**6.** If a pencil of lines and a conic in the plane of the pencil are projective, but not perspective, not more than three lines of the pencil pass through their homologous points on the conic. (*Hint.* Consider the points of intersection of the given conic with the conic generated by the given pencil and a pencil of lines perspective with the given conic.) Dualize.

**7.** The homologous lines of a line conic and a projective pencil of lines in the same plane intersect in points of a "curve of the third order" such that any line of the plane has at most three points in common with it. (This follows readily from the last exercise.)

**8.** The homologous elements of a cone of lines and a projective pencil of planes meet in a "space curve of the third order" such that any plane has at most three points in common with it.

**9.** Dualize the last two propositions.

**77. Projectivities on a conic.** We have seen that two projective conics may coincide (Theorems 8–10), in which case we obtain a projective correspondence among the points or the tangents of the

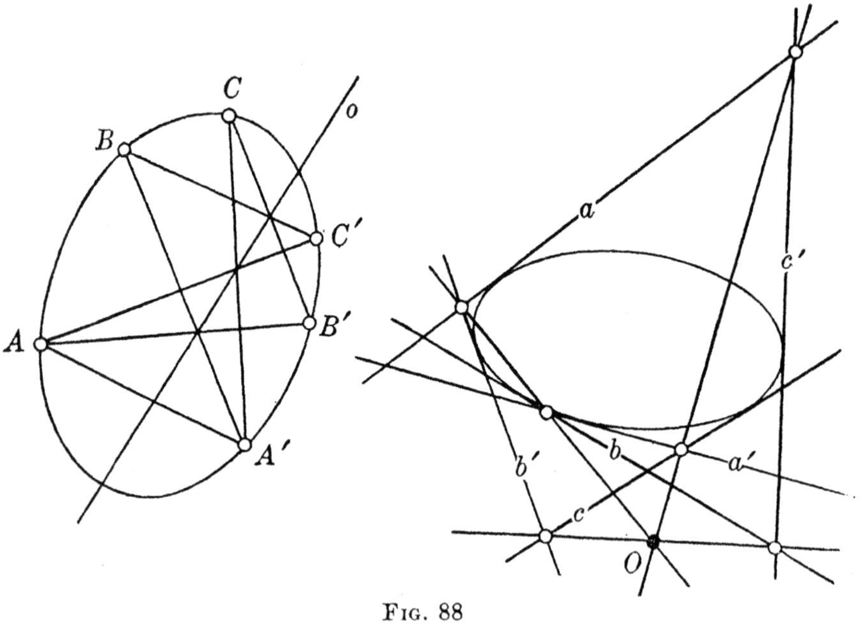


FIG. 88

conic. The construction of the projectivity in this case is very simple, and leads to many important results. It results from the following theorems:

THEOREM 12. *If $A, A'$ are any two distinct homologous points of a projectivity on a conic, and $B, B'$; $C, C'$; etc., are any other pairs of*

THEOREM 12′. *If $a, a'$ are any two distinct homologous tangents of a projectivity on a conic, and $b, b'$; $c, c'$; etc., are any other pairs* *homologous points, the lines $A'B$ and $AB'$, $A'C$ and $AC'$, etc., meet in points of the same line; and this line is independent of the pair $AA'$ chosen.*

*of homologous tangents, the points $a'b$ and $ab'$, $a'c$ and $ac'$, etc., are collinear with the same point; and this point is independent of the pair $aa'$ chosen.*

*Proof.* The pencils of lines $A'(ABC \cdots)$ and $A(A'B'C' \cdots)$ are projective (Theorem 10), and since they have a self-corresponding line $AA'$, they are perspective, and the pairs of homologous lines of these two pencils therefore meet in the points of a line (fig. 88). This proves the first part of the theorem on the left. That the line thus determined is independent of the homologous pair $AA'$ chosen then follows at once from the fact this line is the Pascal line of the simple hexagon $AB'CA'BC'$, so that the lines $B'C$ and $BC'$ and all other analogously formed pairs of lines meet on it. The theorem on the right follows by duality.

DEFINITION. The line and the point determined by the above dual theorems are called the *axis* and the *center* of the projectivity respectively.

COROLLARY 1. *A (nonidentical) projectivity on a conic is uniquely determined when the axis of homology and one pair of distinct homologous points are given.*

COROLLARY 1'. *A (nonidentical) projectivity on a conic is uniquely determined when the center and one pair of distinct homologous tangents are given.*

These corollaries follow directly from the construction of the projectivity arising from the above theorem. This construction is as follows: Given the axis $o$ and a pair of distinct homologous points $AA'$, to get the point $P'$ homologous with any point $P$ on the conic; join $P$ to $A'$; the point $P'$ is then on the line joining $A$ to the point of intersection of $A'P$ with $o$. Or, given the center $O$ and a pair of distinct homologous tangents $aa'$, to construct the tangent $p'$ homologous with any tangent $p$; the line joining the point $a'p$ to the center meets $a$ in a point of $p'$.

COROLLARY 2. *Every double point of a projectivity on a conic is on the axis of the projectivity; and, conversely, every point common to the axis and the conic is a double point.*

COROLLARY 2'. *Every double line of a projectivity on a conic contains the center of the projectivity; and, conversely, every tangent of a conic through the center is a double line of the projectivity.*

COROLLARY 3. *A projectivity among the points on a conic is parabolic if and only if the axis is tangent to the conic.*

COROLLARY 3′. *A projectivity among the tangents to a conic is parabolic if and only if the center is a point of the conic.*

THEOREM 13. *A projectivity among the points of a conic determines a projectivity of the tangents in which the tangents at pairs of homologous points are homologous.*

*Proof.* This follows at once from the fact that the collineation in the plane of the conic which generates the projectivity transforms the tangent at any point of the conic into the tangent at the homologous point, and hence also generates a projectivity between the tangents.

THEOREM 14. *The center of a projectivity of tangents on a conic and the axis of the corresponding projectivity of points are pole and polar with respect to the conic.*

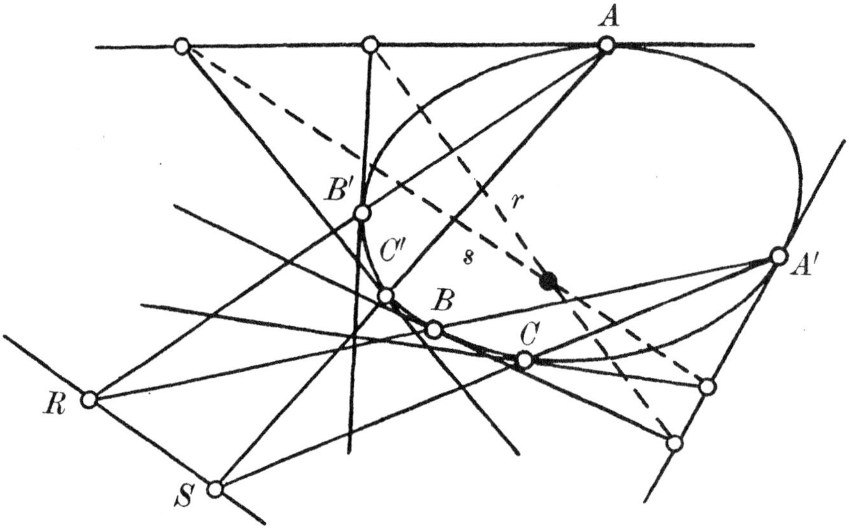


FIG. 89

*Proof.* Let $AA'$, $BB'$, $CC'$ (fig. 89) be three pairs of homologous points ($AA'$ being distinct), and let $A'B$ and $AB'$, $A'C$ and $AC'$, meet in points $R$ and $S$ respectively, which determine the axis of the projectivity of points. Now the polar of $R$ with respect to the conic is determined by the intersections of the pairs of tangents at $A'$, $B$ and $A$, $B'$ respectively; and the polar of $S$ is determined by the pairs of tangents at $A'$, $C$ and $A$, $C'$ respectively (Theorem 13, Chap. V). The pole of the axis $RS$ is then determined as the intersection of these two polars (Theorem 17, Chap. V). But by definition these two polars also determine the center of the projectivity of tangents.

This theorem is obvious if the projectivity has double elements; the proof given, however, applies to all cases.

The collineation generating the projectivity on the conic transforms the conic into itself and clearly leaves the center and axis invariant. The set of all collineations in the plane leaving the conic invariant form a group (cf. p. 67). In determining a transformation of this group, any point or any line of the plane may be chosen arbitrarily as a double point or a double line of the collineation; and any two points or lines of the conic may be chosen as a homologous pair of the collineation. The collineation is then, however, uniquely determined. In fact, we have already seen that the projectivity on the conic is uniquely determined by its center and axis and one pair of homologous elements (Theorem 12, Cor. 1); and the theorem just proved shows that if the center of the projectivity is given, the axis is uniquely determined, and conversely.

COROLLARY 1. *A plane projective collineation which leaves a nondegenerate conic in its plane invariant is of Type I if it has two double points on the conic, unless it is of period two, in which case it is of Type IV; and is of Type III if the corresponding projectivity on the conic is parabolic.*

COROLLARY 2. *An elation or a collineation of Type II transforms every nondegenerate conic of its plane into a different conic.*

COROLLARY 3. *A plane projective collineation which leaves a conic in its plane invariant and has no double point on the conic has one and only one double point in the plane.*

THEOREM 15. *The group of projective collineations in a plane leaving a nondegenerate conic invariant is simply isomorphic* with the general projective group on a line.*

*Proof.* Let $A$ be any point of the invariant conic. Any projectivity on the conic then gives rise to a projectivity in the flat pencil at $A$ in which two lines are homologous if they meet the conic in a pair of homologous points. And, conversely, any projectivity in the flat

* Two groups are said to be *simply isomorphic* if it is possible to establish a (1,1) correspondence between the elements of the two groups such that to the product of any two elements of one of the groups corresponds the product of the two corresponding elements of the other.

pencil at $A$ gives rise to a projectivity on the conic. The group of all projectivities on a conic is therefore simply isomorphic with the group of all projectivities in a flat pencil, since it is clear that in the correspondence described between the projectivities in the flat pencil and on the conic, the products of corresponding pairs of projectivities will be corresponding projectivities. Hence the group of plane collineations leaving the conic invariant is simply isomorphic with the general projective group in a flat pencil and hence with the general projective group on a line.

**78. Involutions.** An involution was defined (p. 102) as any projectivity in a one-dimensional form which is of period two, i.e. by the relation $\mathrm{I}^2 = \mathbf{1}\,(\mathrm{I} \neq \mathbf{1})$, where I represents an involution. This relation is clearly equivalent to the other, $\mathrm{I} = \mathrm{I}^{-1}(\mathrm{I} \neq \mathbf{1})$, so that any projectivity (not the identity) in a one-dimensional form, which is identical with its inverse, is an involution. It will be recalled that since an involution makes every pair of homologous elements correspond doubly, i.e. $A$ to $A'$ and $A'$ to $A$, an involution may also be considered as a pairing of the elements of a one-dimensional form; any such pair is then called a *conjugate pair* of the involution. We propose now to consider this important class of projectivities more in detail. To this end it seems desirable to collect the fundamental properties of involutions which have been obtained in previous chapters. They are as follows:

1. *If the relation $\pi^2(A) = A$ holds for a single element $A$ (not a double element of $\pi$) of a one-dimensional form, the projectivity $\pi$ is an involution, and the relation holds for every element of the form* (Theorem 26, Chap. IV).

2. *An involution is uniquely determined when two pairs of conjugate elements are given* (Theorem 26, Cor., Chap. IV).

3. *The opposite pairs of any quadrangular set are three pairs of an involution* (Theorem 27, Chap. IV).

4. *If $M$, $N$ are distinct double elements of any projectivity in a one-dimensional form and $A$, $A'$ and $B$, $B'$ are any two pairs of homologous elements of the projectivity, the pairs of elements $MN$, $AB'$, $A'B$ are three pairs of an involution* (Theorem 27, Cor. 3, Chap. IV).

5. *If $M$, $N$ are double elements of an involution, they are distinct, and every conjugate pair of the involution is harmonic with $M$, $N$* (Theorem 27, Cor. 1, Chap. IV).

6. *An involution is uniquely determined, if two double elements are given, or if one double element and another conjugate pair are given.* (This follows directly from the preceding.)

7. *An involution is represented analytically by a bilinear form* $cxx' - a(x + x') - b = 0$, *or by the transformation*

$$x' = \frac{ax + b}{cx - a} \qquad a^2 + bc \neq 0$$

(Theorem 12, Cor. 3, Chap. VI).

8. *An involution with double elements* $m, n$ *may be represented by the transformation*

$$\frac{x' - m}{x' - n} = -\frac{x - m}{x - n}$$

(Theorem 1, Cors. 2, 3, Chap. VIII).

We recall, finally, the Second Theorem of Desargues and its various modifications (§ 46, Chap. V), which need not be repeated at this place. It has been seen in the preceding sections that any projectivity in a one-dimensional primitive form may be transformed into a projectivity on a conic. We shall find that the construction of an involution on a conic is especially simple, and may be used to advantage in deriving further properties of involutions. Under duality we may confine our consideration to the case of an involution of points on a conic.

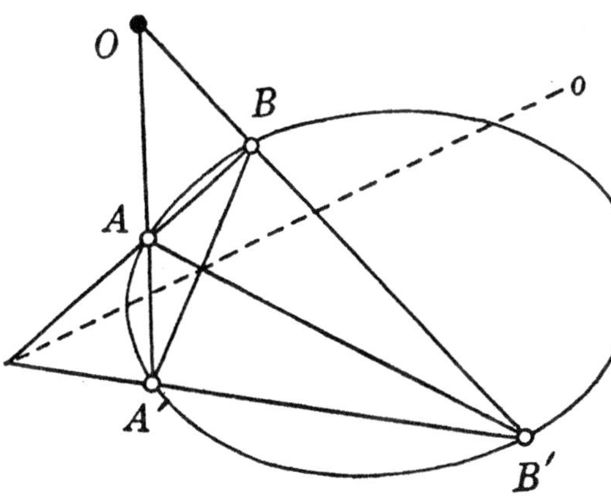

FIG. 90

THEOREM 16. *The lines joining the conjugate points of an involution on a conic all pass through the center of the involution.*

*Proof.* Let $A, A'$ (fig. 90) be any conjugate pair ($A$ not a double point) of an involution of points on a conic $C^2$. The line $AA'$ is then an invariant line of the collineation generating the involution. Every line joining a pair of distinct conjugate points of the involution is therefore invariant, and the generating collineation must be a perspective collineation, since any collineation leaving four lines invariant is either perspective or the identity

(Theorem 9, Cor. 3, Chap. III). It remains only to show that the center of this perspective collineation is the center of the involution. Let $B$, $B'$ ($B$ not a double point) be any other conjugate pair of the involution, distinct from $A$, $A'$. Then the lines $AB'$ and $A'B$ intersect on the axis of the involution. But since $B$, $B'$ correspond to each other doubly, it follows that the lines $AB$ and $A'B'$ also intersect on the axis. This axis then joins two of the diagonal points of the quadrangle $AA'BB'$. The center of the perspective collineation is determined as the intersection of the lines $AA'$ and $BB'$, i.e. it is the third diagonal point of the quadrangle $AA'BB'$. The center of the collineation is therefore the pole of the axis of the involution (Theorem 14, Chap. V) and is therefore (Theorem 14, above) the center of the involution.

Since this center of the involution is clearly not on the conic, the generating collineation of any involution of the conic is a homology, whose center $O$ and axis $o$ are pole and polar with respect to the conic. A homology of period two is sometimes called a *harmonic homology*, since it transforms any point $P$ of the plane into its harmonic conjugate with respect to $O$ and the point in which $OP$ meets the axis. It is also called a *projective reflection* or a *point-line reflection*. Clearly this is the only kind of homology that can leave a conic invariant.

The construction of the pairs of an involution on a conic is now very simple. If two conjugate pairs $A, A'$ and $B, B'$ are given, the lines $AA'$ and $BB'$ determine the center of the involution. The conjugate of any other point $C$ on the conic is then determined as the intersection with the conic of the line joining $C$ to the center. If the involution has double points, the tangents at these points pass through the center of the involution; and, conversely, if tangents can be drawn to the conic from the center of the involution, the points of contact of these tangents are double points of the involution.

The great importance of involutions is in part due to the following theorem:

THEOREM 17. *Any projectivity in a one-dimensional form may be obtained as the product of two involutions.*

*Proof.* Let $\Pi$ be the projectivity in question, and let $A$ be any point of the one-dimensional form which is not a double point.

Further, let $\Pi(A) = A'$ and $\Pi(A') = A''$. Then, if $I_1$ is the involution of which $A'$ is a double point and of which $AA''$ is a conjugate pair (Prop. 6, p. 222), we have

$$I_1 \cdot \Pi(AA') = A'A,$$

so that in the projectivity $I_1 \cdot \Pi$ the pair $AA'$ corresponds to itself doubly. $I_1 \cdot \Pi$ is therefore an involution (Prop. 1, p. 221). If it be denoted by $I_2$, we have $I_1 \cdot \Pi = I_2$, or $\Pi = I_1 \cdot I_2$, which was to be proved.

This proof gives at once:

COROLLARY 1. *Any projectivity* $\Pi$ *may be represented as the product of two involutions,* $\Pi = I_1 \cdot I_2$, *either of which (but not both) has an arbitrary point (not a double point of* $\Pi$*) for a double point.*

*Proof.* We have seen above that the involution $I_1$ may have an arbitrary point ($A'$) for a double point. If in the above argument we let $I_2$ be the involution of which $A'$ is a double point and $AA''$ is a conjugate pair, we have $\Pi \cdot I_2(A'A'') = A''A'$; whence $\Pi \cdot I_2$ is an involution, say $I_1$. We then have $\Pi = I_1 \cdot I_2$, in which $I_2$ has the arbitrary point $A'$ for a double point.

The argument given above for the proof of the theorem applies without change when $A = A''$, i.e. when the projectivity $\Pi$ is an involution. This leads readily to the following important theorem:

COROLLARY 2. *If* $A\,A'$ *is a conjugate pair of an involution* I, *the involution of which* $A, A'$ *are double points transforms* I *into itself, and the two involutions are commutative.*

*Proof.* The proof of Theorem 17 gives at once $I = I_1 \cdot I_2$, where $I_1$ is determined as the involution of which $A, A'$ are double points. We have then $I_1 \cdot I = I_2$, from which follows, by taking the inverse of both sides of the equality, $I \cdot I_1 = I_2^{-1} = I_2$, or $I_1 \cdot I = I \cdot I_1$, or $I_1 \cdot I \cdot I_1 = I$.

As an immediate corollary of the preceding we have

COROLLARY 3. *The product of two involutions with double points* $A, A'$ *and* $B, B'$ *respectively transforms into itself the involution in which* $A\,A'$ *and* $B\,B'$ *are two conjugate pairs.*

Involutions related as are the two in Cor. 2 above are worthy of special attention.

DEFINITION. Two involutions are said to be *harmonic* if their product is an involution.

THEOREM 18. *Two harmonic involutions are commutative.*

*Proof.* If $I_1, I_2$ are harmonic, we have, by definition, $I_1 \cdot I_2 = I_3$, where $I_3$ is an involution. This gives at once the relations $I_1 \cdot I_2 \cdot I_3 = \mathbf{1}$ and $I_1 \cdot I_2 = I_2 \cdot I_1$.

COROLLARY. *Conversely, if two distinct involutions are commutative, they are harmonic.*

For from the relation $I_1 \cdot I_2 = I_2 \cdot I_1$ follows $(I_1 \cdot I_2)^2 = \mathbf{1}$; i.e. $I_1 \cdot I_2$ is an involution, since $I_1 \cdot I_2 \neq \mathbf{1}$.

DEFINITION. The set of involutions harmonic with a given involution is called a *pencil* of involutions.

It follows then from Theorem 17, Cor. 2, that the set of all involutions in which two given elements form a conjugate pair is a pencil. Thus the double points of the involutions of such a pencil are the pairs of an involution.

**79. Involutions associated with a given projectivity.** In deriving further theorems on involutions we shall find it desirable to suppose the projectivities in question to be on a conic.

THEOREM 19. *If a projectivity on a conic is represented as the product of two involutions, the axis of the projectivity is the line joining the centers of the two involutions.*

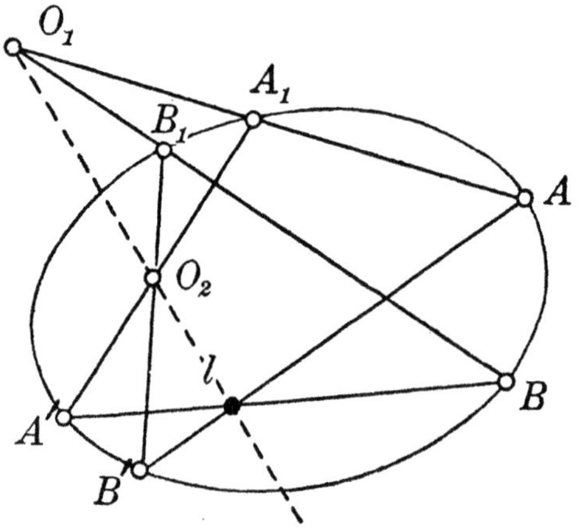

FIG. 91

*Proof.* Let the given projectivity be $\Pi = I_2 \cdot I_1$; $I_1, I_2$ being two involutions. Let $O_1, O_2$ be the centers of $I_1, I_2$ respectively (fig. 91), and let $A$ and $B$ be any two points on the conic which are not double points of either of the involutions $I_1$ or $I_2$ and which are not a conjugate pair of $I_1$ or $I_2$. If, then, we have $\Pi(AB) = A'B'$, we have, by hypothesis, $I_1(AB) = A_1B_1$ and $I_2(A_1B_1) = A'B'$; $A_1, B_1$ being uniquely determined points of the conic, such that the lines $AA_1$, $BB_1$ intersect in $O_1$ and the lines $A_1A'$, $B_1B'$ intersect in $O_2$. The Pascal line of the hexagon $AA_1A'BB_1B'$ then passes through $O_1$, $O_2$ and the intersection of the lines $AB'$ and $A'B$. But the latter point is a point on the axis of $\Pi$. This proves the theorem.

COROLLARY. *A projectivity on a conic is the product of two involutions, the center of one of which may be any arbitrary point (not a double point) on the axis of the projectivity; the center of the other is then uniquely determined.*

*Proof.* Let the projectivity $\Pi$ be determined by its axis $l$ and any pair of homologous points $A, A'$ (fig. 91). Let $O_1$ be any point on the axis not a double point of $\Pi$, and let $\mathrm{I}_1$ be the involution of which $O_1$ is the center. If, then, $\mathrm{I}_1(A) = A_1$, the center $O_2$ of the involution $\mathrm{I}_2$ such that $\Pi = \mathrm{I}_2 \cdot \mathrm{I}_1$ is clearly determined as the intersection of the line $A_1A'$ with the axis. For by the theorem the product $I_2 \cdot I_1$ is a projectivity having $l$ for an axis, and it has the points $A, A'$ as a homologous pair. This shows that the center of the first involution may be any point on the axis (not a double point). The modification of this argument in order to show that the center of the second involution may be chosen arbitrarily (instead of the center of the first) is obvious.

THEOREM 20. *There is one and only one involution commutative with a given nonparabolic noninvolutoric projectivity. If the projectivity is represented on a conic, the center of this involution is the center of the projectivity.*

*Proof.* Let the given nonparabolic projectivity $\Pi$ be on a conic, and let I be any involution commutative with $\Pi$; i.e. such that we have $\Pi \cdot \mathrm{I} = \mathrm{I} \cdot \Pi$. This is equivalent to $\Pi \cdot \mathrm{I} \cdot \Pi^{-1} = \mathrm{I}$. That is to say, I is transformed into itself by $\Pi$. Hence the center of I is transformed into itself by the collineation generating $\Pi$. But by hypothesis the only invariant points of this collineation are its center and the points (if existent) in which its axis meets the conic. Since the center of I cannot be on the conic, it must coincide with the center of $\Pi$. Moreover, if the center of I is the same as the center of $\Pi$, I is transformed into itself by the collineation generating $\Pi$, $\Pi \cdot \mathrm{I} \cdot \Pi^{-1} = \mathrm{I}$. Hence $\Pi \cdot \mathrm{I} = \mathrm{I} \cdot \Pi$. Hence I is the one and only involution commutative with $\Pi$.

COROLLARY 1. *There is no involution commutative with a parabolic projectivity.*

DEFINITION. The involution commutative with a given nonparabolic noninvolutoric projectivity is called the *involution belonging to the given projectivity.* An involution *belongs to itself.*

COROLLARY 2. *If a nonparabolic projectivity has double points, the involution belonging to the projectivity has the same double points.*

For if the axis of the projectivity meets the conic in two points, the tangents to the conic at these points meet in the pole of the axis.

It is to be noted that the involution I belonging to a given projectivity $\Pi$ transforms $\Pi$ into itself, and is transformed into itself by $\Pi$. Indeed, from the relation $\Pi \cdot I = I \cdot \Pi$ follow at once the relations $I \cdot \Pi \cdot I = \Pi$ and $\Pi \cdot I \cdot \Pi^{-1} = I$. Conversely, from the equation $\Pi \cdot I \cdot \Pi^{-1}$ follows $\Pi \cdot I = I \cdot \Pi$.

THEOREM 21. *The necessary and sufficient condition that two involutions on a conic be harmonic is that their centers be conjugate with respect to the conic.*

*Proof.* The condition is sufficient. For let $I_1$, $I_2$ be two involutions on the conic whose centers $O_1$, $O_2$ respectively are conjugate with respect to the conic (fig. 92). Let $A$ be any point of the conic not a double point of either involution, and let $I_1(A) = A_1$ and $I_2(A_1) = A'$. If, then, $I_1(A') = A_1'$, the center $O_1$ is a diagonal point of the quadrangle $AA_1A'A_1'$, and the center $O_2$ is on the side $A_1A'$. Since, by hypothesis, $O_2$ is conjugate to $O_1$ with respect to the conic, it must be the diagonal point on $A_1A'$, i.e. it must be collinear with $AA_1'$. We have then $I_2 \cdot I_1(AA') = A'A$, i.e. the projectivity $I_2 \cdot I_1$ is an involution $I_3$. The center $O_3$ of the involution $I_3$ is then the pole of the line $O_1O_2$ with respect to the conic (Theorem 19). The triangle $O_1O_2O_3$ is therefore self-polar with respect to the conic. It follows readily also that the condition is necessary. For the relation $I_1 \cdot I_2 = I_3$ leads at once to the relation $I_2 = I_1 \cdot I_3$. If $O_1$, $O_2$, $O_3$ are the centers respectively of the involutions $I_1$, $I_2$, $I_3$, the former of these two relations shows (Theorem 19) that $O_3$ is the pole of the line $O_1O_2$; while the latter shows that $O_2$ is the pole of the line $O_1O_3$. The triangle $O_1O_2O_3$ is therefore self-polar.

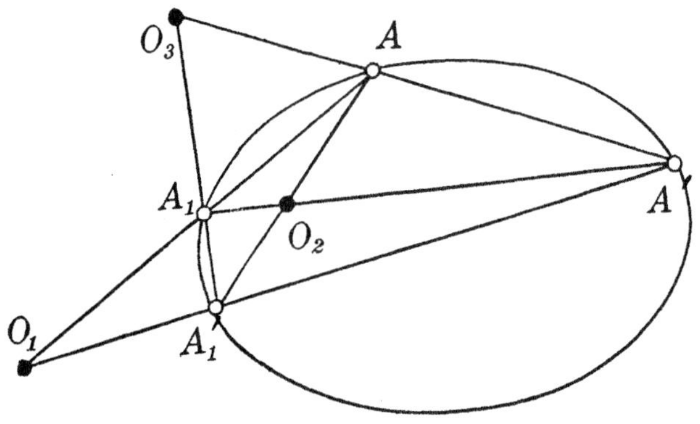


FIG. 92

COROLLARY 1. *Given any two involutions, there exists a third involution which is harmonic with each of the given involutions.*

For if we take the two involutions on a conic, the involution whose center is the pole with respect to the conic of the line joining the centers of the given involutions clearly satisfies the condition of the theorem for each of the latter.

COROLLARY 2. *Three involutions each of which is harmonic to the other two constitute, together with the identity, a group.*

COROLLARY 3. *The centers of all involutions in a pencil of involutions are collinear.*

THEOREM 22. *The set of all projectivities to which belongs the same involution* I *forms a commutative group.*

*Proof.* If $\Pi$, $\Pi_1$ are two projectivities to each of which belongs the involution I, we have the relations $I \cdot \Pi \cdot I = \Pi$ and $I \cdot \Pi_1 \cdot I = \Pi_1$, from which follows $I \cdot \Pi^{-1} \cdot I = \Pi^{-1}$ and, by multiplication, the relation $I \cdot \Pi \cdot I \cdot I \cdot \Pi_1 \cdot I = I \cdot \Pi \cdot \Pi_1 \cdot I = \Pi \cdot \Pi_1$, which shows that the set forms a group. To show that any two projectivities of this group are commutative, we need only suppose the projectivities given on a conic. Let $A$ be any point on this conic, and let $\Pi(A) = A'$ and $\Pi_1(A') = A_1'$, so that $\Pi_1 \cdot \Pi(A) = A_1'$. Since the same involution I belongs, by hypothesis, both to $\Pi$ and $\Pi_1$, these two projectivities have the same axis; let it be the line $l$ (fig. 93). The point $\Pi_1(A) = A_1$ is now readily determined (Theorem 12) as the intersection with the conic of the line joining $A'$ to the intersection of the line $AA_1'$ with the axis $l$. In like manner, $\Pi(A_1)$ is determined as the intersection with the conic of the line joining $A$ to the intersection of the line $A_1A'$ with the axis $l$. Hence $\Pi(A_1) = A_1'$, and hence $\Pi \cdot \Pi_1(A) = A_1'$.

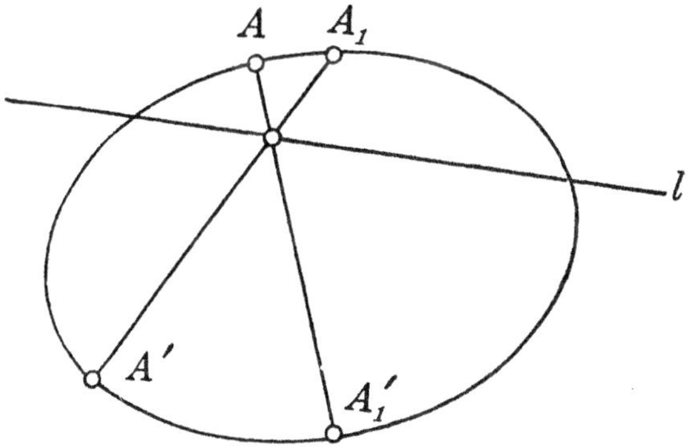

FIG. 93

It is noteworthy that when the common axis of the projectivities of this group meets the conic in two points, which are then common double points of all the projectivities of the group, the group is the same as the one listed as Type 2, p. 209. If, however, our geometry admits of a line in the plane of a conic but not meeting the conic, the argument just given proves the existence of a commutative group none of the projectivities of which have a double point.

THEOREM 23. *Two involutions have a conjugate pair (or a double point) in common if and only if the product of the two involutions has two double points (or is parabolic).*

*Proof.* This follows at once if the involutions are taken on a conic. For a common conjugate pair (or double point) must be on the line joining the centers of the two involutions. This line must then meet the conic in two points (or be tangent to it) in order that the involutions may have a conjugate pair (or a double point) in common.

### EXERCISES

**1.** Dualize all the theorems and corollaries of the last two sections.

**2.** The product of two involutions on a conic is parabolic if and only if the line joining the centers of the involutions is tangent to the conic. Dualize.

**3.** Any involution of a pencil is uniquely determined when one of its conjugate pairs is given.

**4.** Let $\Pi$ be a noninvolutoric projectivity, and let $\mathrm{I}$ be the involution belonging to $\Pi$; further, let $\Pi(AA') = A'A''$, $A$ being any point on which the projectivity operates which is not a double point, and let $\mathrm{I}(A') = A_1'$. Show, by taking the projectivity on a conic, that the points $A'A_1'$ are harmonic with the points $AA''$.

**5.** Derive the theorem of Ex. 4 directly as a corollary of Prop. 4, p. 221, assuming that the projectivity $\Pi$ has two distinct double points.

**6.** From the theorem of Ex. 4 show how to construct the involution belonging to a projectivity $\Pi$ on a line without making use of any double points the projectivity may have.

**7.** A projectivity is uniquely determined if the involution belonging to it and one pair of homologous points are given.

**8.** The product of two involutions $\mathrm{I}_1$, $\mathrm{I}_2$ is a projectivity to which belongs the involution which is harmonic with each of the involutions $\mathrm{I}_1$, $\mathrm{I}_2$.

**9.** Conversely, every projectivity to which a given involution $\mathrm{I}$ belongs can be obtained as the product of two involutions harmonic with $\mathrm{I}$.

**10.** Show that any two projectivities $\Pi_1$, $\Pi_2$ may be obtained as the product of involutions in the form $\Pi_1 = \mathrm{I}\cdot\mathrm{I}_1$, $\Pi_2 = \mathrm{I}_2\cdot\mathrm{I}$; and hence that the product of the two projectivities is given by $\Pi_2\cdot\Pi_1 = \mathrm{I}_2\cdot\mathrm{I}_1$.

**11.** Show that a projectivity $\Pi = \mathrm{I}\cdot\mathrm{I}_1$ may also be written $\Pi = \mathrm{I}_2\cdot\mathrm{I}$, $\mathrm{I}_2$ being a uniquely determined involution; and that in this case the two involutions $\mathrm{I}_1$, $\mathrm{I}_2$ are distinct unless $\Pi$ is involutoric.

**12.** Show that if $I_1$, $I_2$, $I_3$ are three involutions of the same pencil, the relation $(I_1 \cdot I_2 \cdot I_3)^2 = 1$ must hold.

**13.** If $aa'$, $bb'$, $cc'$ are the coördinates of three pairs of points in involution, show that $\frac{a'-b}{a'-c} \cdot \frac{b'-c}{b'-a} \cdot \frac{c'-a}{c'-b} = 1$.

**80. Harmonic transformations.** The definition of harmonic involutions in the section above is a special case of a more general notion which can be defined for (1, 1) transformations of any kind whatever.

DEFINITION. Two distinct transformations A and B are said to be *harmonic* if they satisfy the relation $(AB^{-1})^2 = 1$ or the equivalent relation $(BA^{-1})^2 = 1$, provided that $AB^{-1} \neq 1$.

A number of theorems which are easy consequences of this definition when taken in conjunction with the two preceding sections are stated in the following exercises. (Cf. C. Segre, Note sur les homographies binaires et leur faisceaux, Journal für die reine und angewandte Mathematik, Vol. 100 (1887), pp. 317–330, and H. Wiener, Ueber die aus zwei Spiegelungen zusammengesetzten Verwandtschaften, Berichte d. K. sächsischen Gesellschaft der Wissenschaften, Leipzig, Vol. 43 (1891), pp. 644–673.)

## EXERCISES

**1.** If A and B are two distinct involutoric transformations, they are harmonic to their product AB.

**2.** If three involutoric transformations A, B, Γ satisfy the relations $(AB\Gamma)^2 = 1$, $AB\Gamma \neq 1$, they are all three harmonic to the transformation AB.

**3.** If a transformation Σ is the product of two involutoric transformations A, B (i.e. $\Sigma = AB$) and Γ is an involutoric transformation harmonic to Σ, then we have $(AB\Gamma)^2 = 1$.

**4.** If $A$, $B$, $C$, $A'$, $B'$, $C'$ are six points of a line, the involutions A, B, Γ, such that $\Gamma(AA') = B'B$, $A(BB') = C'C$, $B(CC') = A'A$, are all harmonic to the same projectivity. Show that if the six points are taken on a conic, this proposition is equivalent to Pascal's theorem (Theorem 3, Chap. V).

**5.** The set of involutions of a one-dimensional form which are harmonic to a given nonparabolic projectivity form a pencil. Hence, if an involution with double points is harmonic to a projectivity with two double points, the two pairs of double points form a harmonic set.

**6.** Let $O$ be a fixed point of a line $l$, and let $C$ be called the *mid-point* of a pair of points $A$, $B$, provided that $C$ is the harmonic conjugate of $O$ with respect to $A$ and $B$. If $A$, $B$, $C$, $A'$, $B'$, $C'$ are any six points of $l$ distinct from $O$, and $AB'$ have the same mid-point as $A'B$, and $BC'$ have the same mid-point as $B'C$, then $CA'$ will have the same mid-point as $C'A$.

**7.** Any two involutions of the same one-dimensional form determine a pencil of involutions. Given two involutions A, B and a point $M$, show how to construct the other double point of that involution of the pencil of which one double point is $M$.

**8.** The involutions of conjugate points on a line $l$ with regard to the conics of any pencil of conics in a plane with $l$ form a pencil of involutions.

**9.** If two nonparabolic projectivities are commutative, the involutions belonging to them coincide, unless both projectivities are involutions, in which case the involutions may be harmonic.

**10.** If $[\Pi]$ is the set of projectivities to which belongs an involution I and $A$ and $B$ are two given points, then we have $[\Pi(A)] \barwedge [\Pi(B)]$.

**11.** A conic through two of the four common points of a pencil of conics of Type $I$ meets the conics of the pencil in pairs of an involution. Extend this theorem to the other types of pencils of conics. Dualize.

**12.** The pairs of second points of intersection of the opposite sides of a complete quadrangle with a conic circumscribed to its diagonal triangle are in involution (Sturm, Die Lehre von den Geometrischen Verwandtschaften, Vol. I, p. 149).

**81. Scale on a conic.** The notions of a point algebra and a scale which we have developed hitherto only for the elements of one-dimensional primitive forms may also be studied to advantage on a conic. The constructions for the sum and the product of two points (numbers) on a conic are remarkably simple. As in the case on the line, let 0, 1, $\infty$ be any three arbitrary distinct points on a conic $C^2$. Regarding these as the *fundamental points* of our *scale* on the conic, the sum and the product of any two points $x$, $y$ on the conic (which are distinct from $\infty$) are defined as follows:

Definition. The conjugate of 0 in the involution on the conic having $\infty$ for a double point and $x$, $y$ for a conjugate pair is called the *sum* of the two points $x$, $y$ and is denoted by $x + y$ (fig. 94, left). The conjugate of 1 in the involution determined on the conic by the conjugate pairs 0, $\infty$ and $x$, $y$ is called the *product* of the points $x$, $y$ and is denoted by $x \cdot y$ (fig. 94, right).

It will be noted that under Assumption P this definition is entirely equivalent to the definitions of the sum and product of two points on a line, previously given (Chap. VI). To construct the point $x + y$ on the conic (fig. 94), we need only determine the center of the involution in question as the intersection of the tangent at $\infty$ with the line joining the points $x, y$. The point $x + y$ is then determined as the intersection with the conic of the line joining the center to the point 0. Similarly, to obtain the product of the points $x, y$ we determine the center of the involution as the intersection of the lines $0\infty$ and $xy$. The point $x \cdot y$ is then the intersection with the conic of the line joining this center to

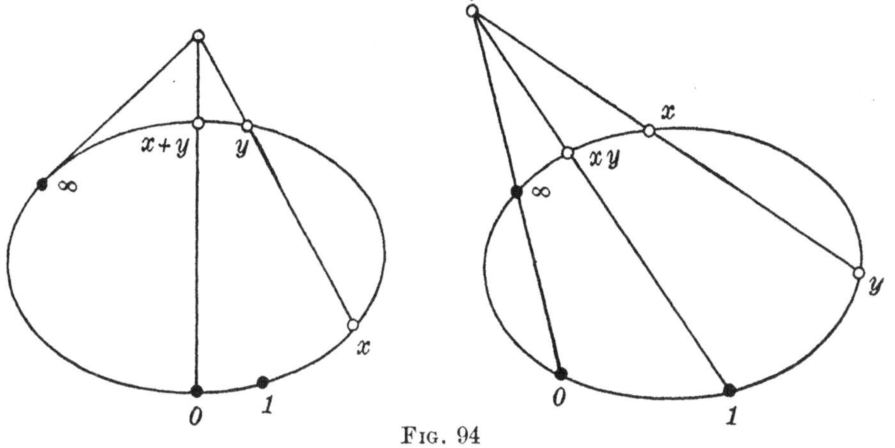


FIG. 94

the point 1. The inverse operations (subtraction and division) lead to equally simple constructions. Since the scale thus defined is obviously projective with the scale on a line, it is not necessary to derive again the fundamental properties of addition and subtraction, multiplication and division. It is clear from this consideration that *the points of a conic form a field with reference to the operations just defined.* This fact will be found of use in the analytic treatment of conics.

At this point we will make use of it to discuss the existence of the square root of a number in the field of points. It is clear from the

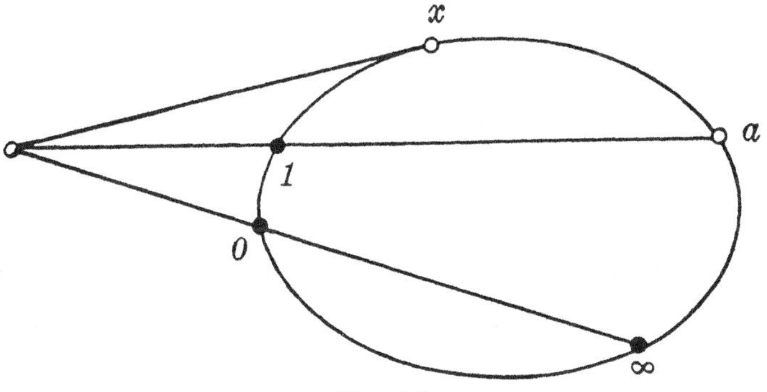


FIG. 95

preceding discussion that if a number $x$ satisfies the equation $x^2 = a$, the tangent to the conic at the point $x$ must pass through the intersection of the lines $0\infty$ and $1\,a$ (fig. 95). *A number $a$ will therefore have a square root in the field if and only if a tangent can be drawn to* *the conic from the intersection of the lines* $0\infty$ *and* $1a$; *and, conversely, if the number* $a$ *has a square root in the field, a tangent can be drawn to the conic from this point of intersection.* It follows at once that if a number $a$ has a square root $x$, it also has another which is obtained by drawing the second tangent to the conic from the point of intersection of the lines $0\infty$ and $1a$. Since this tangent meets the conic in a point which is the harmonic conjugate of $x$ with respect to $0\infty$, it follows that this second square root is $-x$. It follows also from this construction that the point 1 has the two square roots 1 and $-1$ in any field in which 1 and $-1$ are distinct, i.e. whenever $H_0$ is satisfied.

We may use these considerations to derive the following theorem, which will be used later.

THEOREM 24. *If* $AA'$, $BB'$ *are any two distinct pairs of an involution, there exists one and only one pair* $CC'$ *distinct from* $BB'$ *such that the cross ratios* $R(AA', BB')$ *and* $R(AA', CC')$ *are equal.*

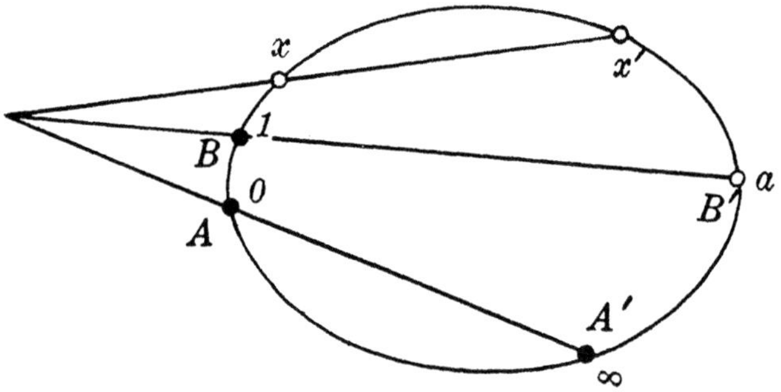


FIG. 96

*Proof.* Let the involution be taken on a conic, and let the pairs $AA'$ and $BB'$ be represented by the points $0\infty$ and $1a$ respectively (fig. 96). Let $xx'$ be any other pair of the involution. We then have, clearly from the above, $xx' = a$. Further, the cross ratios in question give

$$R(0\infty, 1a) = \frac{1}{a}, \quad R(0\infty, xx') = \frac{x}{x'}.$$

These are equal, if and only if $x' = ax$, or if $xx' = ax^2$. But this implies the relation $a = ax^2$, and since we have $a \neq 0$, this gives $x^2 = 1$. The only pair of the involution satisfying the conditions of the theorem is therefore the pair $CC' = -1, -a$.

## EXERCISES

**1.** Show that an involution which has two harmonic conjugate pairs has double points if and only if $-1$ has a square root in the field.

**2.** Show that any involution may be represented by the equation $x'x = a$.

**3.** The equation of Ex. 13, p. 230, is the condition that the lines joining the three pairs of points $aa'$, $bb'$, $cc'$ on a conic are concurrent.

**4.** Show that if the involution $x'x = a$ has a conjugate pair $bb'$ such that the cross ratio $\text{R}(0\infty, bb')$ has the value $\lambda$, the number $a\lambda$ has a square root in the field.

**82. Parametric representation of a conic.** Let a scale be established on a conic $C^2$ by choosing three distinct points of the conic as the fundamental points, say, $O = 0$, $M = \infty$, $A = 1$. Then let us establish a system of nonhomogeneous point coördinates in the plane of the conic as follows: Let the line $OM$ be the $x$-axis, with $O$ as origin and $M$ as $\infty_x$ (fig. 97). Let the tangents at $O$ and $M$ to the conic meet in a point $N$, and let the tangent $ON$ be the $y$-axis, with $N$ as $\infty_y$. Finally, let the point $A$ be the point $(1, 1)$, so that the line $AN$ meets the $x$-axis in the point for which $x = 1$, and $AM$ meets the $y$-axis in the point for which $y = 1$. Now let $P = \lambda$ be any point on the conic. The coördinates $(x, y)$ of $P$ are determined by the intersections of the lines $PN$ and $PM$ with the $x$-axis and the $y$-axis respectively. We have at once the relation

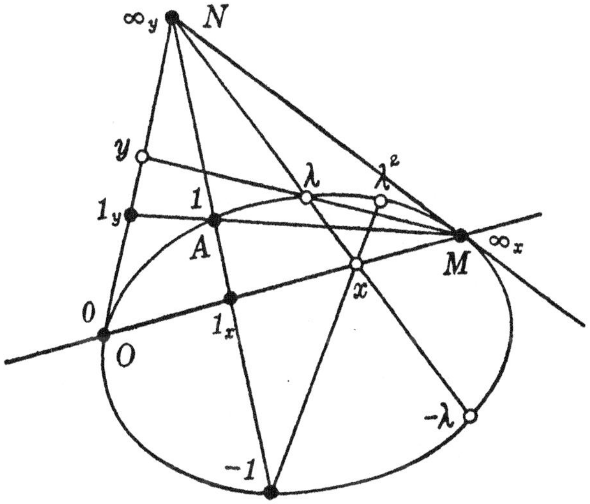

FIG. 97

$$y = \lambda,$$

since the points $0, \infty, 1, \lambda$ on the conic are perspective from $M$ with points $0, \infty, 1, y$ on the $y$-axis. To determine $x$ in terms of $\lambda$, we note, first, that from the constructions given, any line through $N$ meets the conic (if at all) in two points whose sum in the scale is 0. In particular, the points $1, -1$ on the conic are collinear with $N$ and the point 1 on the $x$-axis, and the points $\lambda, -\lambda$ on the conic are collinear with $N$ and the point $x$ on the $x$-axis. Since the latter point is also on the line joining 0 and $\infty$ on the conic, the construction for multiplication on the conic shows that any line through the point $x$ on the $x$-axis meets the conic (if at all) in two points whose product is constant, and hence equal to $-\lambda^2$. The line joining the point $x$ on the $x$-axis to the point $-1$ on the conic therefore meets the conic again in the point $\lambda^2$. But now we have $0, \infty, 1, \lambda^2$ on the conic perspective from the point $-1$ on the conic with the points $0, \infty, 1, x$ on the $x$-axis. This gives the relation

$$x = \lambda^2.$$

We may now readily express these relations in homogeneous form. If the triangle $OMN$ is taken as triangle of reference, $ON$ being $x_1 = 0$, $OM$ being $x_2 = 0$, and the point $A$ being the point $(1, 1, 1)$, we pass from the nonhomogeneous to the homogeneous by simply placing $x = x_1/x_3$, $y = x_2/x_3$. *The points of the conic $C^2$ may then be represented by the relations*

$$(1) \qquad x_1 : x_2 : x_3 = \lambda^2 : \lambda : 1.$$

This agrees with our preceding results, since the elimination of $\lambda$ between these equations gives at once

$$x_2^2 - x_1 x_3 = 0,$$

which we have previously obtained as the equation of the conic.

It is to be noted that the point $M$ on the conic, which corresponds to the value $\lambda = \infty$, is exceptional in this equation. This exceptional character is readily removed by writing the parameter $\lambda$ homogeneously $\lambda = \lambda_1 : \lambda_2$. Equations (1) then readily give

THEOREM 25. *A conic may be represented analytically by the equations $x_1 : x_2 : x_3 = \lambda_1^2 : \lambda_1\lambda_2 : \lambda_2^2$.*

This is called a *parametric representation of a conic.*

### EXERCISES

**1.** Show that the equation of the line joining two points $\lambda_1, \lambda_2$ on the conic (1) above is $x_1 - (\lambda_1 + \lambda_2)\, x_2 + \lambda_1\lambda_2 x_3 = 0$; and that the equation of the tangent to the conic at a point $\lambda_1$ is $x_1 - 2\,\lambda_1 x_2 + \lambda_1^2 x_3 = 0$. Dualize.

**2.** Show that any collineation leaving the conic (1) invariant is of the form

$x_1' : x_2' : x_3' = \alpha^2 x_1 + 2\,\alpha\beta x_2 + \beta^2 x_3 : \alpha\gamma x_1 + (\alpha\delta + \beta\gamma)\, x_2 + \beta\delta x_3 : \gamma^2 x_1 + 2\,\gamma\delta x_2 + \delta^2 x_3$. (*Hint.* Use the parametric representation of the conic and let the projectivity generated on the conic by the collineation be $\lambda_1' = \alpha\lambda_1 + \beta\lambda_2$, $\lambda_2' = \gamma\lambda_1 + \delta\lambda_2$.)

# CHAPTER IX

## GEOMETRIC CONSTRUCTIONS. INVARIANTS

**83. The degree of a geometric problem.** The specification of a line by two of its points may be regarded as a *geometric operation.** The plane dual of this operation is the specification of a point by two lines. In space we have hitherto made use of the following geometric operations: the specification of a line by two planes (this is the space dual of the first operation mentioned above); the specification of a plane by two intersecting lines (the space dual of the second operation above); the specification of a plane by three of its points or by a point and a line; the specification of a point by three planes or by a plane and a line. These operations are known as *linear* operations or operations of the *first degree,* and the elements determined by them from a set of given elements are said to be obtained by *linear constructions,* or by *constructions of the first degree.* The reason for this terminology is found in the corresponding analytic formulations. Indeed, it is at once clear that each of the two linear operations in a plane corresponds analytically to the solution of a pair of linear equations; and the linear operations in space clearly correspond to the solution of systems of three equations, each of the first degree. Any problem which can be solved by a finite sequence of linear constructions is said to be a *linear problem* or a *problem of the first degree.* Any such problem has, if determinate, one and only one solution.

In the usual representation of the ordinary real projective geometry in a plane by means of points and lines drawn, let us say, with a pencil on a sheet of paper, the linear constructions are evidently those that can be carried out by the use of a straightedge alone. There is no familiar mechanical

* An *operation* on one or more elements is defined as a correspondence whereby to the set of given elements corresponds an element of some sort (cf. § 48). If the latter element is uniquely defined by the set of given elements (in general, the order of the given elements is an essential factor of this determination), the operation is said to be *one-valued.* The operation referred to in the text is then a one-valued operation defined for any two distinct points and associating with any such pair (the order of the points is in this case immaterial) a new element, viz. a line.

device for drawing lines and planes in space. But a picture (which is the section by a plane of a projection from a point) of the lines and points of intersection of linearly constructed planes may be constructed with a straight-edge (cf. the definition of a plane).

As examples of linear problems we mention: (*a*) the determination of the point homologous with a given point in a projectivity on a line of which three pairs of homologous points are given; (*b*) the determination of the sixth point of a quadrangular set of which five points are given; (*c*) the determination of the second double point of a projectivity on a line of which one double point and two pairs of homologous points are given (this is equivalent to (*b*)); (*d*) the determination of the second point of intersection of a line with a conic, one point of intersection and four other points of the conic being given, etc.

The analytic relations existing between geometric elements offer a convenient means of classifying geometric problems.* Confining ourselves, for the sake of brevity, to problems in a plane, a geometric problem consists in constructing certain points, lines, etc., which bear given relations to a certain set of points, lines, etc., which are supposed given in advance. In fact, we may suppose that the elements sought are points only; for if a line is to be determined, it is sufficient to determine two points of this line; or if a conic is sought, it is sufficient to determine five points of this conic, etc. Similar considerations may also be applied to the given elements of the problem, to the effect that we may assume these given elements all to be points. This merely involves replacing any given elements that are not points by certain sets of points having the property of uniquely determining these elements. Confining our discussion to problems in which this is possible, any geometric problem may be reduced to one or more problems of the following form: *Given in a plane a certain finite number of points, to construct a point which shall bear to the given points certain given relations.*

In the analytic formulation of such a problem the given points are supposed to be determined by their coördinates (homogeneous or nonhomogeneous), referred to a certain frame of reference. The vertices of this frame of reference are either points contained among the given points, or some or all of them are additional points which we

* The remainder of this section follows closely the discussion given in Castelnuovo, Lezioni di geometria, Rome-Milan, Vol. I (1904), pp. 467 ff.

suppose added to the given points. The set of all given points then gives rise to a certain set of coördinates, which we will denote by $1, a, b, c, \cdots$,* and which are supposed known. These numbers together with all numbers obtainable from them by a finite number of rational operations constitute a set of numbers,

$$\mathsf{K} = [1, a, b, c, \cdots],$$

which we will call *the domain of rationality defined by the data.*† In addition to the coördinates of the known points (which, for the sake of simplicity, we will suppose given in nonhomogeneous form), the coördinates $(x, y)$ of the point sought must be considered. The conditions of the problem then lead to certain analytic relations which these coördinates $x$, $y$ and $a, b, c \cdots$ must satisfy. Eliminating one of the variables, say $y$, we obtain two equations,

$$f_1(x) = 0, \quad f_2(x, y) = 0,$$

the first containing $x$ but not $y$; the second, in general, containing both $x$ and $y$. The problem is thus replaced by two problems: the first depending on the solution of $f_1(x) = 0$ to determine the abscissa of the unknown point; the second to determine the ordinate, assuming the abscissa to be known.

In view of this fact we may confine ourselves to the discussion of problems depending on a single equation with one unknown. Such problems may be classified according to the equation to which they give rise. A problem is said to be *algebraic* if the equation on which its solution depends is algebraic, i.e. if this equation can be put in the form

$$(1) \qquad x^n + a_1x^{n-1} + a_2x^{n-2} + \cdots + a_n = 0,$$

in which the coefficients $a_1, a_2, \cdots, a_n$ are numbers of the domain of rationality defined by the data. Any problem which is not algebraic is said to be *transcendental.* Algebraic problems (which alone will be considered) may in turn be classified according to the degree $n$ of

* In case homogeneous coördinates are used, $a, b, c, \cdots$ denote the mutual ratios of the coördinates of the given elements.

† A moment's consideration will show that the points whose coördinates are numbers of this domain are the points obtainable from the data by linear constructions. Geometrically, any domain of rationality on a line may be defined as any class of points on a line which is closed under harmonic constructions; i.e. such that if $A$, $B$, $C$ are any three points of the class, the harmonic conjugate of $A$ with respect to $B$ and $C$ is a point of the class.

the equation on which their solutions depend. We have thus problems of the first degree (already referred to), depending merely on the solution of an equation of the first degree; problems of the second degree, depending on the solution of an equation of the second degree, etc.

Account must however be taken of the fact that equation (1) may be *reducible* within the domain K; in other words, that the left member of this equation may be the product of two or more polynomials whose coefficients are numbers of K. In fact, let us suppose, for example, that this equation may be written in the form

$$\phi_1(x) \cdot \phi_2(x) = 0,$$

where $\phi_1$, $\phi_2$ are two polynomials of the kind indicated, and of degrees $n_1$ and $n_2$ respectively ($n_1 + n_2 = n$). Equation (1) is then equivalent to the two equations

$$\phi_1(x) = 0, \quad \phi_2(x) = 0.$$

Then either it happens that one of these two equations, e.g. the first, furnishes all the solutions of the given problem, in which case $\phi_1$ being assumed irreducible in K, the problem is not of degree $n$, but of degree $n_1 < n$; or, both equations furnish solutions of the problem, in which case $\phi_2$ also being assumed irreducible in K, the problem reduces to two problems, one of degree $n_1$ and one of degree $n_2$. In speaking of a problem of the $n$th degree we will therefore always assume that the associated equation of degree $n$ is irreducible in the domain of rationality defined by the data. Moreover, we have tacitly assumed throughout this discussion that equation (1) has a root; we shall see presently that this assumption can always be satisfied by the introduction, if necessary, of so-called *improper elements*. It is important to note, however, since our Assumptions A, E, P do not in any way limit the field of numbers to which the coördinates of all elements of our space belong, and since equations of degree greater than one do not always have a root in a given field when the coefficients of the equation belong to this field, there exist spaces in which problems of degree higher than the first may have no solutions. Thus in the ordinary real projective geometry a problem of the second degree will have a (real) solution only if the quadratic equation on which it depends has a (real) root.

The example of a problem of the second degree given in the next section will serve to illustrate the general discussion given above.

**84. The intersection of a given line with a given conic.** Given a conic defined, let us say, by three points $A$, $B$, $C$ and the tangents at $A$ and $B$; to find the points of intersection of a given line with this conic. Using nonhomogeneous coördinates and choosing as $x$-axis one of the given tangents to the conic, as $y$-axis the line joining the points $A$ and $B$, and as the point (1, 1) the point $C$, the equation of the conic may be assumed to be of the form

$$x^2 - y = 0.$$

The equation of the given line may then be assumed to be of the form

$$y = px + q.*$$

The domain of rationality defined by the data is in this case

$$\mathsf{K} = [1, p, q].$$

The elimination of $y$ between the two equations above then leads to the equation

(1) $$x^2 - px - q = 0.$$

This equation is not in general reducible in the domain $\mathsf{K}$. The problem of determining the points of intersection of an arbitrary line in a plane with a given conic in this plane is then a problem of the second degree. If equation (1) has a root in the field of the geometry, it is clear that this root gives rise to a solution of the problem proposed; if this equation has no root in the field, the problem has no solution.

If, on the other hand, one point of intersection of the line with the conic is given, so that one root of equation (1), say $x = r$, is known, the domain given by the data is

$$\mathsf{K}' = [1, p, q, r],$$

and in this domain (1) is reducible; in fact, it is equivalent to the equation

$$(x + r - p)\ (x - r) = 0.$$

The problem of finding the remaining point of intersection then depends merely on the solution of the linear equation

$$x + r - p = 0;$$

* There is no loss in generality in assuming this form; for if in the choice of coördinates the equation of the given line were of the form $x = c$, we should merely have to choose the other tangent as $x$-axis to bring the problem into the form here assumed.

that is, the problem is of the first degree, as already noted among the examples of linear problems.

It is important to note that equation (1) is the most general form of equation of the second degree. It follows that *every problem of the second degree in a plane can be reduced to the construction of the points of intersection of an arbitrary line with a particular conic.* We shall return to this later (§ 86).

**85. Improper elements. Proposition $K_2$.** We have called attention frequently to the fact that the nature of the field of points on a line is not completely determined by Assumptions A, E, P, under which we are working. We have seen in particular that this field may be finite or infinite. The example of an analytic space discussed in the Introduction shows that the theory thus far developed applies equally well whether we assume the field of points on a line to consist of all the ordinary rational numbers, or of all the ordinary real numbers, or of all the ordinary complex numbers. According to which of these cases we assume, our space may be said to be the *ordinary rational space*, or the *ordinary real space*, or the *ordinary complex space*. Now, in the latter we know that every number has a square root. Moreover, each of the former spaces (the rational and the real) are clearly contained in the complex space as subspaces. Suppose now that our space S is one in which not every number has a square root. In such a case it is often convenient to be able to think of our space S as forming a subspace in a more extensive space S′, in which some or all of these numbers do have square roots.

We have seen that the ordinary rational and ordinary real spaces are such that they may be regarded as subspaces of a more extensive space in the number system associated with which the square root of any number always exists. In fact, they may be regarded as subspaces of the ordinary complex space which has this property. For a general field it is easy to prove that *if $a_1, a_2, \cdots, a_n$ are any finite set of elements of a field* F, *there exists a field* F′, *containing all the elements of* F, *such that each of the elements $a_1, a_2, \cdots, a_n$ is a square in* F′. This is, of course, less general than the theorem that a field F′ exists in which every element of F is a square, but it is sufficiently general for many geometric purposes. In the presence of Assumptions A, E, P, $H_0$ it is equivalent (cf. § 54) to the following statement:

PROPOSITION $K_2$. *If any finite number of involutions are given in a space* S *satisfying Assumptions* A, E, P, *there exists a space* S′ *of which* S *is a subspace,* such that all the given involutions have double points in* S′.

A proof of this theorem will be found at the end of the chapter. The proposition is, from the analytic point of view, that the domain of rationality determined by a quadratic problem may be extended so as to include solutions of that problem. The space S′ may be called an *extended space.* The elements of S may be called *proper elements,* and those of S′ which are not in S may be called *improper.* A projective transformation which changes every proper element into a proper element is likewise a *proper transformation;* one which transforms proper elements into improper elements, on the other hand, is called an *improper transformation.* Taking Proposition $K_2$ for the present as an assumption like A, E, P, and $H_0$, and noting that it is consistent with these other assumptions because they are all satisfied by the ordinary complex space, we proceed to derive some of its consequences.

THEOREM 1. *A proper one-dimensional projectivity without proper double elements may always be regarded in an extended space as having two improper double elements.* (A, E, P, $H_0$, $K_2$)†

*Proof.* Suppose the projectivity given on a conic. If the involution which belongs to this projectivity had two proper double points, they would be the intersections of the axis of the projectivity with the conic, and hence the given projectivity would have proper double points. Let S′ be the extended space in which ($K_2$) the involution has double points. There are then two points of S′ in which the axis of the projectivity meets the conic, and these are, by Theorem 20, Chap. VIII, the double points of the given projectivity.

COROLLARY 1. *If a line does not meet a conic in proper points, it may be regarded in an extended space as meeting it in two improper points.* (A, E, P, $H_0$, $K_2$)

COROLLARY 2. *Every quadratic equation with proper coefficients has two roots which, if distinct, are both proper or both improper.* (A, E, P, $H_0$, $K_2$)

* We use the word *subspace* to mean any space, every point of which is a point of the space of which it is a subspace. With this understanding the subspace may be identical with the space of which it is a subspace. The ordinary complex space then satisfies Proposition $K_2$.

† Cf. Ex., p. 261.

For the double points of any projectivity satisfy an equation of the form $cx^2 + (d - a)x - b = 0$ (Theorem 11, Cor. 4, Chap. VI), and any quadratic equation may be put into this form.

THEOREM 2. *Any two involutions in the same one-dimensional form have a conjugate pair in common, which may be proper or improper.* (A, E, P, $H_0$, $K_2$)

This follows at once from the preceding and Theorem 23, Chap. VIII.

COROLLARY. *In any involution there exists a conjugate pair, proper or improper, which is harmonic with any given conjugate pair.* (A, E, P, $H_0$, $K_2$)

For the involution which has the given pair for double elements has (by the theorem) a pair, proper or improper, in common with the given involution. The latter pair satisfies the condition of the theorem (Theorem 27, Cor. 1, Chap. IV).

We have seen earlier (Theorem 4, Cor., Chap. VIII) that any two involutions with double points are conjugate. Under Proposition $K_2$ we may remove the restriction and say that *any two involutions are conjugate in an extended space dependent on the two involutions.* If the involutions are on coplanar lines, we have the following:

THEOREM 3. *Two involutions on distinct lines in the same plane are perspective (the center of perspectivity being proper or improper), provided the point of intersection of the lines is a double point for both or for neither of the involutions.* (A, E, P, $K_2$)

*Proof.* If the point of intersection $O$ of the two lines be a double point of each of the involutions, let $Q$ and $R$ be an arbitrary pair of one involution and $Q'$ and $R'$ an arbitrary pair of the other involution. The point of intersection of the lines $QQ'$ and $RR'$ is then a center of a perspectivity which transforms elements which determine the first involution into elements which determine the second. If the point $O$ is a double point of neither of the two involutions, let $M$ be a double point of one and $M'$ of the other (these double points are proper or else exist in an extended space $S'$ which exists by Proposition $K_2$). Also let $N$ and $N'$ be the conjugates of $O$ in the two involutions. Then by the same argument as before, the point of intersection of the lines $MM'$, $NN'$ may be taken as the center of the perspectivity.

It was proved in § 66, Chap. VII, that the equation of any point conic is of the form

(1) $a_{11}x_1^2 + a_{22}x_2^2 + a_{33}x_3^2 + 2\,a_{12}x_1x_2 + 2\,a_{13}x_1x_3 + 2\,a_{23}x_2x_3 = 0$;

but it was not shown that every equation of this form represents a conic. The line $x_1 = 0$ contains the point $(0, x_2, x_3)$ satisfying (1), provided the ratio $x_2 : x_3$ satisfies the quadratic equation

$$a_{22}x_2^2 + 2\,a_{23}x_2x_3 + a_{33}x_3^2 = 0.$$

Similarly, the lines $x_2 = 0$ and $x_3 = 0$ contain points of the locus defined by (1), provided two other quadratic equations are satisfied. By Proposition $K_2$ there exists an extended space in which these three quadratic equations are solvable. Hence (1) is satisfied by the coördinates of at least two distinct points $P$, $Q$ (proper or improper).*

A linear transformation

$$(2)\qquad \begin{aligned} \rho x_1' &= b_{11}x_1 + b_{12}x_2 + b_{13}x_3 \\ \rho x_2' &= b_{21}x_1 + b_{22}x_2 + b_{23}x_3 \\ \rho x_3' &= b_{31}x_1 + b_{32}x_2 + b_{33}x_3 \end{aligned}$$

evidently transforms the points satisfying (1) into points satisfying another equation of the second degree. If, then, (2) is so chosen as to transform $P$ and $Q$ into the points (0, 0, 1) and (0, 1, 0) respectively, (1) will be transformed into an equation which is satisfied by the latter pair of points, and which is therefore of the form

$$(3)\qquad ax_1^2 + c_1x_2x_3 + c_2x_1x_3 + c_3x_1x_2 = 0.$$

If $c_1 = 0$, the points satisfying (3) lie on the two lines

$$x_1 = 0, \quad ax_1 + c_2x_3 + c_3x_2 = 0;$$

and hence (1) is satisfied by the points on the lines into which these lines are transformed by the inverse of (2). If $c_1 \neq 0$, the transformation

$$(4)\qquad \begin{aligned} x_1 &= x_1' \\ x_2 &= -\frac{c_2}{c_1}x_1' + x_2' \\ x_3 &= x_3' \end{aligned}$$

* Proposition $K_2$ has been used merely to establish the existence of points satisfying (1). In case there are proper points satisfying (1), the whole argument can be made without $K_2$.

transforms the points $(x_1, x_2, x_3)$ satisfying (3) into points $(x_1', x_2', x_3')$ satisfying

$$(5) \qquad \left(a - \frac{c_2 c_3}{c_1}\right) x_1'^2 + (c_1 x_3' + c_3 x_1')\, x_2' = 0.$$

But (5) is in the form which was proved in Theorem 7, Chap. VII, to be the equation of a conic. As the points which satisfy (5) are transformed by the inverse of the product of the collineations (2) and (4) into points which satisfy (1), we see that in all cases (1) represents a point conic (proper or improper, degenerate or nondegenerate).

This gives rise to the two following dual theorems:

THEOREM 4. *Every equation of the form*

$$a_{11}x_1^2 + a_{22}x_2^2 + a_{33}x_3^2 + 2\, a_{12}x_1x_2 + 2\, a_{13}x_1x_3 + 2\, a_{23}x_2x_3 = 0$$

*represents a point conic (proper or improper) which may, however, degenerate; and, conversely, every point conic may be represented by an equation of this form.* (A, E, P, $H_0$, $K_2$)

THEOREM 4'. *Every equation of the form*

$$A_{11}u_1^2 + A_{22}u_2^2 + A_{33}u_3^2 + 2\, A_{12}u_1u_2 + 2\, A_{13}u_1u_3 + 2\, A_{23}u_2u_3 = 0$$

*represents a line conic (proper or improper) which may, however, degenerate; and, conversely, every line conic may be represented by an equation of this form.* (A, E, P, $H_0$, $K_2$)

**86. Problems of the second degree.** We have seen in § 83 that any problem of the first degree can be solved completely by means of linear constructions; but that a problem of degree higher than the first cannot be solved by linear constructions alone. In regard to problems of the second degree in a plane, however, it was seen in § 84 that any such problem may be reduced to the problem of finding the points of intersection of an arbitrary line in the plane with a particular conic in the plane. This result we may state in the following form:

THEOREM 5. *Any problem of the second degree in a plane may be solved by linear constructions if the intersections of every line in the plane with a single conic in this plane are assumed known.* (A, E, P, $H_0$, $K_2$)

In the usual representation of the projective geometry of a real plane by means of points, lines, etc., drawn with a pencil, say, on a sheet of paper, the linear constructions, as has already been noted, are those that can be performed with the use of a straightedge alone. It will be shown later that any conic in the real geometry is equivalent projectively to a circle. The instrument usually employed to draw circles is the compass. It is then clear that in this representation *any problem of the second degree can be solved by means of a straightedge and compass alone.* The theorem just stated, however, shows that if *a single circle* is drawn once for all in the plane, the straightedge alone suffices for the solution of any problem of the second degree in this plane. The discussion immediately following serves to indicate briefly how this may be accomplished.

We proceed to show how this theorem may be used in the solution of problems of the second degree. Any such problem may be reduced more or less readily to the first of the following:

PROBLEM 1. *To find the double points of a projectivity on a line of which three pairs of homologous points are given.* We may assume

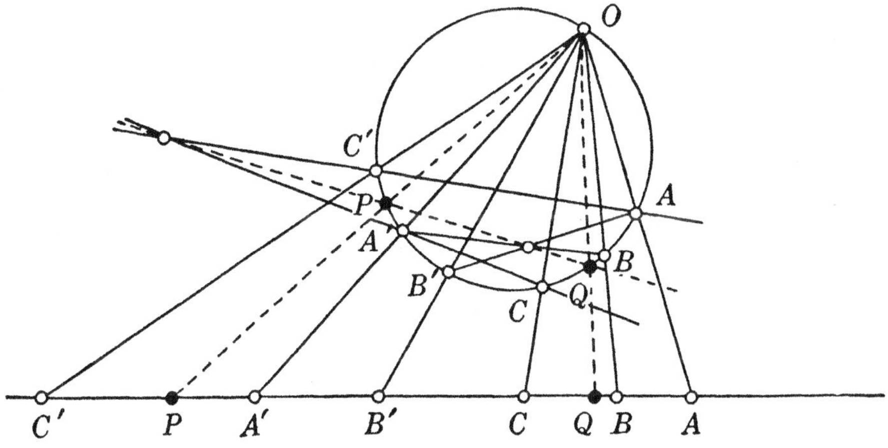


FIG. 98

that the given pairs of homologous points all consist of distinct points (otherwise the problem is linear). In accordance with Theorem 5, we suppose given a conic (in a plane with the line) and assume known the intersections of any line of the plane with this conic. Let $O$ be any point of the given conic, and with $O$ as center project the given pairs of homologous points on the conic (fig. 98). These define a projectivity on the conic. Construct the axis of this projectivity and let it meet the conic in the points $P, Q$. The lines $OP$, $OQ$ then meet the given line in the required double points.

PROBLEM 2. *To find the points of intersection of a given line with a conic of which five points are given.* Let $A, B, C, D, E$ be the given points of the conic. The conic is then defined by the projectivity $D(A, B, C) \barwedge E(A, B, C)$ between the pencils of lines at $D$ and $E$.

This projectivity gives rise to a projectivity on the given line of which three pairs of homologous points are known. The double points of the latter projectivity are the points of intersection of the line with the conic. The problem is thus reduced to Problem 1.

PROBLEM 3. We have seen that it is possible for two triangles in a plane to be perspective from four different centers (cf. Ex. 8, p. 105). The maximum number of ways in which it is conceivable that two triangles may be perspective is clearly equal to the number of permutations of three things three at a time, i.e. six. The question then arises, *Is it possible to construct two triangles that are perspective from six different centers?* Let the two triangles be $ABC$ and $A'B'C'$, and let

$$x_1 = 0, \quad x_2 = 0, \quad x_3 = 0$$

be the sides of the first opposite to $A$, $B$, $C$ respectively. Let the sides of the second opposite to $A'$, $B'$, $C'$ respectively be

$$x_1 + x_2 + x_3 = 0, \quad x_1 + k'x_2 + k''x_3 = 0, \quad x_1 + l'x_2 + l''x_3 = 0.$$

The condition for $ABC \underset{\wedge}{=} A'B'C'$ is that the points of intersection of corresponding sides be collinear, i.e.

$$(1) \qquad \begin{vmatrix} 0 & 1 & -1 \\ -k'' & 0 & 1 \\ -l' & 1 & 0 \end{vmatrix} = k'' - l' = 0.$$

In like manner, the condition for $BCA \underset{\wedge}{=} A'B'C'$ is

$$(2) \qquad \begin{vmatrix} 0 & -l'' & l' \\ -1 & 0 & 1 \\ -k' & 1 & 0 \end{vmatrix} = k'l'' - l' = 0.$$

From these two conditions follows

$$\begin{vmatrix} 0 & -k'' & k' \\ -l'' & 0 & 1 \\ 1 & -1 & 0 \end{vmatrix} = k'l'' - k'' = 0,$$

which is the condition for $CAB \underset{\wedge}{=} A'B'C'$. Hence, *if two triangles are in the relations $ABC \underset{\wedge}{=} A'B'C'$ and $BCA \underset{\wedge}{=} A'B'C'$, they are also in the relation $CAB \underset{\wedge}{=} A'B'C'$.* Two triangles in this relation are said to be *triply perspective* (cf. Ex. 2, p. 100). The domain of rationality defined by the data of our problem is clearly

$$\mathsf{K} = [1].$$

Since numbers in this domain may be found which satisfy equations (1) and (2), the problem of constructing two triply perspective triangles is linear.

The condition for $ACB \underset{\wedge}{=} A'B'C'$ is

$$(3) \qquad k' - l'' = 0.$$

If relations (1), (2), and (3) are satisfied, the triangles will be perspective from four centers. Let $k$ be the common value of $k'$ and $l''$ (3), and let $l$ be the common value of $l'$ and $k''$ (1). Relation (2) then gives the condition $k^2 - l = 0$. The relations

$$k' = l'' = k, \quad l' = k'' = k^2$$

then define two quadruply perspective triangles. The problem of constructing two such triangles is therefore still linear.

If now we add the condition for $CBA \underset{\wedge}{=} A'B'C'$, the two triangles will, by what precedes, be perspective from six different centers. The latter condition is

$$(4) \qquad k''l' - l'' = 0.$$

With the preceding conditions (1), (2), (3) and the notation adopted above, this leads to the condition

$$k^3 = l^3 = 1.$$

The equation $k^3 - 1 = 0$ is, however, reducible in K; indeed, it is equivalent to

$$k - 1 = 0, \quad k^2 + k + 1 = 0.$$

The first of these equations leads to the condition that $A'$, $B'$, $C'$ are collinear, and does not therefore give a solution of the problem. The problem of constructing two triangles that are sextuply perspective is therefore of the second degree. The equation

$$k^2 + k + 1 = 0$$

has two roots $w$, $w^2$ (proper or improper and, in general,* distinct). Hence our problem has two solutions. One of these consists of the triangles

$$x_1 = 0, \quad x_2 = 0, \quad x_3 = 0;$$
$$x_1 + x_2 + x_3 = 0, \quad x_1 + wx_2 + w^2x_3 = 0, \quad x_1 + w^2x_2 + wx_3 = 0.$$

* They can coincide only if the number system is such that $1 + 1 + 1 = 0$; e.g. in a finite space involving the modulus 3.

Two of the sides of the second triangle may be improper.* The points of intersection of the sides of one of these triangles with the sides of the other are the following nine points:

$$(5)\quad \begin{matrix} (\ 0, & -1, & 1) & (\ 0, & w^2, & -w) & (\ 0, & w, & -w^2) \\ (-1, & 0, & 1) & (-w^2, & 0, & 1\ ) & (-w, & 0, & 1\ ) \\ (-1, & 1, & 0) & (\ w, & -1, & 0\ ) & (\ w^2, & -1, & 0\ ) \end{matrix}$$

They form a configuration

| | |
|---|---|
| 9 | 4 |
| 3 | 12 |

which contains four configurations

| | |
|---|---|
| 9 | 3 |
| 3 | 9 |

of the kind studied in § 36, Chap. IV. All triples of points in the same row or column or term of the determinant expansion of their matrix are collinear.† If one line is omitted from a finite plane (in the sense of § 72, Chap. VII) having four points on each line, the remaining nine points and twelve lines are isomorphic with this configuration.

## EXERCISES

*The problems in a plane given below that are of the second degree are to be solved by linear constructions, with the assumption that the points of intersection of any line in the plane with a given fixed conic in the plane are known; i.e. "with a straight-edge and a given circle in the plane."*

**1.** Construct the points of intersection of a given line with a conic determined by (i) four points and a tangent through one of them; (ii) three points and the tangents through two of them; (iii) five tangents.

**2.** Construct the conjugate pair common to two involutions on a line.

**3.** Given a conic determined by five points, construct a triangle inscribed in this conic whose sides pass through three given points of the plane.

* It may be noted that in the ordinary real geometry two sides of the second triangle are necessarily improper, so that in this geometry our problem has no *real* solution.

† They all lie on any cubic curve of the form $x_1^3 + x_2^3 + x_3^3 + 3\lambda x_1x_2x_3 = 0$ for any value of $\lambda$, and are, in fact, the points of inflexion of the cubic. This configuration forms the point of departure for a variety of investigations leading into many different branches of mathematics.

**4.** Given a triangle $A_2B_2C_2$ inscribed in a triangle $A_1B_1C_1$. In how many ways can a triangle $A_3B_3C_3$ be inscribed in $A_2B_2C_2$ and circumscribed to $A_1B_1C_1$? Show that in one case, in which one vertex of $A_3B_3C_3$ may be chosen arbitrarily, the problem is linear (cf. § 36, Chap. IV); and that in another case the problem is quadratic. Show that this problem gives all configurations of the symbol $\boxed{\begin{matrix} 9 & 3 \\ 3 & 9 \end{matrix}}$. Give the constructions for all cases (cf. S. Kantor, Sitzungsberichte der mathematisch-naturwissenschaftlichen Classe der Kaiserlichen Akademie der Wissenschaften zu Wien, Vol. LXXXIV (1881), p. 915).

**5.** If opposite vertices of a simple plane hexagon $P_1P_2P_3P_4P_5P_6$ are on three concurrent lines, and the lines $P_1P_2$, $P_3P_4$, $P_5P_6$ are concurrent, then the lines $P_2P_3$, $P_4P_5$, $P_6P_1$ are also concurrent, and the figure thus formed is a configuration of Pappus.

**6.** Show how to construct a simple $n$-point inscribed in a given simple $n$-point and circumscribed to another given simple $n$-point.

**7.** Show how to inscribe in a given conic a simple $n$-point whose sides pass respectively through $n$ given points.

**8.** Construct a conic through four points and tangent to a line not meeting any of the four points.

**9.** Construct a conic through three points and tangent to two lines not meeting any of the points.

**10.** Construct a conic through four given points and meeting a given line in two points harmonic with two given points on the line.

**11.** If $A$ is a given point of a conic and $X$, $Y$ are two variable points of the conic such that $AX$, $AY$ always pass through a conjugate pair of a given involution on a line $l$, the line $XY$ will always pass through a fixed point $B$. The line $AB$ and the tangent to the conic at $A$ pass through a conjugate pair of the given involution.

**12.** Given a collineation in a plane and a line which does not contain a fixed point of the collineation; show that there is one and only one point on the line which is transformed by the collineation into another point on the line.

**13.** Given four skew lines, show that there are in general two lines which meet each of the given four lines; and that if there are three such lines, there is one through every point on one of the lines.

**14.** Given in a plane two systems of five points $A_1A_2A_3A_4A_5$ and $B_1B_2B_3B_4B_5$; given also a point $X$ in the plane, determine a point $Y$ such that we have $X(A_1A_2A_3A_4A_5) \barwedge Y(B_1B_2B_3B_4B_5)$. In general, there is one and only one such point $Y$. Under what condition is there more than one? (R. Sturm, Mathematische Annalen, Vol. I (1869), p. 533.*)

* This is a special case of the so-called problem of projectivity. For references and a systematic treatment see Sturm, Die Lehre von den geometrischen Verwandtschaften, Vol. I, p. 348.

**87. Invariants of linear and quadratic binary forms.** An expression of the form $a_1x_1 + a_2x_2$ is called a *linear binary form* in the two variables $x_1$, $x_2$. The word *linear* refers to the degree in the variables, the word *binary* to the number (two) of the variables. A convenient notation for such a form is $a_x$. The equation

$$a_x = a_1x_1 + a_2x_2 = 0$$

defines a unique element $A$ of a one-dimensional form in which a scale has been established, viz. the element whose homogeneous coördinates are $(x_1, x_2) = (a_2, -a_1)$. If $b_x = b_1x_1 + b_2x_2$ is another linear binary form determining the element $B$, say, the question arises as to the condition under which the two elements $A$ and $B$ coincide. This condition is at once obtained as the vanishing of the determinant $\Delta$ formed by the coefficients of the two forms; i.e. the elements $A$ and $B$ will coincide if and only if we have

$$\Delta = \begin{vmatrix} a_1 & a_2 \\ b_1 & b_2 \end{vmatrix} = 0.$$

Now suppose the two elements $A$ and $B$ are subjected to any projective transformation $\Pi$:

$$\Pi: \begin{aligned} x_1 &= \alpha x_1' + \beta x_2', \\ x_2 &= \gamma x_1' + \delta x_2', \end{aligned} \qquad \begin{vmatrix} \alpha & \beta \\ \gamma & \delta \end{vmatrix} \neq 0.$$

The forms $a_x$ and $b_x$ will be transformed into two forms $a'_{x'}$ and $b'_{x'}$ respectively, which, when equated to 0, define the points $A'$, $B'$ into which the points $A$, $B$ are transformed by $\Pi$. The coefficients of the forms $a'_{x'}$, $b'_{x'}$ in terms of those of $a_x$, $b_x$ are readily calculated as follows:

$$\begin{aligned} a_1x_1 + a_2x_2 &= a_1(\alpha x_1' + \beta x_2') + a_2(\gamma x_1' + \delta x_2') \\ &= (\alpha a_1 + \gamma a_2)\,x_1' + (\beta a_1 + \delta a_2)\,x_2', \end{aligned}$$

which gives

$$a_1' = \alpha a_1 + \gamma a_2, \quad a_2' = \beta a_1 + \delta a_2.$$

Similarly, we find

$$b_1' = \alpha b_1 + \gamma b_2, \quad b_2' = \beta b_1 + \delta b_2.$$

Now it is clear that if the elements $A$, $B$ coincide, so also will the new elements $A'$, $B'$ coincide. If we have $\Delta = 0$, therefore we should also have $\Delta' = \begin{vmatrix} a_1' & a_2' \\ b_1' & b_2' \end{vmatrix} = 0$. That this is the case is readily verified. We have

$$\Delta' = \begin{vmatrix} \alpha a_1 + \gamma a_2 & \beta a_1 + \delta a_2 \\ \alpha b_1 + \gamma b_2 & \beta b_1 + \delta b_2 \end{vmatrix} = \begin{vmatrix} a_1 & a_2 \\ b_1 & b_2 \end{vmatrix} \cdot \begin{vmatrix} \alpha & \beta \\ \gamma & \delta \end{vmatrix},$$

by a well-known theorem in determinants. This relation may also be written

$$\Delta' = \begin{vmatrix} \alpha & \beta \\ \gamma & \delta \end{vmatrix} \cdot \Delta.$$

The determinant $\Delta$ is then a function of the coefficients of the forms $a_x$, $b_x$, with the property that, if the two forms are subjected to a linear homogeneous transformation of the variables (with nonvanishing determinant), the same function of the coefficients of the new forms is equal to the function of the coefficients of the old forms multiplied by an expression which is a function of the coefficients of the transformation only. Such a function of the coefficients of two forms is called a (simultaneous) *invariant* of the forms.

Suppose, now, we form the product $a_x \cdot b_x$ of the two forms $a_x$, $b_x$. If multiplied out, this product is of the form

$$a_x^2 = a_{11}x_1^2 + 2\,a_{12}x_1x_2 + a_{22}x_2^2.$$

Any such form is called a *quadratic binary form*. Under Proposition $K_2$ every such form may be factored into two linear factors (proper or improper), and hence any such form represents two elements (proper or improper) of a one-dimensional form. These two elements will coincide, if and only if the discriminant $D_a \doteq a_{12}^2 - a_{11} \cdot a_{22}$ of the quadratic form vanishes. The condition $D_a = 0$ therefore expresses a property which is invariant under any projectivity. If, then, the form $a_x^2$ be subjected to a projective transformation, the discriminant $D_{a'}$ of the new form $a_x'$ must vanish whenever $D_a$ vanishes. There must accordingly be a relation of the form $D_{a'} = k \cdot D_a$. If $a_x^2$ be subjected to the transformation $\Pi$ given above, the coefficients $a_{11}'$, $a_{12}'$, $a_{22}'$ of the new form $a_x'$ are readily found to be

$$(1) \qquad \begin{aligned} a_{11}' &= a_{11}\alpha^2 + 2\,a_{12}\alpha\gamma + a_{22}\gamma^2, \\ a_{12}' &= a_{11}\alpha\beta + a_{12}(\alpha\delta + \beta\gamma) + a_{22}\gamma\delta, \\ a_{22}' &= a_{11}\beta^2 + 2\,a_{12}\beta\delta + a_{22}\delta^2. \end{aligned}$$

By actual computation the reader may then verify the relation

$$D_{a'} = a_{12}'^2 - a_{11}'a_{22}' = \begin{vmatrix} \alpha & \beta \\ \gamma & \delta \end{vmatrix}^2 \cdot (a_{12}^2 - a_{11}a_{22}) = (\alpha\delta - \beta\gamma)^2 \cdot D_a.$$

The discriminant $D_a$ of a quadratic form $a_x^2$ is therefore called an *invariant* of the form.

Suppose, now, we consider two binary quadratic forms

$$a_x^2 = a_{11}x_1^2 + 2\,a_{12}x_1x_2 + a_{22}x_2^2,$$
$$b_x^2 = b_{11}x_1^2 + 2\,b_{12}x_1x_2 + b_{22}x_2^2.$$

Each of these (under $K_2$) represents a pair of points (proper or improper). Let us seek the condition that these two pairs be harmonic. This property is invariant under projective transformations; we may therefore expect the condition sought to be an invariant of the two forms. We know that if $a_1$, $a_2$ are the *nonhomogeneous* coördinates of the two points represented by $a_x^2 = 0$, we have relations

$$a_1 \cdot a_2 = \frac{a_{22}}{a_{11}}, \quad a_1 + a_2 = -\frac{2\,a_{12}}{a_{11}},$$

with similar relations for the nonhomogeneous coördinates $b_1$, $b_2$ of the points represented by $b_x^2 = 0$. The two pairs of points $a_1$, $a_2$; $b_1$, $b_2$ will be harmonic if we have (Theorem 13, Cor. 2, Chap. VI)

$$\frac{a_1 - b_1}{a_1 - b_2} \cdot \frac{a_2 - b_2}{a_2 - b_1} = -1.$$

This relation may readily be changed into the following:

$$a_1a_2 + b_1b_2 - \tfrac{1}{2}(a_1 + a_2)(b_1 + b_2) = 0,$$

which, on substituting from the relations just given, becomes

$$D_{ab} = a_{11}b_{22} + a_{22}b_{11} - 2\,a_{12}b_{12} = 0.$$

This is the condition sought. If we form the same function of the coefficients of the two forms $a_x'^2$, $b_x'^2$ obtained from $a_x^2$, $b_x^2$ by subjecting them to the transformation $\Pi$, and substitute from equations (1), we obtain the relation

$$D_{a'b'} = (\alpha\delta - \beta\gamma)^2 \cdot D_{ab}.$$

In the three examples of invariants of binary forms thus far obtained, the function of the new coefficients was always equal to the function of the old coefficients multiplied by a power of the determinant of the transformation. This is a general theorem regarding invariants to which we shall refer again in § 90, when a formal definition of an invariant will be given. Before closing this section, however, let us consider briefly the cross ratio R $(a_1a_2, b_1b_2)$ of the two pairs of points represented by $a_x^2 = 0$, $b_x^2 = 0$. This cross ratio is entirely unchanged when the two forms are subjected to a projective transformation. If, therefore, this cross ratio be calculated in terms of the coefficients of the two forms, the resulting function of the coefficients must be exactly equal to the same function of the coefficients of the forms $a'_x$, $b'_x$; the power of the determinant referred to above is in this case zero. Such an invariant is called an *absolute invariant;* for purposes of distinction the invariants which when transformed are multiplied by a power $\neq 0$ of the determinant of the transformation are then called *relative invariants.*

## EXERCISES

**1.** Show that the cross ratio $\text{R}(a_1a_2,\ b_1b_2)$ referred to at the end of the last section is

$$\text{R}(a_1a_2,\ b_1b_2) = \frac{D_{ab} + 2\sqrt{D_aD_b}}{D_{ab} - 2\sqrt{D_aD_b}};$$

and hence show, by reference to preceding results, that it is indeed an absolute invariant.

**2.** Given three pairs of points defined by the three binary quadratic forms $a_x^2 = 0$, $b_x^2 = 0$, $c_x^2 = 0$; show that the three will be in involution if we have

$$\begin{vmatrix} a_{11} & a_{12} & a_{22} \\ b_{11} & b_{12} & b_{22} \\ c_{11} & c_{12} & c_{22} \end{vmatrix} = 0.$$

Hence show that the above determinant is a simultaneous invariant of the three forms (cf. Ex. 13, p. 230).

**88. Proposition $\text{K}_n$.** If we form the product of $n$ linear binary forms $a_x \cdot a'_x \cdot a''_x \cdot \quad \cdots a_x^{(n-1)}$, we obtain an expression of the form

$$a_x^n = a_0x_1^n + na_1x_1^{n-1}x_2 + \frac{n(n-1)}{2}a_2x_1^{n-2}x_2^2 + \cdots + na_{n-1}x_1x_2^{n-1} + a_nx_2^n.$$

An expression of this form is called a *binary homogeneous form or quantic of the nth degree.* If it is obtained as the product of $n$ linear forms, it will represent a set of $n$ points on a line (or a set of $n$ elements of some one-dimensional form).

If it is of the second degree, we have, by Proposition $\text{K}_2$, that there exists an extended space in which it represents a pair of points. At the end of this chapter there will be proved the following generalization of $\text{K}_2$:

PROPOSITION $K_n$. *If $a_x^k$, $a_x^l$, $\cdots$ are a finite number of binary homogeneous forms whose coefficients are proper in a space* S *which satisfies Assumptions* A, E, P, *there exists a space* S′, *of which* S *is a subspace, in the number system of which each of these forms is a product of linear factors.*

As in § 85, S′ is called an *extended space*, and elements in S′ but not in S are called *improper elements.* Proposition $K_n$ thus implies that an equation of the form $a_x^n = 0$ can always be thought of as representing $n$ (distinct or partly coinciding) improper points in an extended space in case it does not represent any proper points.

Proposition $K_n$ could be introduced as an (not independent) assumption in addition to A, E, P, and $H_0$. Its consistency with the other assumptions would be shown by the example of the ordinary complex space in which it is equivalent to the fundamental theorem of algebra.

**89. Taylor's theorem. Polar forms.** It is desirable at this point to borrow an important theorem from elementary algebra.

DEFINITION. Given a term $Ax_i^n$ of any polynomial, the expression $nAx_i^{n-1}$ is called the *derivative of $Ax_i^n$ with respect to $x_i$* in symbols

$$\frac{\partial}{\partial x_i} A\, x_i^n = nAx_i^{n-1}.$$

The *derivative of a polynomial* with respect to $x_i$ is, by definition, the sum of the derivatives of its respective terms.

This definition gives at once $\frac{\partial}{\partial x_i} A = 0$, if $A$ is independent of $x_i$. Applied to a term of a binary form it gives

$$\frac{\partial}{\partial x_1} kx_1^n x_2^m = nkx_1^{n-1}x_2^m, \quad \frac{\partial}{\partial x_2} kx_1^n x_2^m = mkx_1^n x_2^{m-1}.$$

With this definition it is possible to derive *Taylor's theorem* for the expansion of a polynomial. *We state it for a binary form as follows:

Given the binary form

$$f(x_1, x_2) = a_x^n = a_0 x_1^n + na_1 x_1^{n-1} x_2 + \frac{n(n-1)}{2} a_2 x_1^{n-2} x_2^2 + \cdots + na_{n-1} x_1 x_2^{n-1} + a_n x_2^n.$$

* For the proof of this theorem on the basis of the definition just given, cf. Fine, College Algebra, pp. 460–462.

If herein we substitute for $x_1$, $x_2$ respectively the expressions $x_1 + \lambda y_1$, $x_2 + \lambda y_2$, we obtain,

$$f(x_1 + \lambda y_1, x_2 + \lambda y_2) = f(x_1, x_2) + \lambda\left(y_1 \frac{\partial}{\partial x_1} + y_2 \frac{\partial}{\partial x_2}\right) f(x_1, x_2)$$
$$+ \frac{\lambda^2}{2!}\left(y_1 \frac{\partial}{\partial x_1} + y_2 \frac{\partial}{\partial x_2}\right)^2 f(x_1, x_2)$$
$$+ \cdots + \frac{\lambda^n}{n!}\left(y_1 \frac{\partial}{\partial x_1} + y_2 \frac{\partial}{\partial x_2}\right)^n f(x_1, x_2).$$

Here the parentheses are *differential operators*. Thus

$$\left(y_1 \frac{\partial}{\partial x_1} + y_2 \frac{\partial}{\partial x_2}\right)^2 f = y_1^2 \frac{\partial^2 f}{\partial x_1^2} + 2\, y_1 y_2 \frac{\partial^2 f}{\partial x_2 \partial x_1} + y_2^2 \frac{\partial^2 f}{\partial x_2^2},$$

where $\frac{\partial^2 f}{\partial x_1^2}$ means $\frac{\partial}{\partial x_1}\left[\frac{\partial f}{\partial x_1}\right]$, $\frac{\partial^2 f}{\partial x_2 \partial x_1}$ means $\frac{\partial}{\partial x_2}\left[\frac{\partial f}{\partial x_1}\right]$, etc. It is readily proved for any term of a polynomial (and hence for the polynomial itself) that the value of such a higher derivative as $\partial^2 f/\partial x_2 \partial x_1$ is independent of the order of differentiation; i.e. that we have

$$\frac{\partial^2 f}{\partial x_2 \partial x_1} = \frac{\partial^2 f}{\partial x_1 \partial x_2}.$$

DEFINITION. The coefficient of $\lambda$ in the above expansion, viz. $y_1 \partial f/\partial x_1 + y_2 \partial f/\partial x_2$ is called the *first polar form of* $(y_1, y_2)$ *with respect to* $f(x_1, x_2)$; the coefficient of $\lambda^2$ is called the *second*; the coefficient of $\lambda^n$ is called the *nth polar form of* $(y_1, y_2)$ *with respect to the form* $f$. If any polar form be equated to 0, it represents a set of points which is called the *first, second, $\cdots$, nth polar of the point* $(y_1, y_2)$ *with respect to the set of points represented by* $f(x_1, x_2) = 0$.

Consider now a binary form $f(x_1, x_2) = 0$ and the effect upon it of a projective transformation

$$\Pi: \begin{array}{l} x_1' = \alpha x_1 + \beta x_2, \\ x_2' = \gamma x_1 + \delta x_2. \end{array} \qquad (\alpha\delta - \beta\gamma \neq 0)$$

If we substitute these values in $f(x_1, x_2)$, we obtain a new form $F(x_1', x_2')$. A point $(x_1, x_2)$ represented by $f(x_1, x_2) = 0$ will be transformed into a point $(x_1', x_2')$ represented by the form $F(x_1', x_2') = 0$. Moreover, if the point $(y_1, y_2)$ be subjected to the same projectivity, it is evident from the nature of the expansion given above that the polars of $(y_1, y_2)$ with respect to $f(x_1, x_2) = 0$ are transformed into the polars of $(y_1', y_2')$ with respect to $F(x_1', x_2') = 0$.

We may summarize the results thus obtained as follows:

THEOREM 6. *If a binary form $f$ is transformed by a projective transformation into the form $F$, the set of points represented by $f=0$ is transformed into the set represented by $F=0$. Any polar of a point $(y_1, y_2)$ with respect to $f=0$ is transformed into the corresponding polar of the point $(y_1', y_2')$ with respect to $F=0$.*

The following is a simple illustration of a polar of a point with respect to a set of points on a line.

The form $x_1x_2=0$ represents the two points whose nonhomogeneous coördinates are 0 and $\infty$ respectively. The first polar of any point $(y_1, y_2)$ with respect to this form is clearly $y_1x_2+y_2x_1=0$, and represents the point $(-y_1, y_2)$; in other words, the first polar of a point $P$ with respect to the pair of points represented by the given form is the harmonic conjugate of this point with respect to the pair.

### EXERCISE

Determine the geometrical construction of the $(n-1)$th polar of a point with respect to a set of $n$ distinct points on a line (cf. Ex. 3, p. 51).

**90. Invariants and covariants of binary forms.** DEFINITION. If a binary form $a_x^n = a_0x_1^n + na_1x_1^{n-1}x_2 + \cdots + a_nx_2^n$ be changed by the transformation

$$\Pi: \begin{matrix} x_1' = \alpha x_1 + \beta x_2, \\ x_2' = \gamma x_1 + \delta x_2 \end{matrix} \qquad (\alpha\delta - \beta\gamma \neq 0)$$

into a new form $A_{x'}^n = A_0x_1'^n + A_1x_1'^{n-1}x_2' + \cdots + A_nx_2'^n$, any rational function $I(a_0, a_1, \cdots, a_n)$ of the coefficients such that we have

$$I(A_0, A_1, \cdots, A_n) = \phi(\alpha, \beta, \gamma, \delta) \cdot I(a_0, a_1, \cdots, a_n)$$

is called an *invariant* of the form $a_x^n$. A function

$$C(a_0, a_1, \cdots, a_n;\ x_1, x_2)$$

of the coefficients and the variables such that we have

$$C(A_0, A_1, \cdots, A_n;\ x_1', x_2') = \psi(\alpha, \beta, \gamma, \delta) \cdot C(a_0, a_1, \cdots, a_n;\ x_1, x_2)$$

is called a *covariant* of the form $a_x^n$. The same terms apply to functions $I$ and $C$ of the coefficients and variables of any finite number of binary forms with the property that the same function of the coefficients and variables of the new forms is equal to the original function multiplied by a function of $\alpha, \beta, \gamma, \delta$ only; they are then called *simultaneous invariants* or *covariants*.

In § 87 we gave several examples of invariants of binary forms, linear and quadratic. It is evident from the definition that the *condition obtained by equating to 0 any invariant of a form (or of a system of forms) must determine a property of the set of points represented by the form (or forms) which is invariant under a projective transformation.* Hence the complete study of the projective geometry of a single line would involve the complete theory of invariants and covariants of binary forms. It is not our purpose in this book to give an account of this theory. But we will mention one theorem which we have already seen verified in special cases:

*The functions $\phi(\alpha, \beta, \gamma, \delta)$ and $\psi(\alpha, \beta, \gamma, \delta)$ occurring in the definition above are always powers of the determinant $\alpha\delta - \beta\gamma$ of the projective transformation in question.**

Before closing this section we will give a simple example of a covariant. Consider two binary quadratic forms $a_x^2$, $b_x^2$ and form the new quantic

$$C_{ab} = (a_0b_1 - a_1b_0)\,x_1^2 + (a_0b_2 - a_2b_0)\,x_1x_2 + (a_1b_2 - a_2b_1)\,x_2^2.$$

By means of equations (1), § 87, the reader may then verify without difficulty that the relation

$$C_{a'b'} = (\alpha\delta - \beta\gamma)\cdot C_{ab}$$

holds, which proves $C_{ab}$ to be a covariant. The two points represented by $C_{ab} = 0$ are the double points (proper or improper) of the involution of which the pairs determined by $a_x^2 = 0$, $b_x^2 = 0$ are conjugate pairs. This shows why the form should be a covariant.

### EXERCISE

Prove the statement contained in the next to the last sentence.

**91. Ternary and quaternary forms and their invariants.** The remarks which have been made above regarding binary forms can evidently be generalized. A *p-ary form of the nth degree* is a polynomial of the nth degree homogeneous in $p$ variables. When the number of variables is three or four, the form is called *ternary* or *quaternary* respectively. The general ternary form of the second degree when equated to zero has been shown to be the equation of a conic. In general, the set of points (proper and improper) in a plane which satisfy an equation

$$a_x^n = a_1x_1^n + a_2x_2^n + a_3x_3^n + \cdots = 0$$

* For proof, cf., for example, Grace and Young, Algebra of Invariants, pp. 21, 22.

obtained by equating to zero a ternary form of the $n$th degree is called an *algebraic curve of the nth degree (order)*. Similarly, the set of points determined in space by a quaternary form of the $n$th degree equated to zero is called an *algebraic surface of the nth degree.*

The definitions of invariants and covariants of $p$-ary forms is precisely the same as that given above for binary forms, allowance being made for the change in the number of variables. Just as in the binary case, if an invariant of a ternary or quaternary form vanishes, the corresponding function of the coefficients of any projectively equivalent form also vanishes, and consequently *it represents a property of the corresponding algebraic curve or surface which is not changed when the curve or surface undergoes a projective transformation.* Similar remarks apply to covariants of systems of ternary and quaternary forms.

Invariants and covariants as defined above are with respect to the group of all projective collineations. The geometric properties which they represent are properties unaltered by any projective collineation. Like definitions can of course be made of invariants with respect to any subgroup of the total group. Evidently any function of the coefficients of a form which is invariant under the group of all collineations will also be an invariant under any subgroup. But there will in general be functions which remain invariant under a subgroup but which are not invariant under the total group. These correspond to properties of figures which are invariant under the subgroup without being invariant under the total group. We thus arrive at the fundamental notion of *a geometry as associated with a given group*, a subject to which we shall return in detail in a later chapter.

## EXERCISES

**1.** Define by analogy with the developments of § 89, the $n-1$ polars of a ternary or quaternary form of the $n$th degree.

**2.** Regarding a triangle as a curve of the third degree, show that the second polar of a point with regard to a triangle is the polar line defined on page 46.

**3.** Generalize Ex. 2 in the plane and in space, and dualize.

**4.** Prove that the discriminant $\begin{vmatrix} a_{11} & a_{12} & a_{13} \\ a_{12} & a_{22} & a_{23} \\ a_{13} & a_{23} & a_{33} \end{vmatrix}$ of the ternary quadratic form

$$a_{11}x_1^2 + a_{22}x_2^2 + a_{33}x_3^2 + 2\,(a_{12}x_1x_2 + a_{13}x_1x_3 + a_{23}x_2x_3) = 0$$

is an invariant. What is its geometrical interpretation? Cf. Ex., p. 187.

**92. Proof of Proposition $K_n$.** *Given a rational integral function*

$$\phi(x) = a_0x^n + a_1x^{n-1} + \cdots + a_n, \quad a_0 \neq 0,$$

*whose coefficients belong to a given field* $F$, *and which is irreducible in* $F$, *there exists a field* $F'$, *containing* $F$, *in which the equation* $\phi(x) = 0$ *has a root.*

Let $f(x)$ be any rational integral function of $x$ with coefficients in $F$, and let $j$ be an arbitrary symbol not an element of $F$. Consider the class $F_j = [f(j)]$ of all symbols $f(j)$, where $[f(x)]$ is the class of all rational integral functions with coefficients in $F$. We proceed to define laws of combination for the elements of $F_j$ which render the latter a field. The process depends on the theorem * that any polynomial $f(x)$ can be represented uniquely in the form

$$f(x) = q(x)\phi(x) + r(x),$$

where $q(x)$ and $r(x)$ are polynomials belonging to $F$, — i.e. with coefficients in $F$, — and where $r(x)$ is of degree lower than the degree $n$ of $\phi(x)$. If two polynomials $f_1$, $f_2$ belonging to $F$ are such that their difference is exactly divisible by $\phi(x)$, then they are said to be *congruent* modulo $\phi(x)$, in symbols $f_1 \equiv f_2$, mod. $\phi(x)$.

1. Two elements $f_1(j)$, $f_2(j)$ of $F_j$ are said to be *equal*, if and only if $f_1(x)$ and $f_2(x)$ are congruent mod. $\phi(x)$. By virtue of the theorem referred to above, every element $f(j)$ of $F_j$ is equal to one and only one element $f'(j)$ of degree less than $n$. We need hence consider only those elements $f(j)$ of degree less than $n$. Further, it follows from this definition that $\phi(j) = 0$.

2. If $f_1(x) + f_2(x) \equiv f_3(x)$, mod. $\phi(x)$, then $f_1(j) + f_2(j) = f_3(j)$.

3. If $f_1(x) \cdot f_2(x) \equiv f_3(x)$, mod. $\phi(x)$, then $f_1(j) \cdot f_2(j) = f_3(j)$.

Addition and multiplication of the elements of $F_j$ having thus been defined, the associative and distributive laws follow as immediate consequences of the corresponding laws for the polynomials $f(x)$. It remains merely to show that the inverse operations exist and are unique. That addition has a unique inverse is obvious. To prove that the same holds for multiplication (with the exception of 0) we need only recall † that, since $\phi(x)$ and any polynomial $f(x)$ have no common factors, there exist two polynomials $h(x)$ and $k(x)$ with coefficients in $F$ such that

$$h(x) \cdot f(x) + k(x) \cdot \phi(x) = 1.$$

* Fine, College Algebra, p. 156. † Fine, loc. cit., p. 208.

This gives at once $$h(j) \cdot f(j) = 1,$$
so that every element $f(j)$ distinct from 0 has a reciprocal. The class $F_j$ is therefore a field with respect to the operations of addition and mutiplication defined above (cf. § 52), such that $\phi(j) = 0$. It follows at once* that $x - j$ is a factor of $\phi(x)$ in the field $F_j$, which is therefore the required field $F'$. The quotient $\phi(x)/(x-j)$ is either irreducible in $F_j$, or, if reducible, has certain irreducible factors. If the degree of one of the latter is greater than unity, the above process may be repeated leading to a field $F_{j, j'}$, $j'$ being a zero of the factor in question. Continuing in this way, it is possible to construct a field $F_{j, j', \dots, j^{(m)}}$, where $m \leqq n - 1$, in which $\phi(x)$ is completely reducible, i.e. in which $\phi(x)$ may be decomposed into $n$ linear factors. This gives the following corollary:

*Given a polynomial $\phi(x)$ belonging to a given field $F$, there exists a field $F'$ containing $F$ in which $\phi(x)$ is completely reducible.*

Finally, an obvious extension of this argument gives the corollary:

*Given a finite number of polynomials each of which belongs to a given field $F$, there exists a field $F'$, containing $F$, in which each of the given polynomials is completely reducible.*

This corollary is equivalent to Proposition $K_n$. For if $S$ be any space, let $F$ be the number system on one of its lines. Then, as in the Introduction (p. 11), $F'$ determines an analytic space which is the required space $S'$ of Proposition $K_n$.

The more general question at once presents itself: Given a field $F$, does there exist a field $F'$, containing $F$, in which *every* polynomial belonging to $F$ is completely reducible? The argument used above does not appear to offer a direct answer to this question. The question has, however, recently been answered in the affirmative by an extension of the above argument which assumes the possibility of "well ordering" any class.†

## EXERCISE

Many theorems of this and other chapters are given as dependent on A, E, P, $H_0$, whereas they are provable without the use of $H_0$. Determine which theorems are true in those spaces for which $H_0$ is false.

* Fine, College Algebra, p. 169.

† Cf. E. Steinitz, Algebraische Theorie der Körper, Journal für reine u. angewandte Mathematik, Vol. CXXXVII (1909), p. 167; especially pp. 271–286.

# CHAPTER X*

## PROJECTIVE TRANSFORMATIONS OF TWO-DIMENSIONAL FORMS

**93. Correlations between two-dimensional forms.** DEFINITION. A projective correspondence between the elements of a plane of points and the elements of a plane of lines (whether they be on the same or on different bases) is called a *correlation*. Likewise, a projective correspondence between the elements of a bundle of planes and the elements of a bundle of lines is called a *correlation*.†

Under the principle of duality we may confine ourselves to a consideration of correlations between planes. In such a correlation, then, to every point of the plane of points corresponds a unique line of the plane of lines; and to every pencil of points in the plane of points corresponds a unique projective pencil of lines in the plane of lines. In particular, if the plane of points and the plane of lines are on the same base, we have a correlation in a planar field, whereby to every point $P$ of the plane corresponds a unique line $p$ of the same plane, and in which, if $P_1$, $P_2$, $P_3$, $P_4$ are collinear points, the corresponding lines $p_1$, $p_2$, $p_3$, $p_4$ are concurrent and such that

$$R(P_1P_2, P_3P_4) = R(p_1p_2, p_3p_4).$$

That a correlation $\Gamma$ transforms the points $[P]$ of a plane into the lines $[p]$ of the plane, we indicate as usual by the functional notation

$$\Gamma(P) = p.$$

The points on a line $l$ are transformed by $\Gamma$ into the lines on a point $L$. This determines a transformation of the lines $[l]$ into the points $[L]$, which we may denote by $\Gamma'$, thus:

$$\Gamma'(l) = L.$$

That $\Gamma'$ is also a correlation is evident (the formal proof may be supplied by the reader). The transformation $\Gamma'$ is called the correlation *induced* by $\Gamma$. If a correlation $\Gamma$ transforms the lines $[l]$ of a

* All developments of this chapter are on the basis of Assumptions A, E, P, and $H_0$. Cf. the exercise at the end of the last chapter.

† The terms *reciprocity* and *duality* are sometimes used in place of *correlation*.

plane into the points $[L]$ of the plane, the correlation which transforms the points $[ll']$ into the lines $[LL']$ is the correlation induced by $\Gamma$. If $\Gamma'$ is induced by $\Gamma$, it is clear that $\Gamma$ is induced by $\Gamma'$. For if we have

$$\Gamma(P_1P_2P_3\cdots)=p_1p_2p_3\cdots,$$

we have also

$$\Gamma'((P_1P_2)(P_2P_3)\cdots)=(p_1p_2)(p_2p_3)\cdots,$$

and hence the induced correlation of $\Gamma'$ transforms $P_2$ into $p_2$, etc.

That correlations in a plane exist follows from the existence of the polar system of a conic. The latter is in fact a projective transformation in which to every point in the plane of the conic corresponds a unique line of the plane, to every line corresponds a unique point, and to every pencil of points (lines) corresponds a projective pencil of lines (points) (Theorem 18, Cor., Chap. V). This example is, however, of a special type having the peculiarity that, if a point $P$ corresponds to a line $p$, then in the induced correlation the line $p$ will correspond to the point $P$; i.e. in a polar system the points and lines correspond doubly. This is by no means the case in every correlation.

DEFINITION. A correlation in a plane in which the points and lines correspond doubly is called a *polarity.*

It has been found convenient in the case of a polarity defined by a conic to study a transformation of points into lines and the induced transformation of lines into points simultaneously. Analogously, in studying collineations we have regarded a transformation T of points $P_1$, $P_2$, $P_3$, $P_4$ into points $P_1'$, $P_2'$, $P_3'$, $P_4'$, and the transformation T$'$ of the lines $P_1P_2$, $P_2P_3$, $P_3P_4$, $P_4P_1$ into the lines $P_1'P_2'$, $P_2'P_3'$, $P_3'P_4'$, $P_4'P_1'$ as the same collineation. In like manner, when considering a transformation of the points and lines of a plane into its lines and points respectively, a correlation $\Gamma$ operating on the points and its induced correlation $\Gamma'$ operating on the lines constitute one transformation of the points and lines of the plane. For this sort of transformation we shall also use the term *correlation.* In the first instance a correlation in a plane is a correspondence between a plane of points (lines) and a plane of lines (points). In the extended sense it is a transformation of a planar field either into itself or into another planar field, in which an element of one kind (point or line) corresponds to an element of the other kind.

The following theorem is an immediate consequence of the definition and the fact that the resultant of any two projective correspondences is a projective correspondence.

THEOREM 1. *The resultant of two correlations is a projective collineation, and the resultant of a correlation and a projective collineation is a correlation.*

We now proceed to derive the *fundamental theorem* for correlations between two-dimensional forms.

THEOREM 2. *A correlation between two two-dimensional primitive forms is uniquely defined when four pairs of homologous elements are given, provided that no three elements of either form are on the same one-dimensional primitive form.*

*Proof.* Let the two forms be a plane of points $\alpha$ and a plane of lines $\alpha'$. Let $C^2$ be any conic in $\alpha'$, and let the four pairs of homologous elements be $A$, $B$, $C$, $D$ in $\alpha$ and $a'$, $b'$, $c'$, $d'$ in $\alpha'$. Let $A'$, $B'$, $C'$, $D'$ be the poles of $a'$, $b'$, $c'$, $d'$ respectively with respect to $C^2$. If the four points $A$, $B$, $C$, $D$ are the vertices of a quadrangle and the four points $A'$, $B'$, $C'$, $D'$ are likewise the vertices of a quadrangle (and this implies that no three of the lines $a'$, $b'$, $c'$, $d'$ are concurrent), there exists one and only one collineation transforming $A$ into $A'$, $B$ into $B'$, $C$ into $C'$, and $D$ into $D'$ (Theorem 18, Chap. IV). Let this collineation be denoted by T, and let the polarity defined by the conic $C^2$ be denoted by P. Then the projective transformation $\Gamma$ which is the resultant of these two transforms $A$ into $a'$, $B$ into $b'$, etc. Moreover, there cannot be more than one correspondence effecting this transformation. For, suppose there were two, $\Gamma$ and $\Gamma_1$. Then the projective correspondence $\Gamma_1^{-1} \cdot \Gamma$ would leave each of the four points $A$, $B$, $C$, $D$ fixed; i.e. would be the identity (Theorem 18, Chap. IV). But this would imply $\Gamma_1 = \Gamma$.

THEOREM 3. *A correlation which interchanges the vertices of a triangle with the opposite sides is a polarity.*

*Proof.* Let the vertices of the given triangle be $A$, $B$, $C$, and let the opposite sides be respectively $a$, $b$, $c$. Let $P$ be any point of the plane $ABC$ which is not on a side of the triangle. The line $p$ into which $P$ is transformed by the given correlation $\Gamma$ does not, then, pass through a vertex of the triangle $ABC$. The correlation $\Gamma$ is determined by the equation $\Gamma(ABCP) = abcp$, and, by hypothesis, is such that $\Gamma(abc) = ABC$. The points $[Q]$ of $c$ are transformed into the lines $[q]$ on $C$, and these meet $c$ in a pencil $[Q']$ projective with $[Q]$ (fig. 99). Since $A$ corresponds to $B$ and $B$ to $A$ in the projectivity $[Q] \barwedge [Q']$, this projectivity is an involution I. The point $Q_0$ in which

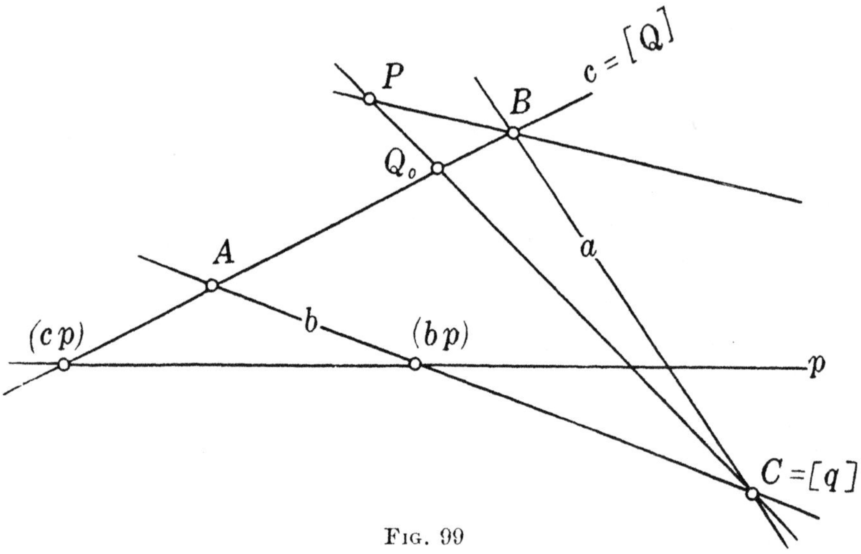


FIG. 99

$CP$ meets $c$ is transformed by $\Gamma$ into a line on the point $cp$; and since $Q_0$ and $cp$ are paired in I, it follows that $cp$ is transformed into the line $CQ_0 = CP$. In like manner, $bp$ is transformed into $BP$. Hence $p = (cp, bp)$ is transformed into $P = (CP, BP)$.

THEOREM 4. *Any projective collineation, $\Pi$, in a plane, $\alpha$, is the product of two polarities.*

*Proof.* Let $Aa$ be a lineal element of $\alpha$, and let

$$\Pi(Aa) = A'a', \quad \Pi(A'a') = A''a''.$$

Unless $\Pi$ is perspective, $Aa$ may be so chosen that $A, A', A''$ are not collinear, $aa'a''$ are not concurrent, and no line of one of the three lineal elements passes through the point of another. In this case there exists a polarity P such that $\mathrm{P}(AA'A'') = a''a'a$, namely the polarity defined by the conic with regard to which $AA''(aa'')$ is a self-polar triangle and to which $a'$ is tangent at $A'$. If $\Pi$ is perspective, the existence of P follows directly on choosing $Aa$, so that neither $A$ nor $a$ is fixed. We then have

$$\mathrm{P}\Pi(AA'aa') = a'aA'A,$$

and hence the triangle $AA'(aa')$ is self-reciprocal. Hence (Theorem 3) $\mathrm{P}\Pi = \mathrm{P}_1$ is a polarity, and therefore $\Pi = \mathrm{PP}_1$.

**94. Analytic representation of a correlation between two planes. Bilinear forms.** Let a system of simultaneous point-and-line coördinates be established in a planar field. We then have

THEOREM 5. *Any correlation in a plane is given as a transformation of points into lines by equations of the form*

$$(1)\qquad \begin{aligned} \rho u_1' &= a_{11}x_1 + a_{12}x_2 + a_{13}x_3, \\ \rho u_2' &= a_{21}x_1 + a_{22}x_2 + a_{23}x_3, \\ \rho u_3' &= a_{31}x_1 + a_{32}x_2 + a_{33}x_3, \end{aligned}$$

*where the determinant $A$ of the coefficients $a_{ij}$ is different from zero. Conversely, every transformation of this form in which the determinant $A$ is different from zero represents a correlation.*

The proof of this theorem is completely analogous to the proof of Theorem 8, Chapter VII, and need not be repeated here.

As a corollary we have

COROLLARY 1. *The transformation $\rho u_1' = x_1$, $\rho u_2' = x_2$, $\rho u_3' = x_3$ in a plane represents a polarity in which to every side of the triangle of reference corresponds the opposite vertex.*

Also, if $(u_1', u_2', u_3')$ be interpreted as line coördinates in a plane different from that containing the points $(x_1, x_2, x_3)$ (and if the number systems are so related that the correspondence $X' = X$ between the two planes is projective), we have at once

COROLLARY 2. *The equations of Theorem 5 also represent a correlation between the plane of $(x_1, x_2, x_3)$ and the plane of $(u_1', u_2', u_3')$.*

Returning now to the consideration of a correlation in a plane (planar field), we have seen that the equations (1) give the coördinates $(u_1', u_2', u_3')$ of the line $u' = \Gamma(X)$, which corresponds to the point $X = (x_1, x_2, x_3)$. By solving these equations for $x_i$,

$$(2)\qquad \begin{aligned} \sigma x_1 &= A_{11}u_1' + A_{21}u_2' + A_{31}u_3', \\ \sigma x_2 &= A_{12}u_1' + A_{22}u_2' + A_{32}u_3', \\ \sigma x_3 &= A_{13}u_1' + A_{23}u_2' + A_{33}u_3', \end{aligned}$$

we obtain the coördinates of $X = \Gamma^{-1}(u')$ in terms of the coördinates $u_i'$ of the line to which $X$ is homologous in the inverse correlation $\Gamma^{-1}$. If, however, we seek the coördinates of the point $X' = \Gamma(u)$ which corresponds to any line $u$ in the correlation $\Gamma$, we may proceed as follows:

Let the equation of the point $X' = (x_1', x_2', x_3')$ in line coördinates be

$$u_1'x_1' + u_2'x_2' + u_3'x_3' = 0.$$

Substituting in this equation from (1) and arranging the terms as a linear expression in $x_1, x_2, x_3$,

$$u_1x_1 + u_2x_2 + u_3x_3 = 0,$$

we readily find

$$(3)\qquad \begin{aligned} \tau u_1 &= a_{11}x_1' + a_{21}x_2' + a_{31}x_3', \\ \tau u_2 &= a_{12}x_1' + a_{22}x_2' + a_{32}x_3', \\ \tau u_3 &= a_{13}x_1' + a_{23}x_2' + a_{33}x_3'. \end{aligned}$$

The coördinates of $X'$ in terms of the coördinates of $u$ are then given by

$$(4)\qquad \begin{aligned} \upsilon x_1' &= A_{11}u_1 + A_{12}u_2 + A_{13}u_3, \\ \upsilon x_2' &= A_{21}u_1 + A_{22}u_2 + A_{23}u_3, \\ \upsilon x_3' &= A_{31}u_1 + A_{32}u_2 + A_{33}u_3. \end{aligned}$$

This is the analytic expression of the correlation as a transformation of lines into points; i.e. of the induced correlation of $\Gamma$. These equations clearly apply also in the case of a correlation between two different planes.

It is perhaps well to emphasize the fact that Equations (1) express $\Gamma$ as a transformation of points into lines, while Equations (4) represent the induced correlation of lines into points. Since we consider a correlation as a transformation of points into lines and lines into points, $\Gamma$ is completely represented by (1) and (4) taken together. Equations (2) and (3) taken together represent the inverse of $\Gamma$.

Another way of representing $\Gamma$ analytically is obtained by observing that the point $(x_1, x_2, x_3)$ is transformed by $\Gamma$ into the line whose equation in current coördinates $(x_1', x_2', x_3')$ is

$$u_1'x_1' + u_2'x_2' + u_3'x_3' = 0,$$

or,

$$(5)\quad (a_{11}x_1 + a_{12}x_2 + a_{13}x_3)\,x_1' + (a_{21}x_1 + a_{22}x_2 + a_{23}x_3)\,x_2' + (a_{31}x_1 + a_{32}x_2 + a_{33}x_3)\,x_3' = 0.$$

The left-hand member of (5) is a general *ternary bilinear form*. We have then

Corollary 3. *Any ternary bilinear form in which the determinant $A$ is different from zero represents a correlation in a plane.*

**95. General projective group. Representation by matrices.** The *general projective group* of transformations in a plane (which, under duality, we take as representative of the two-dimensional primitive forms) consists of all projective collineations (including the identity) and all correlations in the plane. Since the product of two collineations is a collineation, the set of all projective collineations forms a subgroup of the general group. Since, however, the product of two correlations is a collineation, there exists no subgroup consisting entirely of correlations.*

According to the point of view developed in the last chapter, the projective geometry of a plane is concerned with theorems which state properties invariant under the general projective group in the plane. In particular, *the principle of duality may be regarded as a consequence of the presence of correlations in this group.*

Analytically, collineations and correlations may be regarded as aspects of the theory of matrices. The collineation

$$x_i' = \sum_{j=1}^{3} a_{ij} x_j \qquad (i = 1, 2, 3)$$

may be conveniently represented by the matrix A of the coefficients $a_{ij}$:

$$A = (a_{ij}) = \begin{pmatrix} a_{11} & a_{12} & a_{13} \\ a_{21} & a_{22} & a_{23} \\ a_{31} & a_{32} & a_{33} \end{pmatrix}.$$

The product of two collineations $A = (a_{ij})$ and $B = (b_{ij})$ is then given by the product of their matrices:

$$BA = (b_{ij})(a_{ij}) = \begin{pmatrix} b_{11} & b_{12} & b_{13} \\ b_{21} & b_{22} & b_{23} \\ b_{31} & b_{32} & b_{33} \end{pmatrix} \begin{pmatrix} a_{11} & a_{12} & a_{13} \\ a_{21} & a_{22} & a_{23} \\ a_{31} & a_{32} & a_{33} \end{pmatrix}$$

$$= \begin{pmatrix} b_{11}a_{11} + b_{12}a_{21} + b_{13}a_{31} & b_{11}a_{12} + b_{12}a_{22} + b_{13}a_{32} & b_{11}a_{13} + b_{12}a_{23} + b_{13}a_{33} \\ b_{21}a_{11} + b_{22}a_{21} + b_{23}a_{31} & b_{21}a_{12} + b_{22}a_{22} + b_{23}a_{32} & b_{21}a_{13} + b_{22}a_{23} + b_{23}a_{33} \\ b_{31}a_{11} + b_{32}a_{21} + b_{33}a_{31} & b_{31}a_{12} + b_{32}a_{22} + b_{33}a_{32} & b_{31}a_{13} + b_{32}a_{23} + b_{33}a_{33} \end{pmatrix},$$

the element of the $i$th row and the $j$th column of the matrix BA being obtained by multiplying each element of the $i$th row of B by the corresponding element of the $j$th column of A and adding the products thus obtained. It is clear that *two collineations are not in general commutative.*

* A polarity and the identity form a group; but this forms no exception to the statement just made, since the identity must be regarded as a collineation.

Of the two matrices

$$\begin{pmatrix} a_{11} & a_{12} & a_{13} \\ a_{21} & a_{22} & a_{23} \\ a_{31} & a_{32} & a_{33} \end{pmatrix} \text{ and } \begin{pmatrix} a_{11} & a_{21} & a_{31} \\ a_{12} & a_{22} & a_{32} \\ a_{13} & a_{23} & a_{33} \end{pmatrix},$$

either of which is obtained from the other by interchanging rows and columns, one is called the *conjugate* or *transposed* matrix of the other. The matrix

$$\begin{pmatrix} A_{11} & A_{21} & A_{31} \\ A_{12} & A_{22} & A_{32} \\ A_{13} & A_{23} & A_{33} \end{pmatrix}$$

is called the adjoint matrix of the matrix A. The *adjoint* matrix is clearly obtained by replacing each element of the transposed matrix by its cofactor. Equations (2) of § 67 show that *the adjoint of a given matrix represents the inverse of the collineation represented by the given matrix.* Indeed, by direct multiplication,

$$\begin{pmatrix} a_{11} & a_{12} & a_{13} \\ a_{21} & a_{22} & a_{23} \\ a_{31} & a_{32} & a_{33} \end{pmatrix} \begin{pmatrix} A_{11} & A_{21} & A_{31} \\ A_{12} & A_{22} & A_{32} \\ A_{13} & A_{23} & A_{33} \end{pmatrix} = \begin{pmatrix} A & 0 & 0 \\ 0 & A & 0 \\ 0 & 0 & A \end{pmatrix};$$

and the matrix just obtained clearly represents the identical collineation. Since, when a matrix is thought of as representing a collineation, we may evidently remove any common factor from all the elements of the matrix, the latter matrix is equivalent to the so-called identical matrix,*

$$\begin{pmatrix} 1 & 0 & 0 \\ 0 & 1 & 0 \\ 0 & 0 & 1 \end{pmatrix}.$$

Furthermore, Equations (3), § 67, show that *if a given matrix represents a collineation in point coördinates, the conjugate of the adjoint matrix represents the same collineation in line coördinates.* Also from the representation of the product of two matrices just derived, follows the important result:

*The determinant of the product of two matrices (collineations) is equal to the product of the determinants of the two matrices (collineations).*

* In the general theory of matrices these two matrices are not, however, regarded as the same. It is only the interpretation of them as collineations which renders them equivalent.

From what has just been said it is clear that a matrix does not completely define a collineation, unless the nature of the coördinates is specified. If it is desired to exhibit the coördinates in the notation, we may write the collineation $x_i' = \Sigma a_{ij} x_j$ in the symbolic form

$$x' = (a_{ij})\, x.$$

The matrix $(a_{ij})$ may then be regarded as an *operator* transforming the coördinates $x = (x_1, x_2, x_3)$ into the coördinates $x' = (x_1', x_2', x_3')$. If we place $\bar{a}_{ij} = a_{ji}$, the matrix conjugate to $(a_{ij})$ is $(\bar{a}_{ij})$. Also by placing $\bar{A}_{ij} = A_{ji}$, the adjoint matrix of $(a_{ij})$ is $(\bar{A}_{ij})$. The inverse of the above collineation is then written

$$x = (\bar{A}_{ij})\, x'.$$

Furthermore, the collineation $x' = (a_{ij})x$ is represented in line coördinates by the equation

$$u' = (A_{ij})\, u.$$

This more complete notation will not be found necessary in general in the analytic treatment of collineations, when no correlations are present, but it is essential in the representation of correlations by means of matrices.

The correlation (1) of § 94 may clearly be represented symbolically by the equation

$$u' = (a_{ij})\, x,$$

where the matrix $(a_{ij})$ is to be regarded as an operator transforming the point $x$ into the line $u'$. This correlation is then expressed as a transformation of lines into points by

$$x' = (A_{ij})\, u.$$

The product of two correlations $u' = (a_{ij})\, x$ and $u' = (b_{ij})\, x$ is therefore represented by

$$x' = (B_{ij})\,(a_{ij})\, x$$

(cf. Equations (4), § 94), or by

$$u' = (b_{ij})\,(A_{ij})\, u.$$

Also, the inverse of the correlation $u' = (a_{ij})\, x$ is given by

$$x = (\bar{A}_{ij})\, u',$$

or by

$$u = (\bar{a}_{ij})\, x'.$$

**EXERCISE**

Show that if $[\Pi]$ is the set of all collineations in a plane and $\Gamma_1$ is any correlation, the set of all correlations in the plane is $[\Pi\Gamma_1]$, so that the two sets of transformations $[\Pi]$ and $[\Pi\Gamma_1]$ comprise the general projective group in the plane. By virtue of this fact the subgroup of all projective collineations is said to be of *index 2* in the general projective group.*

**96. Double points and double lines of a collineation in a plane.** Referring to Equations (1) of § 67 we see that a point $(x_1, x_2, x_3)$ which is transformed into itself by the collineation (1) must satisfy the equations

$$\rho x_1 = a_{11}x_1 + a_{12}x_2 + a_{13}x_3,$$
$$\rho x_2 = a_{21}x_1 + a_{22}x_2 + a_{23}x_3,$$
$$\rho x_3 = a_{31}x_1 + a_{32}x_2 + a_{33}x_3,$$

which, by a simple rearrangement, may be written

$$(1) \qquad \begin{aligned} (a_{11}-\rho)x_1 + \quad a_{12}x_2 \quad + \quad a_{13}x_3 \quad &= 0,\\ a_{21}x_1 \quad + (a_{22}-\rho)x_2 + \quad a_{23}x_3 \quad &= 0,\\ a_{31}x_1 \quad + \quad a_{32}x_2 \quad + (a_{33}-\rho)x_3 &= 0. \end{aligned}$$

If a point $(x_1, x_2, x_3)$ is to satisfy these three equations, the determinant of this system of equations must vanish; i.e. $\rho$ must satisfy the equation

$$(2) \qquad \begin{vmatrix} a_{11}-\rho & a_{12} & a_{13} \\ a_{21} & a_{22}-\rho & a_{23} \\ a_{31} & a_{32} & a_{33}-\rho \end{vmatrix} = 0.$$

This is an equation of the third degree in $\rho$, which cannot have more than three roots in the number system of our geometry.

Suppose that $\rho_1$ is a root of this equation. The system of equations (1) is then consistent (which means geometrically that the three lines represented by them pass through the same point), and the point determined by any two of them (if they are independent, i.e. if they do not represent the same line) is a double point. Solving the first two of these equations, for example, we find as the coördinates $(x_1, x_2, x_3)$ of a double point

$$(3) \qquad x_1 : x_2 : x_3 = \begin{vmatrix} a_{12} & a_{13} \\ a_{22}-\rho_1 & a_{23} \end{vmatrix} : \begin{vmatrix} a_{13} & a_{11}-\rho_1 \\ a_{23} & a_{21} \end{vmatrix} : \begin{vmatrix} a_{11}-\rho_1 & a_{12} \\ a_{21} & a_{22}-\rho_1 \end{vmatrix},$$

* A subgroup $[\Pi]$ of a group is said to be of index $n$, if there exist $n-1$ transformations $\Gamma_i (i = 1, 2, \cdots n-1)$, such that the $n-1$ sets $[\Pi\Gamma_i]$ of transformations together with the set $[\Pi]$ contain all the transformations of the group, while no two transformations within the same set or from any two sets are identical.

which represent a unique point, unless it should happen that all the determinants on the right of this equation vanish. Leaving aside this possibility for the moment, we see that every root of Equation (2), which is called the *characteristic equation* of the collineation (or of the representative matrix), gives rise to a unique double point. Moreover, every double point is obtainable in this way. This is the analytic form of the fact already noted, that *a collineation which is not a homology or an elation cannot have more than three double points*, unless it is the identical collineation.

If, however, all the determinants on the right in Equations (3) vanish, it follows readily that the first two of Equations (1) represent the same line. If the determinants formed analogously from the last two equations do not all vanish, we again get a unique double point; but if the latter also vanish, then all three of the equations above represent the same line. Every point of this line is then a double point, and the collineation must be a homology or an elation. Clearly this can happen only if $\rho_1$ is at least a double root of Equation (2); for we know that a perspective collineation cannot have more than one double point which is not on the axis of the collineation.

A complete enumeration of the possible configurations of double points and lines of a collineation can be made by means of a study of the characteristic equation, making use of the theory of elementary divisors.* It seems more natural in the present connection to start with the existence of one fixed point (Proposition $K_3$) and discuss geometrically the cases that can arise.

By Theorem 4 a collineation is the product of two polarities. Hence any double point has the same polar line in both polarities, and that polar line is a double line. Hence *the invariant figure of double points and lines is self-dual.*

Four points of the plane, no three of which are collinear, cannot be invariant unless the collineation reduces to the identity. If three noncollinear points are invariant, two cases present themselves. If the collineation reduces to the identity on no side of the invariant triangle, the collineation is of Type *I* (cf. § 40, Chap. IV). If the collineation is the identity on one and only one side of the invariant triangle, the collineation is of Type *IV*.† If two distinct points are

* Cf. Bôcher, Introduction to Higher Algebra, Chaps. XX and XXI.

† If it is the identity on more than one side, it is the identical collineation.

invariant, but no point not on the line $l$ joining these two is invariant, two possibilities again arise. If the collineation does not leave every point of this line invariant, there is a unique other line through one of these points that is invariant, since the invariant figure is self-dual. The collineation is then of Type *II*. If every point of the line is invariant, on the other hand, all the lines through a point of the line $l$ must be invariant, since the figure of invariant elements is self-dual. The collineation is then of Type *V*.

If only one point is fixed, only one line can be fixed. The collineation is then parabolic both on the line and on the point, and the collineation is of Type *III*.

We have thus proved that every collineation different from the identity is of one of the five types previously enumerated. Type *I* may be represented by the symbol [1, 1, 1], the three 1's denoting three distinct double points. In Type *IV* there are also three distinct double points, but all points on the line joining two of them are fixed and Equation (1) has one double root. Type *IV* is denoted by [(1, 1), 1]. In Type *II*, as there are only two distinct double points, Equation (1) must have a double root and one simple root. This type is accordingly denoted by the symbol [2, 1], the 2 indicating the double point corresponding to the double root. Type *V* is then naturally represented by [(2, 1)], the parentheses again indicating that every point of the line joining the two points is fixed. Type *III* corresponds to a triple root of (1), and may therefore be denoted by [3]. We have then the following:

THEOREM 6. *Every projective collineation in a plane is of one of the following five types:*

| | | |
|---|---|---|
| [1, 1, 1] | [(1, 1), 1] | |
| | [2, 1] | [(2, 1)] |
| | | [3] |

In this table the first column corresponds to three distinct roots of the characteristic equation, the second column to a double root, the third column to a triple root. The first row corresponds to the cases in which there exist at least three double points which are not collinear; the second row to the case where there exist at least two distinct double points and all such points are on the same line; the third row to the case in which there exists only a single double point.

With every collineation in a plane are associated certain projectivities on the invariant lines and in the pencils on the invariant points. In case the collineation is of Type *I*, it is completely determined if the projectivities on two sides of the invariant triangle are given. There must therefore be a relation between the projectivities on the three sides of the invariant triangle (cf. Ex. 5, p. 276). In a collineation of Type *II* the projectivity is parabolic on one of the invariant lines but not on the other. The point in which the two invariant lines meet may therefore be called *singly parabolic*. The collineation is completely determined if the projectivities on the two invariant lines are given. In a collineation of Type *III* the projectivity on the invariant line is parabolic, as likewise the projectivity on the invariant point. The fixed point may then be called *doubly parabolic*. The projectivities on the invariant lines of a collineation of Type *V* are parabolic except the one on the axis which is the identity. The center is thus a singly parabolic point. In the table of Theorem 6 the symbols 3, 2, and 1 may be taken to indicate doubly and singly and nonparabolic points respectively.*

We give below certain simple, so-called *canonical* forms of the equations defining collineations of these five types.

*Type I.* Let the invariant triangle be the triangle of reference. The collineation is then given by equations of the form

$$\begin{aligned} \rho x_1' &= a_{11}x_1, \\ \rho x_2' &= \qquad a_{22}x_2, \\ \rho x_3' &= \qquad\qquad a_{33}x_3, \end{aligned}$$

in which $a_{11}$, $a_{22}$, $a_{33}$ are the roots of the characteristic equation and must therefore be all distinct.

*Type IV, Homology.* If the vertices of the triangle of reference are taken as invariant points, the equations reduce to the form written above; but since one of the lines $x_1 = 0$, $x_2 = 0$, $x_3 = 0$ is pointwise

* For a more detailed discussion of collineations, reference may be made to Newson, A New Theory of Collineations, etc., American Journal of Mathematics, Vol. XXIV, p. 109.

invariant, we must have either $a_{22}=a_{33}$ or $a_{33}=a_{11}$ or $a_{11}=a_{22}$. Thus the homology may be written

$$\begin{aligned} \rho x_1' &= x_1, \\ \rho x_2' &= \quad x_2, \\ \rho x_3' &= \qquad a_{33}x_3, \qquad (a_{33} \neq 1). \end{aligned}$$

A harmonic homology or reflection is obtained by setting $a_{33}=-1$.

*Type II.* The characteristic equation has one double root, $\rho_1=\rho_2$, say, and a simple root $\rho_3$. Let the double point corresponding to $\rho_1=\rho_2$ be $U_1=(0, 0, 1)$, let the double point corresponding to $\rho_3$ be $U_3=(1, 0, 0)$, and let the third vertex of the triangle of reference be any point on the double line $u_3$ corresponding to $\rho_3$, which line will pass through the point $U_1$. The collineation is then of the form

$$\begin{aligned} \rho x_1' &= a_{11}x_1, \\ \rho x_2' &= \qquad a_{22}x_2, \\ \rho x_3' &= \qquad a_{32}x_2 + a_{33}x_3, \end{aligned}$$

since the lines $x_1=0$ and $x_2=0$ are double lines and $(1, 0, 0)$ is a double point. The characteristic equation of the collineation is clearly

$$(a_{11}-\rho)(a_{22}-\rho)(a_{33}-\rho)=0,$$

and since this must have a double root, it follows that two of the numbers $a_{11}$, $a_{22}$, $a_{33}$ must be equal. To determine which, place $\rho=a_{22}$; using the minors of the second row, we find, as coördinates of the corresponding double point,

$$(0, (a_{11}-a_{22})(a_{22}-a_{33}), a_{32}(a_{11}-a_{22})),$$

which is $U_1$, and hence we have $a_{22}=a_{33}$. The collineation then is of Type *II*, if $a_{11}\neq a_{22}$. Its equations are therefore

$$\begin{aligned} \rho x_1' &= a_{11}x_1, \\ \rho x_2' &= \qquad a_{22}x_2, \\ \rho x_3' &= \qquad a_{32}x_2 + a_{22}x_3, \end{aligned}$$

where $a_{32}\neq 0$ and $a_{11}\neq a_{22}$.

*Type III.* The characteristic equation has a triple root, $\rho_1=\rho_2=\rho_3$, say. Let $U_1=(0, 0, 1)$ be the single double point, and the line $x_1=0$ be the single double line. With this choice of coördinates the collineation has the form

$$\begin{aligned} \rho x_1' &= a_{11}x_1, \\ \rho x_2' &= a_{21}x_1 + a_{22}x_2, \\ \rho x_3' &= a_{31}x_1 + a_{32}x_2 + a_{33}x_3. \end{aligned}$$

By writing the characteristic equation we find, in view of the fact that the equation has a triple root, that $a_{11} = a_{22} = a_{33}$. The form of the collineation is therefore

$$\rho x_1' = x_1,$$
$$\rho x_2' = a_{21}x_1 + x_2,$$
$$\rho x_3' = a_{31}x_1 + a_{32}x_2 + x_3,$$

where the numbers $a_{21}$, $a_{32}$ must be different from 0.

*Type V, Elation.* Choosing (0, 0, 1) as center and $x_1 = 0$ as axis, the equations of the collineation reduce to the form given for Type *III*, where, however, $a_{32}$ must be zero in order that the line $x_1 = 0$ be pointwise invariant. The equations for Type *II* also yield an elation in case $a_{11} = a_{22}$. Thus an elation may be written

$$\rho x_1' = x_1,$$
$$\rho x_2' = \quad x_2,$$
$$\rho x_3' = \quad a_{32}x_2 + x_3.$$

**EXERCISES**

**1.** Determine the collineation which transforms the points $A = (0, 0, 1)$, $B = (0, 1, 0)$, $C = (1, 0, 0)$, $D = (1, 1, 1)$ into the points $B$, $C$, $D$, $A$ respectively. Show that the characteristic equation of this collineation is $(\rho - 1)(\rho^2 + 1) = 0$, which in any field has one root. Determine the double point and double line corresponding to this root. Assuming the field of numbers to be the ordinary complex field, determine the coördinates of the remaining two double points and double lines. Verify, by actually multiplying the matrices, that this collineation is of period 4 (a fact which is evident from the definition of the collineation).

**2.** With the same coördinates for $A$, $B$, $C$, $D$ determine the collineation which transforms these points respectively into the points $B$, $A$, $D$, $C$. The resulting collineation must, from this definition, be a homology. Why? Determine its center and its axis. By actual multiplication of the matrices verify that its square is the identical collineation.

**3.** Express each of the collineations in Exs. 1 and 2 in terms of line coördinates.

**4.** Show that the characteristic cross ratios of the one-dimensional projectivities on the sides of the invariant triangle of the collineation $x_1' = ax_1$, $x_2' = bx_2$, $x_3' = cx_3$ are the ratios of the numbers $a$, $b$, $c$. Hence show that the product of these cross ratios is equal to unity, the double points being taken around the triangle in a given order.

**5.** Prove the latter part of Ex. 4 for the cross ratios of the projectivities on the sides of the invariant triangle of any collineation of Type *I*.

**6.** Write the equations of a collineation of period 3 ; 4 ; 5 ; $\cdots$ ; $n$ ; $\cdots$.

**7.** By properly choosing the system of nonhomogeneous coördinates any collineation of Type *I* may be represented by equations $x' = ax$, $y' = by$. The set of all collineations obtained by giving the parameters $a$, $b$ all possible values forms a group. Show that the collineations $x' = ax$, $y' = a^r y$, where $r$ is constant for all collineations of the set, form a subgroup. Show that every collineation of this subgroup leaves invariant every curve whose equation is $y = cx^r$, where $c$ is any constant. Such curves are called *path curves* of the collineations.

**8.** If $P$ is any point of a given path curve, $p$ the tangent at $P$, and $A, B, C$ the vertices of the invariant triangle, then R $(p, PA, PB, PC)$ is a constant.

**9.** For the values $r = -1, 2, \frac{1}{2}$ the path curves of the collineations of the subgroup described in Ex. 7 are conics tangent to two sides of the invariant triangle at two vertices.

**10.** If $r = 0$, the subgroup of Ex. 7 consists entirely of homologies.

**11.** Prove that any collineation of Type *I* may be expressed in the form

$$x' = k\,(ax + by),$$
$$y' = k\,(bx - ay),$$

with the restriction $a^2 + b^2 = 1$.

**12.** Prove that any collineation can be expressed as a product of collineations of Type *I*.

**13.** Let the invariant figure of a collineation of Type *II* be $A$, $B$, $l$, $m$, where $l = AB$, $B = lm$. The product of such a collineation by another of Type *II* with invariant figure $A'$, $B$, $l$, $m'$ is in general of Type *II*, but may be of Types *III*, *IV*, or *V*. Under what conditions do the latter cases arise?

**14.** Using the notation of Ex. 13, the product of a collineation of Type *II* with invariant figure $A$, $B$, $l$, $m$ by one with invariant figure $A$, $B'$, $l$, $m'$ is in general of Type *II*, but may be of Types *III* or *IV*. Under what conditions do the latter cases arise?

**15.** Prove that any collineation can be expressed as a product of collineations of Type *II*.

**16.** Two collineations of Type *III* with the same invariant figure are not in general commutative.

**17.** Any projective collineation can be expressed as a product of collineations of Type *III*.

**18.** If $\Pi$ is an elation whose center is $C$, and $P$ any point not on the axis, then $P$ and $C$ are harmonically conjugate with respect to $\Pi^{-1}(P)$ and $\Pi(P)$.

**19.** If two coplanar conics are projective, the correspondence between the points of one and the tangents at homologous points of the other determines a correlation.

**20.** If in a collineation between two distinct planes every point of the line of intersection of the planes is self-corresponding, the planes are perspective.

**21.** In nonhomogeneous coördinates a collineation of Type *I* with fixed points $(a_1, a_2)$, $(b_1, b_2)$ $(c_1, c_2)$ may be written

$$x' = \frac{\begin{vmatrix} x & y & 1 & 0 \\ a_1 & a_2 & 1 & a_1 \\ b_1 & b_2 & 1 & kb_1 \\ c_1 & c_2 & 1 & k'c_1 \end{vmatrix}}{\begin{vmatrix} x & y & 1 & 0 \\ a_1 & a_2 & 1 & 1 \\ b_1 & b_2 & 1 & k \\ c_1 & c_2 & 1 & k' \end{vmatrix}}, \qquad y' = \frac{\begin{vmatrix} x & y & 1 & 0 \\ a_1 & a_2 & 1 & a_2 \\ b_1 & b_2 & 1 & kb_2 \\ c_1 & c_2 & 1 & k'c_2 \end{vmatrix}}{\begin{vmatrix} x & y & 1 & 0 \\ a_1 & a_2 & 1 & 1 \\ b_1 & b_2 & 1 & k \\ c_1 & c_2 & 1 & k' \end{vmatrix}}.$$

Type *II* may be written

$$x' = \frac{\begin{vmatrix} x & y & 1 & 0 \\ a_1 & a_2 & 1 & a_1 \\ b_1 & b_2 & 1 & kb_1 \\ s_1 & s_2 & 0 & ta_1 + s_1 \end{vmatrix}}{\begin{vmatrix} x & y & 1 & 0 \\ a_1 & a_2 & 1 & 1 \\ b_1 & b_2 & 1 & k \\ s_1 & s_2 & 0 & t \end{vmatrix}}, \qquad y' = \frac{\begin{vmatrix} x & y & 1 & 0 \\ a_1 & a_2 & 1 & a_2 \\ b_1 & b_2 & 1 & kb_2 \\ s_1 & s_2 & 0 & ta_2 + s_2 \end{vmatrix}}{\begin{vmatrix} x & y & 1 & 0 \\ a_1 & a_2 & 1 & 1 \\ b_1 & b_2 & 1 & k \\ s_1 & s_2 & 0 & t \end{vmatrix}},$$

and Type *III* may be written

$$x' = \frac{\begin{vmatrix} x & y & 1 & 0 \\ a_1 & a_2 & 1 & a_1 \\ s_1 & s_2 & 0 & ta_1 + s_1 \\ w_1 & w_2 & 0 & (\alpha t^2 + 2\beta t)a_1 + 2\alpha s_1 t + w_1 \end{vmatrix}}{\begin{vmatrix} x & y & 1 & 0 \\ a_1 & a_2 & 1 & 1 \\ s_1 & s_2 & 0 & t \\ w_1 & w_2 & 0 & \alpha t^2 + 2\beta t \end{vmatrix}}$$

$$y' = \frac{\begin{vmatrix} x & y & 1 & 0 \\ a_1 & a_2 & 1 & a_2 \\ s_1 & s_2 & 0 & ta_2 + s_2 \\ w_1 & w_2 & 0 & (\alpha t^2 + 2\beta t)a_2 + 2\alpha s_2 t + w_2 \end{vmatrix}}{\begin{vmatrix} x & y & 1 & 0 \\ a_1 & a_2 & 1 & 1 \\ s_1 & s_2 & 0 & t \\ w_1 & w_2 & 0 & \alpha t^2 + 2\beta t \end{vmatrix}}.$$

**97. Double pairs of a correlation.** We inquire now regarding the existence of double pairs of a correlation in a plane. By a *double pair* is meant a point $X$ and a line $u$ such that the correlation transforms $X$ into $u$ and also transforms $u$ into $X$; in symbols, if $\Gamma$ is the correlation, such that $\Gamma(X) = u$ and $\Gamma(u) = X$. We have already seen (Theorem 3) that if the vertices and opposite sides of a triangle are double pairs of a correlation, the correlation is a polarity.

We may note first that the problem of finding the double pairs of a correlation is in one form equivalent to finding the double elements of a certain collineation. In fact, a double pair $X$, $u$ is such that $\Gamma(X)=u$ and $\Gamma^2(X)=\Gamma(u)=X$, so that the point of a double pair of a correlation $\Gamma$ is a double point of the collineation $\Gamma^2$. Similarly, it may be seen that the lines of the double pairs are the double lines of the collineation $\Gamma^2$. It follows also from these considerations that $\Gamma$ is a polarity, if $\Gamma^2$ is the identical collineation.

Analytically, the problem of determining the double pairs of a correlation leads to the question: For what values of $(x_1, x_2, x_3)$ are the coördinates

$$[a_{11}x_1+a_{21}x_2+a_{31}x_3,\quad a_{12}x_1+a_{22}x_2+a_{32}x_3,\quad a_{13}x_1+a_{23}x_2+a_{33}x_3]$$

of the line to which it corresponds proportional to the coördinates

$$[a_{11}x_1+a_{12}x_2+a_{13}x_3,\quad a_{21}x_1+a_{22}x_2+a_{23}x_3,\quad a_{31}x_1+a_{32}x_2+a_{33}x_3]$$

of the line which corresponds to it in the given correlation? If $\rho$ is the unknown factor of proportionality, this condition is expressed by the equations

$$\begin{aligned}(a_{11}-\rho a_{11})\,x_1+(a_{12}-\rho a_{21})\,x_2+(a_{13}-\rho a_{31})\,x_3&=0,\\ (a_{21}-\rho a_{12})\,x_1+(a_{22}-\rho a_{22})\,x_2+(a_{23}-\rho a_{32})\,x_3&=0,\\ (a_{31}-\rho a_{13})\,x_1+(a_{32}-\rho a_{23})\,x_2+(a_{33}-\rho a_{33})\,x_3&=0,\end{aligned} \tag{1}$$

which must be satisfied by the coördinates $(x_1, x_2, x_3)$ of any point of a double pair. The remainder of the treatment of this problem is similar to the corresponding part of the problem of determining the double elements of a collineation (§ 96). The factor of proportionality $\rho$ is determined by the equation

$$\begin{vmatrix} a_{11}-\rho a_{11} & a_{12}-\rho a_{21} & a_{13}-\rho a_{31}\\ a_{21}-\rho a_{12} & a_{22}-\rho a_{22} & a_{23}-\rho a_{32}\\ a_{31}-\rho a_{13} & a_{32}-\rho a_{23} & a_{33}-\rho a_{33}\end{vmatrix}=0, \tag{2}$$

which is of the third degree and has (under Proposition $K_2$) three roots, of which one is 1, and of which the other two may be proper or improper. Every root of this equation when substituted for $\rho$ in (1) renders these equations consistent. The coördinates $(x_1, x_2, x_3)$ are then determined by solving two of these.

If the reciprocity in question is a polarity, Equations (1) must be satisfied identically, i.e. for every set of values $(x_1, x_2, x_3)$. This would imply that all the relations

$$a_{ij}-\rho a_{ji}=0 \qquad (i, j=1, 2, 3)$$

are satisfied.

Let us suppose first that at least one of the diagonal elements of the matrix of the coefficients $(a_{ij})$ be different from 0. If this be $a_{11}$, the relation $a_{11}-\rho a_{11}=0$ gives at once $\rho=1$; and this value leads at once to the further relations

$$a_{ij}=a_{ji}, \qquad (i, j=1, 2, 3).$$

The matrix in question must then be *symmetrical*. If, on the other hand, we have $a_{11}=a_{22}=a_{33}=0$, there must be some coefficient $a_{ij}$ different from 0. Suppose, for example, $a_{12}\neq 0$. Then the relation $a_{12}-ka_{21}=0$ shows that neither $k$ nor $a_{21}$ can be 0. The substitution of one in the other of the relations $a_{12}=ka_{21}$ and $a_{21}=ka_{12}$ then gives $k^2=1$, or $k=\pm 1$. The value $k=1$ again leads to the condition that the matrix of the coefficients be symmetrical. The value $k=-1$ gives $a_{ii}=0$, and $a_{ij}=-a_{ji}$, which would render the matrix skew symmetrical. The determinant of the transformation would on this supposition vanish (since every skew-symmetrical determinant of odd order vanishes), which is contrary to the hypothesis. The value $k=-1$ is therefore impossible. We have thus been led to the following theorem:

THEOREM 7. *The necessary and sufficient condition that a reciprocity in a plane be a polarity is that the matrix of its coefficients be symmetrical.*

If the coördinate system is chosen so that the point which corresponds to $\rho=1$ in Equation (2) is (1, 0, 0), it is clear that we must have $a_{21}=a_{12}$ and $a_{31}=a_{13}$. If the line corresponding doubly to (1, 0, 0) does not pass through it, the coördinates [1, 0, 0] may be assigned to this line. The equations of the correlation thus assume the form

$$(3) \qquad \begin{aligned} \rho u_1' &= a_{11}x_1 \\ \rho u_2' &= \qquad a_{22}x_2 + a_{23}x_3, \\ \rho u_3' &= \qquad a_{32}x_2 + a_{33}x_3, \end{aligned}$$

and Equation (2) reduces to

$$(4) \qquad a_{11}(1-\rho)\begin{vmatrix} a_{22}-\rho a_{22} & a_{23}-\rho a_{32} \\ a_{32}-\rho a_{23} & a_{33}-\rho a_{33} \end{vmatrix}=0.$$

The roots, other than 1, of this equation clearly correspond to points on [1, 0, 0]. Choosing one of these points (Proposition $K_2$) as (0, 0, 1), we have either $a_{23}=a_{32}$, which would lead to a polarity, or $a_{33}=0$.

In the latter case it is evident that (4) has a double root if $a_{32}=-a_{23}$, but that otherwise it has two distinct roots. Therefore a correlation in which (1, 0, 0) and [1, 0, 0] correspond doubly, and which is not a polarity, may be reduced to one of the three forms:

$$I \qquad \begin{aligned} \rho u_1' &= ax_1, \\ \rho u_2' &= \qquad bx_2 + cx_3, \\ \rho u_3' &= \qquad x_2, \end{aligned} \qquad (0 \neq c \neq \pm 1,\ a \neq 0)$$

$$II \qquad \begin{aligned} \rho u_1' &= ax_1, \\ \rho u_2' &= \qquad bx_2 - x_3, \\ \rho u_3' &= \qquad x_2, \end{aligned} \qquad (a \neq 0,\ b \neq 0)$$

$$IV \qquad \begin{aligned} \rho u_1' &= ax_1, \\ \rho u_2' &= \qquad -x_3, \\ \rho u_3' &= \qquad x_2. \end{aligned} \qquad (a \neq 0)$$

The squares of these correlations are collineations of Types *I, II, IV* respectively.

If the line doubly corresponding to (1, 0, 0) does pass through it, the coördinates [0, 1, 0] may be assigned to this line, and the equations of the correlation become

$$\begin{aligned} \rho u_1' &= \qquad x_2 \\ \rho u_2' &= x_1 + a_{22}x_2 + a_{23}x_3, \\ \rho u_3' &= \qquad a_{32}x_2 + a_{33}x_3. \end{aligned} \qquad (a_{33} \neq 0,\ a_{23} \neq a_{32})$$

Equation (2) at the same time reduces to

$$a_{33}(1-\rho)^3 = 0,$$

and the square of the correlation is always of Type *III.* There are thus five types of correlations, the polarity and those whose squares are collineations of Types *I, II, III, IV.*

### EXERCISES *

1. The points which lie upon the lines to which they correspond in a correlation form a conic section $C^2$, and the lines which lie upon the points to which they correspond are the tangents to a conic $K^2$. How are $C^2$ and $K^2$ related, in each of the five types of correlations, to one another and to the doubly corresponding elements?

* On the theory of correlations see Seydewitz, Archiv der Mathematik, 1st series, Vol. VIII (1846), p. 32; and Schröter, Journal für die reine und angewandte Mathematik, Vol. LXXVII (1874), p. 105.

**2.** If a line $a$ does not lie upon the point $A'$ to which it corresponds in a correlation, there is a projectivity between the points of $a$ and the points in which their corresponding lines meet $a$. In the case of a polarity this projectivity is always an involution. In any other correlation the lines upon which this projectivity is involutoric all pass through a unique fixed point $O$. The line $o$ having the dual property corresponds doubly to $O$. The double points of the involutions on the lines through $O$ are on the conic $C^2$, and the double lines of the involutions on the points of $K^2$ are tangent to $K^2$. $O$ and $o$ are polar with respect to $C^2$ and $K^2$. If a correlation determines involutions on three nonconcurrent lines, it is a polarity.

**3.** The lines of $K^2$ through a point $P$ of $C^2$ are the line which is transformed into $P$ and the line into which $P$ is transformed by the given correlation.

**4.** In a polarity $C^2$ and $K^2$ are the same conic.

**5.** A necessary and sufficient condition that a collineation be the product of two reflections is the existence of a correlation which is left invariant by the collineation.*

**98. Fundamental conic of a polarity in a plane.** We have just seen that a polarity in a plane is given by the equations

$$\text{(1)}\qquad \begin{aligned} \rho u_1' &= a_{11}x_1 + a_{12}x_2 + a_{13}x_3, \\ \rho u_2' &= a_{12}x_1 + a_{22}x_2 + a_{23}x_3, \\ \rho u_3' &= a_{13}x_1 + a_{23}x_2 + a_{33}x_3. \end{aligned} \qquad |a_{ij}| \neq 0$$

DEFINITION. Two homologous elements of a polarity in a plane are called *pole* and *polar*, the point being the *pole* of the line and the line being the *polar* of the point. If two points are so situated that one is on the polar of the other, they are said to be *conjugate*.

The condition that two points in a plane of a polarity be conjugate is readily derived. In fact, if two points $P=(x_1, x_2, x_3)$ and $P'=(x_1', x_2', x_3')$ are conjugate, the condition sought is simply that the point $P'$ shall be on the line $p'=[u_1', u_2', u_3']$, the polar of $P$; i.e. $u_1'x_1' + u_2'x_2' + u_3'x_3' = 0$. Substituting for $u_1', u_2', u_3'$ their values in terms of $x_1, x_2, x_3$ from (1), we obtain the desired condition, viz.:

$$\text{(2)}\qquad \begin{aligned} a_{11}x_1x_1' + a_{22}x_2x_2' + a_{33}x_3x_3' + a_{12}(x_1x_2' + x_2x_1') \\ + a_{13}(x_1x_3' + x_3x_1') + a_{23}(x_2x_3' + x_3x_2') = 0. \end{aligned}$$

As was to be expected, this condition is symmetrical in the coördinates of the two points $P$ and $P'$. By placing $x_i' = x_i$ we obtain the

* This is a special case of a theorem of Dunham Jackson, Transactions of the American Mathematical Society, Vol. X (1909), p. 479.

condition that the point $P$ be *self-conjugate*, i.e. that it be on its polar. We thus obtain the result:

THEOREM 8. *The self-conjugate points of the polarity* (1) *are on the conic whose equation is*

$$(3)\quad a_{11}x_1^2 + a_{22}x_2^2 + a_{33}x_3^2 + 2\,a_{12}x_1x_2 + 2\,a_{13}x_1x_3 + 2\,a_{23}\,x_2x_3 = 0\,;$$

*and, conversely, every point of this conic is self-conjugate.*

This conic is called the *fundamental conic* of the polarity. All of its points may be improper, but it can never degenerate, for, if so, the determinant $|a_{ij}|$ would have to vanish (cf. Ex., p. 187). By duality we obtain

THEOREM 8'. *The self-conjugate lines of the polarity* (1) *are lines of the conic*

$$(4)\quad A_{11}u_1^2 + A_{22}u_2^2 + A_{33}u_3^2 + 2\,A_{12}u_1u_2 + 2\,A_{13}u_1u_3 + 2\,A_{23}u_2u_3 = 0\,;$$

*and, conversely, every line of this conic is self-conjugate.*

Every point $X$ of the conic (3) corresponds in the polarity (1) to the tangent to (3) at $X$. For if not, a point $A$ of (3) would be polar to a line $a$ through $A$ and meeting (3) also in a point $B$. $B$ would then be polar to a line $b$ through $B$, and hence the line $a = AB$ would, by the definition of a polarity, be polar to $ab = B$. This would require that $a$ correspond both to $A$ and to $B$.

If now we recall that the polar system of a conic constitutes a polarity (Theorem 18, Cor., Chap. V) in which all the points and lines of the conic, and only these, are self-conjugate, it follows from the above that every polarity is given by the polar system of its fundamental conic. This and other results following immediately from it are contained in the following theorem:

THEOREM 9. *Every polarity is the polar system of a conic, the fundamental conic of the polarity. The self-conjugate points are the points and the self-conjugate lines are the tangents of this conic. Every pole and polar pair are pole and polar with respect to the fundamental conic.*

This establishes that Equation (4) represents the same conic as Equation (3). The last theorem may be utilized to develop the analytic expressions for poles and polars, and tangents to a conic. This we take up in the next section.

**99. Poles and polars with respect to a conic. Tangents.** We have seen that the most general equation of a conic in point coördinates may be written

$$(1)\quad a_{11}x_1^2 + a_{22}x_2^2 + a_{33}x_3^2 + 2\,a_{12}x_1x_2 + 2\,a_{13}x_1x_3 + 2\,a_{23}x_2x_3 = 0.$$

The result of the preceding section shows that the equation of the same conic in line coördinates is

$$(2)\quad A_{11}u_1^2 + A_{22}u_2^2 + A_{33}u_3^2 + 2\,A_{12}u_1u_2 + 2\,A_{13}u_1u_3 + 2\,A_{23}u_2u_3 = 0,$$

where $A_{ij}$ is the cofactor of $a_{ij}$ in the determinant

$$\begin{vmatrix} a_{11} & a_{12} & a_{13} \\ a_{12} & a_{22} & a_{23} \\ a_{13} & a_{23} & a_{33} \end{vmatrix}.$$

This result may also be stated as follows:

THEOREM 10. *The necessary and sufficient condition that the line $u_1x_1 + u_2x_2 + u_3x_3 = 0$ be tangent to the conic* (1) *is that Equation* (2) *be satisfied.*

COROLLARY. *This condition may also be written in the form*

$$\begin{vmatrix} a_{11} & a_{12} & a_{13} & u_1 \\ a_{21} & a_{22} & a_{23} & u_2 \\ a_{31} & a_{32} & a_{33} & u_3 \\ u_1 & u_2 & u_3 & 0 \end{vmatrix} = 0.$$

Equation (2) of the preceding section expresses the condition that the points $(x_1, x_2, x_3)$ and $(x_1', x_2', x_3')$ be conjugate with respect to the conic (1). If in this equation $(x_1', x_2', x_3')$ be supposed given, while $(x_1, x_2, x_3)$ is regarded as variable, this condition is satisfied by all the points of the polar of $(x_1', x_2', x_3')$ with respect to the conic and by no others. It is therefore the equation of this polar. When arranged according to the variable coördinates $x_i$, it becomes

$$(3)\qquad (a_{11}x_1' + a_{12}x_2' + a_{13}x_3')\,x_1 + (a_{12}x_1' + a_{22}x_2' + a_{23}x_3')\,x_2 + (a_{13}x_1' + a_{23}x_2' + a_{33}x_3')\,x_3 = 0;$$

while if we arrange it according to the coördinates $x_i'$, it becomes

$$(4)\qquad (a_{11}x_1 + a_{12}x_2 + a_{13}x_3)\,x_1' + (a_{12}x_1 + a_{22}x_2 + a_{23}x_3)\,x_2' + (a_{13}x_1 + a_{23}x_2 + a_{33}x_3)\,x_3' = 0.$$

Now it is readily verified that the latter of these equations may be derived from the equation (1) of the conic by applying to the left-hand member of this equation the *polar operator*

$$x_1'\frac{\partial}{\partial x_1}+x_2'\frac{\partial}{\partial x_2}+x_3'\frac{\partial}{\partial x_3}$$

(§ 89) and dividing the resulting equation by 2. Furthermore, if we define the symbols $\frac{\partial f}{\partial x_1'}, \frac{\partial f}{\partial x_2'}, \frac{\partial f}{\partial x_3'}$ to be the result of substituting $(x_1', x_2', x_3')$ for $(x_1, x_2, x_3)$ in the expressions $\frac{\partial f}{\partial x_1}, \frac{\partial f}{\partial x_2}, \frac{\partial f}{\partial x_3}$ ($f$ being any polynomial in $x_1, x_2, x_3$), it is readily seen that Equation (3) is equivalent to

$$x_1\frac{\partial f}{\partial x_1'}+x_2\frac{\partial f}{\partial x_2'}+x_3\frac{\partial f}{\partial x_3'}=0,$$

where now $f$ is the left-hand member of (1).

This leads to the following theorem:

THEOREM 11. *If $f=0$ is the equation of a conic in homogeneous point coördinates, the equation of the polar of any point $(x_1', x_2', x_3')$ is given by either of the equations*

$$x_1'\frac{\partial f}{\partial x_1}+x_2'\frac{\partial f}{\partial x_2}+x_3'\frac{\partial f}{\partial x_3}=0 \quad \text{or} \quad x_1\frac{\partial f}{\partial x_1'}+x_2\frac{\partial f}{\partial x_2'}+x_3\frac{\partial f}{\partial x_3'}=0.$$

*If the point $(x_1', x_2', x_3')$ is a point on the conic, either of these equations represents the tangent to the conic $f=0$ at this point.*

**100. Various definitions of conics.** The definition of a (point) conic as the locus of the intersections of homologous lines of two projective flat pencils in the same plane was first given by *Steiner* in 1832 and used about the same time by *Chasles*. The considerations of the preceding sections at once suggest two other methods of definition, one synthetic, the other analytic. The former begins by the synthetic definition of a polarity (cf. p. 263), and then defines *a point conic as the set of all self-conjugate points of a polarity,* and a line conic as the set of all self-conjugate lines of a polarity. This definition was first given by *von Staudt* in 1847. From it he derived the fundamental properties of conics and showed easily that his definition is equivalent to Steiner's. The analytic method is to define a (point) conic as the set of all points satisfying any equation of the second degree, homogeneous in three variables $x_1, x_2, x_3$. This definition (at least in its nonhomogeneous form) dates back to *Descartes* and *Fermat* (1637) and the introduction of the notions of analytic geometry.

The oldest definition of conics is due to the ancient Greek geometers, who defined a conic as the plane section of a circular cone. This definition involves metric ideas and hence does not concern us at this point. We will return to it later. It is of interest to note in passing, however, that from this definition Apollonius (about 200 A.D.) derived a theorem equivalent to the one that the equation of a conic in point coördinates is of the second degree.

The reader will find it a valuable exercise to derive for himself the fundamental properties of polarities synthetically, and thence to develop the theory of conics from von Staudt's definition, at least so far as to show that his definition is equivalent to Steiner's. It may be noted that von Staudt's definition has the advantage over Steiner's of including, without reference to Proposition $K_2$, conics consisting entirely of improper points (since there exist polarities which have no proper self-conjugate points). The reader may in this connection refer to the original work of von Staudt, Die Geometrie der Lage, Nürnberg (1847); or to the textbook of Enriques, Vorlesungen über projective Geometrie, Leipzig (1903).

**EXERCISES**

**1.** Derive the condition of Theorem 10 directly by imposing the condition that the quadratic which determines the intersections of the given line with the conic shall have equal roots. What is the dual of this theorem?

**2.** Verify analytically the fundamental properties of poles and polars with respect to a conic (Theorems 13–18, Chap. V).

**3.** State the dual of Theorem 11.

**4.** Show how to construct the correlation between a plane of points and a plane of lines, having given the homologous pairs $A$, $a'$; $B$, $b'$; $C$, $c'$; $D$, $d'$.

**5.** Show that a correlation between two planes is uniquely determined if two pencils of points in one plane are made projective respectively with two pencils of lines in the other, provided that in this projectivity the point of intersection of the axes of the two pencils of points corresponds to the line joining the two centers of the pencils of lines.

**6.** Show that in our system of homogeneous point and line coördinates the pairs of points and lines with the same coördinates are poles and polars with respect to the conic $x_1^2 + x_2^2 + x_3^2 = 0$.

**7.** On a general line of a plane in which a polarity has been defined the pairs of conjugate points form an involution the double points of which are the (proper or improper) points of intersection of the line with the fundamental conic of the polarity.

**8.** A polarity in a plane is completely defined if a self-polar triangle is given together with one pole and polar pair of which the point is not on a side nor the line on a vertex of the triangle.

**9.** Prove Theorem 3 analytically.

**10.** Given a simple plane pentagon, there exists a polarity in which to each vertex corresponds the opposite side.

**11.** The three points $A'$, $B'$, $C'$ on the sides $BC$, $CA$, $AB$ of a triangle that are conjugate in a polarity to the vertices $A$, $B$, $C$ respectively are collinear (cf. Ex. 13, p. 125).

**12.** Show that a polarity is completely determined when the two involutions of conjugate points on two conjugate lines are given.

**13.** Construct the polarity determined by a self-polar triangle $ABC$ and an involution of conjugate points on a line.

**14.** Construct the polarity determined by two pole and polar pairs $A$, $a$ and $B$, $b$ and one pair of conjugate points $C$, $C'$.

**15.** If a triangle $STU$ is self-polar with regard to a conic $C^2$, and $A$ is any point of $C^2$, there are three triangles having $A$ as a vertex which are inscribed to $C^2$ and circumscribed to $STU$ (Sturm, Die Lehre von den geometrischen Verwandtschaften, Vol. I, p. 147).

**101. Pairs of conics.** If two polarities, i.e. two conics (proper or improper), are given, their product is a collineation which leaves invariant any point or line which has the same polar or pole with regard to both conics. Moreover, any point or line which is not left invariant by this collineation must have different polars or poles with regard to the two conics. Hence the points and lines which have the same polars and poles with regard to two conics in the same plane form one of the five invariant figures of a nonidentical collineation.

*Type I.* If the common self-polar figure of the two conics is of Type *I*, it is a self-polar triangle for both conics. Since any two conics are projectively equivalent (Theorem 9, Chap. VIII), the coördinate system may be so chosen that the equation of one of the conics, $A^2$, is

$$x_1^2 - x_2^2 + x_3^2 = 0. \tag{1}$$

With regard to this conic the triangle (0, 0, 1), (0, 1, 0), (1, 0, 0) is self-polar. The general equation of a conic with respect to which this triangle is self-polar is clearly

$$a_1x_1^2 - a_2x_2^2 + a_3x_3^2 = 0. \tag{2}$$

An equation of the form (2) may therefore be taken as the equation of the other conic, $B^2$, if (1) and (2) have no other common self-polar elements than the fundamental triangle. Consider the set of conics

$$a_1x_1^2 - a_2x_2^2 + a_3x_3^2 + \lambda(x_1^2 - x_2^2 + x_3^2) = 0. \tag{3}$$

The coördinates of any point which satisfy (1) and (2) also satisfy (3). Hence all conics (3) pass through the points common to $A^2$ and $B^2$. For the value $\lambda = -a_3$, (3) gives the pair of lines

$$(4) \qquad (a_1 - a_3)x_1^2 - (a_2 - a_3)x_2^2 = 0,$$

which intersect in (0, 0, 1). The points of intersection of these lines with (1) are common to all the conics (3).

The lines (4) are distinct, unless $a_1 = a_3$ or $a_2 = a_3$. But if $a_1 = a_3$, any point $(x_1', 0, x_3')$ on the line $x_2 = 0$ has the polar $x_1'x_1 + x_3'x_3 = 0$ both with regard to (1) and with regard to (2). The self-polar figure is therefore of Type *IV*. In order that this figure be of Type *I*, the three numbers $a_1$, $a_2$, $a_3$ must all be distinct. If this condition is satisfied, the lines (4) meet the conics (3) in four distinct points.

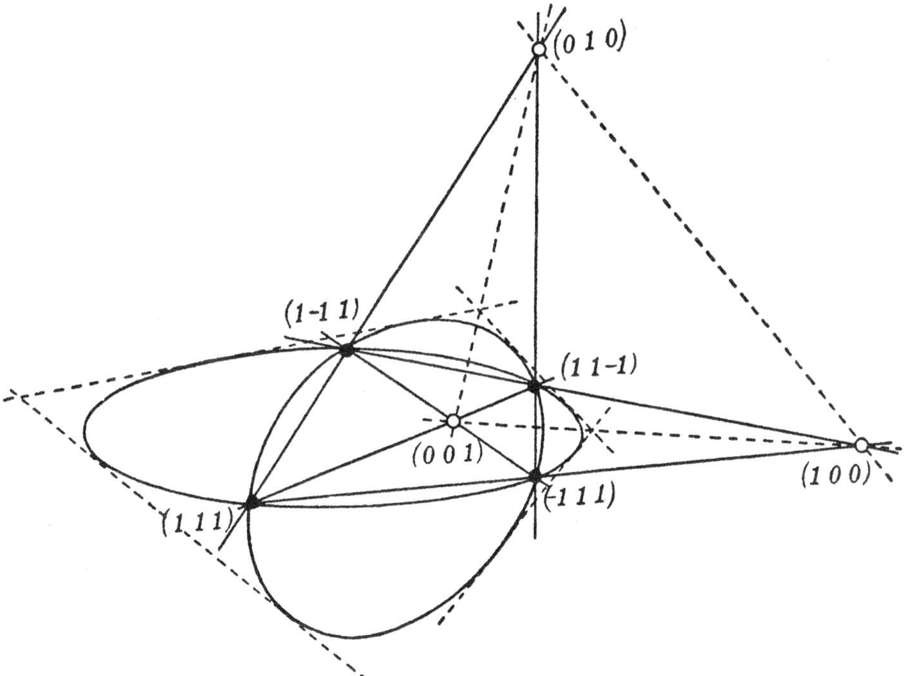


Fig. 100

The actual construction of the points is now a problem of the second degree. We have thus established (fig. 100)

Theorem 12. *If two conics have a common self-polar triangle (and no other common self-polar pair of point and line), they intersect in four distinct points (proper or improper). Any two conics of the pencil determined by these points have the same self-polar triangle. Dually, two such conics have four common tangents, and any two conics of the range determined by these common tangents have the same self-polar triangle.*

COROLLARY. *Any pencil of conics of Type I can be represented by* *

$$(x_1^2 - x_2^2) + \lambda (x_2^2 - x_3^2) = 0, \tag{5}$$

*the four common points being in this case* (1, 1, 1), (1, 1, − 1), (1, − 1, 1), *and* (− 1, 1, 1).

*Type II.* When the common self-polar figure is of Type *II*, one of the points lies on its polar, and therefore this polar is a tangent to each of the conics $A^2$, $B^2$. Since two tangents cannot intersect in a point of contact, the two lines of the self-polar figure are not both tangents. Hence the point $B$ of the self-polar figure which is on only one of the lines is the pole of the line $b$ of the figure which is on only one of the points (fig. 101), and the line $a$ on the two points is tangent to both conics at the point $A$ which is on the two lines.

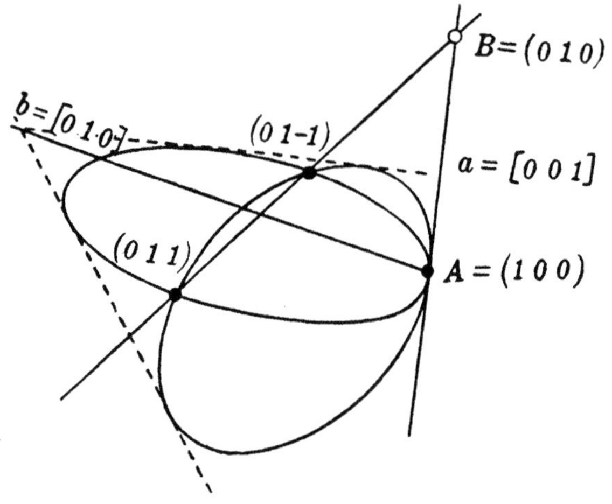

FIG. 101

Choose a system of coördinates with $A = (1, 0, 0)$, $a = [0, 0, 1]$, $B = (0, 1, 0)$, and $b = [0, 1, 0]$. The equation of any conic being

$$a_1 x_1^2 + a_2 x_2^2 + a_3 x_3^2 + 2\, b_1 x_2 x_3 + 2\, b_2 x_1 x_3 + 2\, b_3 x_1 x_2 = 0,$$

the condition that $A$ be on the conic is $a_1 = 0$; that $a$ then be tangent is $b_3 = 0$; that $b$ then be the polar of $B$ is $b_1 = 0$. Hence the general equation of a conic with the given self-polar figure is

$$a_2 x_2^2 + a_3 x_3^2 + 2\, b_2 x_1 x_3 = 0. \tag{6}$$

* Equation (5) is typical for a pencil of conics of Type *I*, and Theorem 12 is a sort of converse to the developments of § 47, Chap. V. The reader will note that if the problem of finding the points of intersection of two conics is set up directly, it is of the fourth degree, but that it is here reduced to a problem of the third degree (the determination of a common self-polar triangle) followed by two quadratic constructions. This corresponds to the well-known solution of the general biquadratic equation (cf. Fine, College Algebra, p. 486). For a further discussion of the analytic geometry of pencils of conics, cf. Clebsch-Lindemann, Vorlesungen über Geometrie, 2d ed., Vol. I, Part I (1906), pp. 212 ff.

Since any two conics are projectively equivalent, $A^2$ may be chosen to be

(7) $$x_2^2 + x_3^2 + 2\,x_1x_3 = 0.$$

The equation of $B^2$ then has the form (6), with the condition that the two conics have no other common self-polar elements. Since the figure in which $a$ is polar to $A$ and $b$ to $B$ can only reduce to Types *IV* or *V*, we must determine under what conditions each point on $a$ or each point on $b$ has the same polar with regard to (6) and (7). The polar of $(x_1', x_2', x_3')$ with regard to (6) is given by

$$a_2x_2'x_2 + a_3x_3'x_3 + b_2x_3'x_1 + b_2x_1'x_3 = 0.$$

Hence the first case can arise only if $a_2 = b_2$; and the second only if $a_3 = b_2$.

Introducing the condition that $a_2$, $a_3$, $b_2$ are all distinct, it is then clear that the set of conics

$$a_2x_2^2 + a_3x_3^2 + 2\,b_2x_1x_3 + \lambda(x_2^2 + x_3^2 + 2\,x_1x_3) = 0$$

contains a line pair for $\lambda = -a_2$, viz. the lines

$$(a_3 - a_2)\,x_3^2 + 2\,(b_2 - a_2)\,x_1x_3 = 0.$$

Hence the conics have in common the points of intersection with (7) of the line

$$(a_3 - a_2)\,x_3 + 2\,(b_2 - a_2)\,x_1 = 0.$$

This gives

THEOREM 13. *If two conics have a common self-polar figure of Type II, they have three points in common and a common tangent at one of them. Dually, they have three common tangents and a common point of contact on one of the tangents. The two conics determine a pencil and also a range of conics of Type II.*

COROLLARY. *Any pencil of conics of Type II may be represented by the equation* $x_2^2 - x_3^2 + \lambda x_3x_1 = 0$. *The conics of this pencil all pass through the points* (0, 1, 1), (0, 1, $-$1), (1, 0, 0) *and are tangent to* $x_3 = 0$.

*Type III.* When the common self-polar figure is of Type *III*, the two conics evidently have a common tangent and a common point of contact, and only one of each. Let the common tangent be $x_3 = 0$, its point of contact be (1, 0, 0), and let $A^2$ be given by

(8) $$x_2^2 + 2\,x_3x_1 = 0.$$

The general equation of a conic tangent to $x_3=0$ at $(1, 0, 0)$ is

$$(9) \qquad a_2x_2^2+a_3x_3^2+2\,b_1x_2x_3+2\,b_2x_1x_3=0,$$

with regard to which the polar of any point $(x_1', x_2', 0)$ on $x_3=0$ is given by

$$(10) \qquad a_2x_2'x_2+b_1x_2'x_3+b_2x_1'x_3=0.$$

This will be identical with the polar of $(x_1', x_2', 0)$ with regard to $A^2$ for all values of $x_1'$, $x_2'$, if $b_2=a_2$ and $b_1=0$. Since $(1, 0, 0)$ only is to have the same polar with regard to both conics, we impose at least one of the conditions $b_2\neq a_2$, $b_1\neq 0$. The line (10) will now be identical with the polar of (8) for any point $(x_1', x_2', 0)$ satisfying the condition

$$\frac{x_2'}{x_1'}=\frac{a_2x_2'}{b_1x_2'+b_2x_1'}.$$

This quadratic equation must have only one root if the self-polar figure is to be of Type *III*. This requires $b_2=a_2$, and as $b_2$, $a_2$ cannot both be 0 unless (9) degenerates, the equation of $B^2$ can be taken as

$$(11) \qquad x_2^2+2\,x_3x_1+a_3x_3^2+2\,b_1x_2x_3=0, \qquad (b_1\neq 0).$$

The conics (8) and (11) now evidently have in common the points of intersection of (8) with the line pair

$$a_3x_3^2+2\,b_1x_2x_3=0,$$

and no other points. Since $x_3=0$ is a tangent, this gives two common points. If the second common point is taken to be $(0, 0, 1)$, the set of conics which have in common the points $(0, 0, 1)$ and $(1, 0, 0)=A$ and the tangent $a$ at $A$, and no other points, may be written (fig. 102)

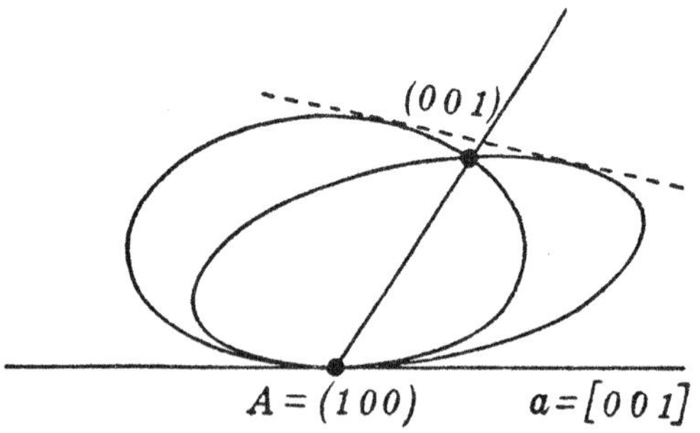


FIG. 102

$$(12) \qquad x_2^2+2\,x_1x_3+\lambda x_2x_3=0.$$

THEOREM 14. *If two conics have a common self-polar figure of Type III, they have two points in common and a common tangent at one of them, and one other common tangent. They determine a pencil and a range of conics of Type III.*

COROLLARY. *A pencil of conics of Type III can be represented by the equation* $x_2^2 + 2x_3x_1 + \lambda x_2x_3 = 0$.

*Type IV.* When the common self-polar figure is of Type *IV*, let the line of fixed points be $x_3 = 0$ and its pole be (0, 0, 1). The coördinates being chosen as they were for Type *I*, the conic $A^2$ has the equation

$$x_1^2 - x_2^2 + x_3^2 = 0;$$

and any other conic having in common with $A^2$ the self-polar triangle (1, 0, 0), (0, 1, 0), (0, 0, 1) has an equation of the form

$$a_1x_1^2 + a_2x_2^2 + a_3x_3^2 = 0.$$

The condition that every point on $x_3 = 0$ shall have the same polar with regard to this conic as with regard to $A^2$ is $a_1 = -a_2$. Hence $B$ may be written

$$x_1^2 - x_2^2 + \lambda x_3^2 = 0.$$

Any conic of this form has the same tangents as $A^2$ at the points (1, 1, 0) and (1, −1, 0) (fig. 103). Hence, if $\lambda$ is a variable parameter, the last equation represents a pencil of conics of Type *IV* according to the classification previously made.

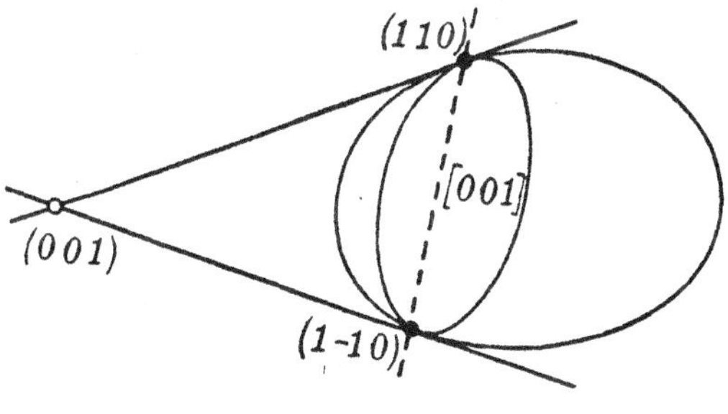

FIG. 103

THEOREM 15. *If two conics have a common self-polar figure of Type IV, they have two points in common and the tangents at these points. They determine a pencil (which is also a range) of conics of Type IV.*

COROLLARY. *A pencil of conics of Type IV may be represented by the equation*

$$x_1^2 - x_2^2 + \lambda x_3^2 = 0;$$

*and also by the equation*

$$x_1^2 + \lambda x_2x_3 = 0.$$

*Type V.* When the common self-polar figure is of Type *V*, let the point of fixed lines be (1, 0, 0) and the line of fixed points be $x_3 = 0$. As in Type *III*, let $A^2$ be given by

$$(8) \qquad x_2^2 + 2x_1x_3 = 0.$$

We have seen, in the discussion of that type, that all points of $x_3 = 0$ have the same polars with respect to (8) and (9), if in (9) we have

$b_2 = a_2$ and $b_1 = 0$. Hence, if $A^2$ and $B^2$ are to have a common self-polar figure of Type *V*, the equation of $B^2$ must have the form

$$(13)\quad a_2(x_2^2 + 2\,x_1x_3) + a_3x_3^2 = 0.$$

From the form of equations (8) and (13) it is evident that the conics have in common only the point (1, 0, 0) and the tangent $x_3 = 0$, and that every point on $x_3 = 0$ has the same polar with respect to both conics (fig. 104). Hence they determine a pencil of Type *V*.

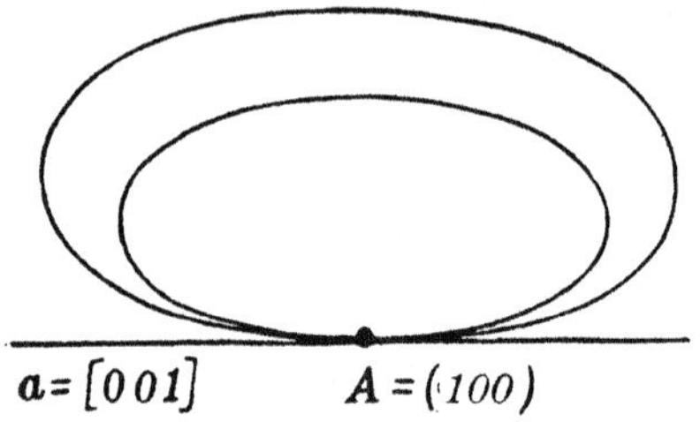


FIG. 104

THEOREM 16. *If two conics have a common self-polar figure of Type V, they have a lineal element (and no other elements) in common and determine a pencil (which is also a range) of conics of Type V according to the classification already given.*

COROLLARY. *A pencil of conics of Type V can be represented by the equation*

$$x_2^2 + 2\,x_1x_3 + \lambda x_3^2 = 0.$$

As an immediate consequence of the corollaries of Theorems 12–16 we have

THEOREM 17. *Any pencil of conics may be written in the form*

$$f + \lambda g = 0,$$

*where $f = 0$ and $g = 0$ are the equations of two conics (degenerate or not) of the pencil.*

## EXERCISES

**1.** Prove analytically that the polars of a point *P* with respect to the conics of a pencil all pass through a point *Q*. The points *P* and *Q* are double points of the involution determined by the conics of the pencil on the line *PQ*. Give a linear construction for *Q* (cf. Ex. 3, p. 136). The correspondence obtained by letting every point *P* correspond to the associated point *Q* is a "quadratic birational transformation." Determine the equations representing this transformation. The point *Q*, which is conjugate to *P* with regard to all the conics of the pencil, is called the *conjugate* of *P* with respect to the pencil. The locus of the conjugates of the points of a line with regard to a pencil of conics is a conic (cf. Ex. 31, p. 140).

**2.** One and only one conic passes through four given points and has two given points as conjugate points, provided the two given points are not conjugate with respect to all the conics of the pencil determined by the given set of four. Show how to construct this conic.

3. One conic in general, or a pencil of conics in a special case, passes through three given points and has two given pairs of points as conjugate points. Give the construction.

4. One conic in general, or a pencil of conics in a special case, passes through two given points and has three pairs of given points as conjugate points; or passes through a given point and has four pairs of given points as conjugate points; or has five given pairs of conjugate points. Give the corresponding constructions for each case.

**102. Problems of the third and fourth degrees.*** The problem of constructing the points of intersection of two conics in the same plane is, in general, of the fourth degree according to the classification of geometric problems described in § 83. Indeed, if one of the coördinates be eliminated between the equations of two conics, the resulting equation is, in general, an irreducible equation of the fourth degree. Moreover, a little consideration will show that any equation of the fourth degree may be obtained in this way. It results that every problem of the fourth degree in a plane may be reduced to the problem of constructing the common points (or by duality the common tangents) of two conics. Further, the problem of finding the remaining intersections of two conics in a plane of which one point of intersection is given, is readily seen to be of the third degree, in general; and any problem of this degree can be reduced to that of finding the remaining intersections of two conics of which one point of intersection is known. It follows that any problem of the third or fourth degree in a plane may be reduced to that of finding the common elements of two conics in the plane.†

A problem of the fourth (or third) degree cannot therefore be solved by the methods sufficient for the solution of problems of the first and second degrees (straight edge and compass).‡ In the case of problems of the second degree we have seen that any such problem could be solved by linear constructions if the intersections of

* In this section we have made use of Amodeo, Lezioni di Geometria Projettiva, pp. 436, 437. Some of the exercises are taken from the same book, pp. 448–451.

† Moreover, we have seen (p. 289, footnote) that any problem of the fourth degree may be reduced to one of the third degree, followed by two of the second degree.

‡ With the usual representation of the ordinary real geometry we should require an instrument to draw conics.

every line in the plane with a fixed conic in that plane were assumed known. Similarly, any problem of the fourth (or third) degree can be solved by linear and quadratic constructions if the intersections of every conic in the plane with a fixed conic in this plane are assumed known. This follows readily from the fact that any conic in the plane can be transformed by linear constructions into the fixed conic. A problem of the third or fourth degree in a plane will then, in the future, be considered solved if it has been reduced to the finding of the intersections of two conics, combined with any linear or quadratic constructions. As a typical problem of the third degree, for example, we give the following:

*To find the double points of a nonperspective collineation in a plane which is determined by four pairs of homologous points.*

*Solution.* When four pairs of homologous elements are given, we can construct linearly the point or line homologous with any given point or line in the plane. Let the collineation be represented by $\Pi$, and let $A$ be any point of the plane which is not on an invariant line. Let $\Pi(A)=A'$ and $\Pi(A')=A''$. The points $A$, $A'$, $A''$ are then not collinear. The pencil of lines at $A$ is projective with the pencil at $A'$, and these two projective pencils generate a conic $C^2$ which passes through all the double points of $\Pi$, and which is tangent at $A'$ to the line $A'A''$ (fig. 105). The conic $C^2$ is transformed by the collineation $\Pi$ into a conic $C_1^2$ generated by the projective pencils of lines at $A'$ and $A''$.

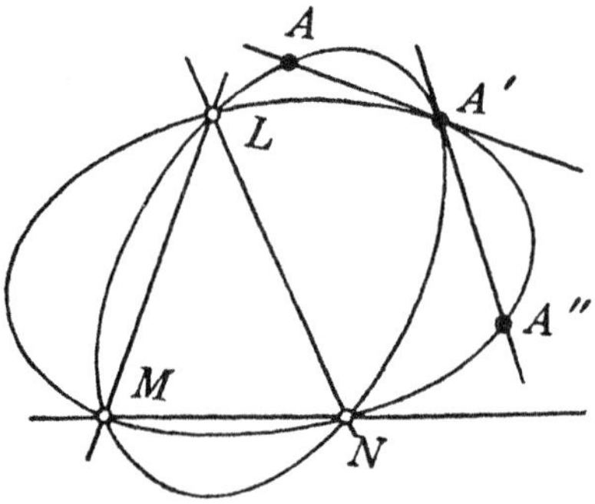

FIG. 105

$C_1^2$ also passes through $A'$ and is tangent at this point to the line $AA'$. The double points of $\Pi$ are also points of $C_1^2$. The point $A'$ is not a double point of $\Pi$ by hypothesis. It is evident, however, that every other point common to the two conics $C^2$ and $C_1^2$ is a double point.

If $C^2$ and $C_1^2$ intersect again in three distinct points $L$, $M$, $N$, the latter form a triangle and the collineation is of Type *I*. If $C^2$ and $C_1^2$ intersect in a point $N$, distinct from $A'$, and are tangent to each other at a third point $L=M$, the collineation has $M$, $N$ for double points and the line $MN$ and the common tangent at $M$ for double lines (fig. 106); it is then of Type *II*. If, finally, the two conics have contact of the second order at a point $L=M=N$, distinct from $A'$, the collineation has the single double line which is tangent to the conics at this point, and is of Type *III* (fig. 107).

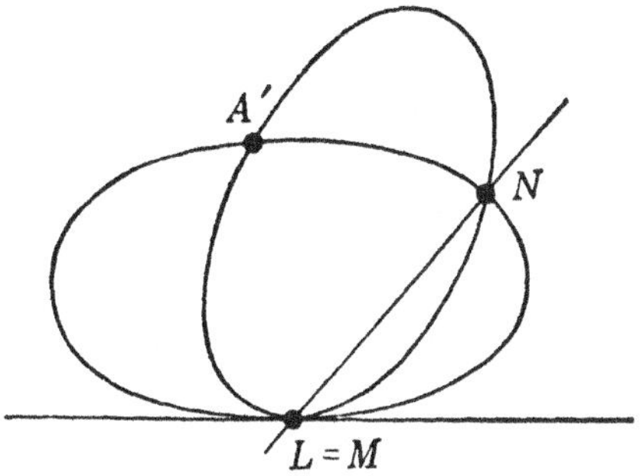

FIG. 106

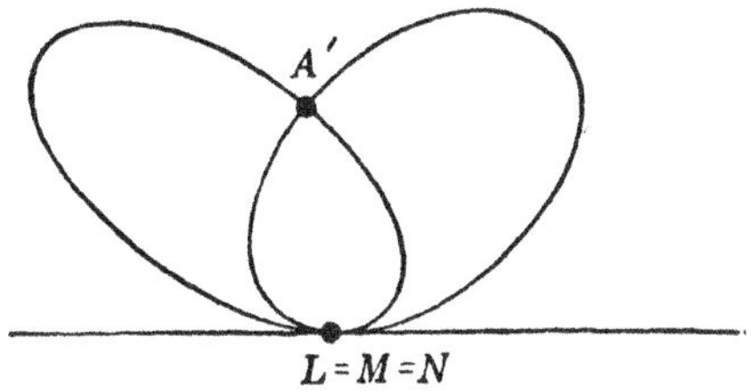

FIG. 107

## EXERCISES

**1.** Give a discussion of the problem above without making at the outset the hypothesis that the collineation is nonperspective.

**2.** Construct the double pairs of a correlation in the plane, which is not a polarity.

**3.** Given two polarities in a plane, construct their common pole and polar pairs.

**4.** On a line tangent to a conic at a point $A$ is given an involution I, and from any pair of conjugates $P$, $P'$ of I are drawn the second tangents $p$, $p'$ to the conic, their points of contact being $Q$, $Q'$ respectively. Show that the locus of the point $pp'$ is a line, $l$, passing through the conjugate, $A'$, of $A$ in the involution I; and that the line $QQ'$ passes through the pole of $l$ with respect to the conic.

**5.** Construct the conic which is tangent at two points to a given conic and which passes through three given points. Dualize.

**6.** The lines joining pairs of homologous points of a noninvolutoric projectivity on a conic $A^2$ are tangent to a second conic $B^2$ which is tangent to $A^2$ at two points, or which hyperosculates $A^2$.

**7.** A pencil of conics of Type *II* is determined by three points $A$, $B$, $C$ and a line $c$ through $C$. What is the locus of the points of contact of the conics of the pencil with the tangents drawn from a given point $P$ of $c$?

**8.** Construct the conics which pass through a given point $P$ and which are tangent at two points to each of two given conics.

**9.** If $f=0$, $g=0$, $h=0$ are the equations of three conics in a plane not belonging to the same pencil, the system of conics given by the equation

$$\lambda f+\mu g+\nu h=0,$$

$\lambda$, $\mu$, $\nu$ being variable parameters, is called a *bundle* of conics. Through every point of the plane passes a pencil of conics belonging to this bundle; through any two distinct points passes in general one and only one conic of the bundle. If the conics $f$, $g$, $h$ have a point in common, this point is common to all the conics of the bundle. Give a nonalgebraic definition of a bundle of conics.

**10.** The set of all conics in a plane passing through the vertices of a triangle form a bundle. If the equations of the sides of this triangle are $l = 0$, $m = 0$, $n = 0$, show that the bundle may be represented by the equation

$$\lambda mn + \mu nl + \nu lm = 0.$$

What are the degenerate conics of this bundle?*

**11.** The set of all conics in a plane which have a given triangle as a self-polar triangle forms a bundle. If the equations of the sides of this triangle are $l = 0$, $m = 0$, $n = 0$, show that the bundle may be represented by the equation

$$\lambda l^2 + \mu m^2 + \nu n^2 = 0.$$

What are the degenerate conics of this bundle?

**12.** The conics of the bundle described in Ex. 11 which pass through a general point $P$ of the plane pass through the other three vertices of the quadrangle, of which one vertex is $P$ and of which the given triangle is the diagonal triangle. What happens when $P$ is on a side of the given triangle? Dualize.

**13.** The reflections whose centers and axes are the vertices and opposite sides of a triangle form a commutative group. Any point of the plane not on a side of the triangle is transformed by the operations of this group into the other three vertices of a complete quadrangle of which the given triangle is the diagonal triangle. If this triangle is taken as the reference triangle, what are the equations of transformation? What conics are transformed into themselves by the group, and how is it associated with the quadrangle-quadrilateral configuration?

**14.** The necessary and sufficient condition that two reflections be commutative is that the center of each shall be on the axis of the other.

**15.** The invariant figure of a collineation may be regarded as composed of two lineal elements, the five types corresponding to various special relations between the two lineal elements.

**16.** A correlation which transforms a lineal element $Aa$ into a lineal element $Bb$ and also transforms $Bb$ into $Aa$ is a polarity.

**17.** How many collineations and correlations are in the group generated by the reflections whose centers and axes are the vertices and opposite sides of a triangle and a polarity with regard to which the triangle is self-polar?

* In connection with this and the two following exercises, cf. Castelnuovo, Lezioni di Geometria Analitica e Projettiva, Vol. I, p. 395.

# CHAPTER XI *

## FAMILIES OF LINES

**103. The regulus.** The following theorem, on which depends the existence of the figures to be studied in this chapter, is logically equivalent (in the presence of Assumptions A and E) to Assumption P. It might have been used to replace that assumption.

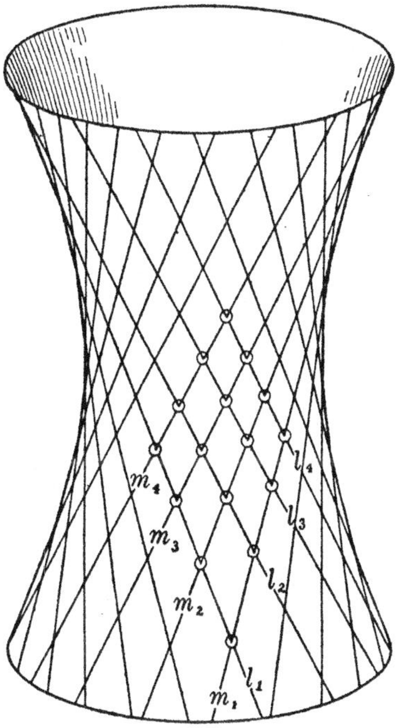

FIG. 108

THEOREM 1. *If* $l_1$, $l_2$, $l_3$ *are three mutally skew lines, and if* $m_1$, $m_2$, $m_3$, $m_4$ *are four lines each of which meets each of the lines* $l_1$, $l_2$, $l_3$, *then any line* $l_4$ *which meets three of the lines* $m_1$, $m_2$, $m_3$, $m_4$ *also meets the fourth.*

*Proof.* The four planes $l_1m_1$, $l_1m_2$, $l_1m_3$, $l_1m_4$ of the pencil with axis $l_1$ are perspective through the pencil of points on $l_3$ with the four planes $l_2m_1$, $l_2m_2$, $l_2m_3$, $l_2m_4$ of the pencil with axis $l_2$ (fig. 108). For, by hypothesis, the lines of intersection $m_1$, $m_2$, $m_3$, $m_4$ of the pairs of homologous planes all meet $l_3$. The set of four points in which the four planes of the pencil on $l_1$ meet $l_4$ is therefore projective with the set of four points in which the four planes of the pencil on $l_2$ meet $l_4$. But $l_4$ meets three of the pairs of homologous planes in points of their lines of intersection, since, by hypothesis, it meets three of the lines $m_1$, $m_2$, $m_3$, $m_4$. Hence in the projectivity on $l_4$ there are three invariant points, and hence (Assumption P) every point is invariant. Hence $l_4$ meets the remaining line of the set $m_1$, $m_2$, $m_3$, $m_4$.

* All the developments of this chapter are on the basis of Assumptions A, E, P, $H_0$. But see the exercise on page 261.

Definition. If $l_1$, $l_2$, $l_3$ are three lines no two of which are in the same plane, the set of all lines which meet each of the three given lines is called a *regulus.* The lines $l_1$, $l_2$, $l_3$ are called *directrices* of this regulus.

It is clear that no two lines of a regulus can intersect, for otherwise two of the directrices would lie in a plane. The next theorem follows at once from the definition.

Theorem 2. *If* $l_1$, $l_2$, $l_3$ *are three lines of a regulus of which* $m_1$, $m_2$, $m_3$ *are directrices,* $m_1$, $m_2$, $m_3$ *are lines of the regulus of which* $l_1$, $l_2$, $l_3$ *are directrices.*

It follows that any three lines no two of which lie in a plane are directrices of one and only one regulus and are lines of one and only one regulus.

Definition. Two reguli which are such that every line of one meets all the lines of the other are said to be *conjugate.* The lines of a regulus are called its *generators* or *rulers;* the lines of a conjugate regulus are called the *directrices* of the given regulus.

Theorem 3. *Every regulus has one and only one conjugate regulus.*

This follows immediately from the preceding. Also from the proof of Theorem 1 we have

Theorem 4. *The correspondence established by the lines of a regulus between the points of two lines of its conjugate regulus is projective.*

Theorem 4′. *The correspondence established by the lines of a regulus between the planes on any two lines of its conjugate regulus is projective.*

Theorem 5. *The set of all lines joining pairs of homologous points of two projective pencils of points on skew lines is a regulus.*

Theorem 5′. *The set of all lines of intersection of pairs of homologous planes of two projective pencils of planes on skew lines is a regulus.*

*Proof.* We may confine ourselves to the proof of the theorem on the left. By Theorem 6, Chap. III, the two pencils of points are perspective through a pencil of planes. Every line joining a pair of homologous points of these two pencils, therefore, meets the axis of the pencil of planes. Hence all these lines meet three (necessarily skew) lines, namely the axes of the two pencils of points and of the pencil of planes, and therefore satisfy the definition of a regulus. Moreover, every line which meets these three lines joins a pair of homologous points of the two pencils of points.

THEOREM 6. *If* $[p]$ *are the lines of a regulus and* $q$ *is a directrix of the regulus, the pencil of points* $q[p]$ *is projective with the pencil of planes* $q[p]$.

*Proof.* Let $q'$ be any other directrix. By Theorem 4 the pencil of points $q[p]$ is perspective with the pencil of points $q'[p]$. But each of the points of this pencil lies on the corresponding plane $qp$. Hence the pencil of points $q'[p]$ is also perspective with the pencil of planes $q[p]$.

### EXERCISES

**1.** Every point which is on a line of a regulus is also on a line of its conjugate regulus.

**2.** A plane which contains one line of a regulus contains also a line of its conjugate regulus.

**3.** Show that a regulus is uniquely defined by two of its lines and three of its points,* provided no two of the latter are coplanar with either of the given lines.

**4.** If four lines of a regulus cut any line of the conjugate regulus in points of a harmonic set, they are cut by every such line in points of a harmonic set. Hence give a construction for the harmonic conjugate of a line of a regulus with respect to two other lines of the regulus.

**5.** Two distinct reguli can have in common at most two distinct lines.

**6.** Show how to construct a regulus having in common with a given regulus one and but one ruler.

**104. The polar system of a regulus.** A plane meets every line of a regulus in a point, unless it contains a line of the regulus, in which case it meets all the other lines in points that are collinear. Since the regulus may be thought of as the lines of intersection of pairs of homologous planes of two projective axial pencils (Theorem 5′), the section by a plane consists of the points of intersection of pairs of homologous lines of two projective flat pencils. Hence the section of a regulus by a plane is a point conic, and the conjugate regulus has the same section. By duality the projection of a regulus and its conjugate from any point is a cone of planes.

The last remark implies that a line conic is the "picture" in a plane of a regulus and its conjugate. For such a picture is clearly a plane section of the projection of the object depicted from the eye of an observer. Fig. 108 illustrates this fact.

* By a point of a regulus is meant any point on a line of the regulus.

The section of a regulus by a plane containing a line of the regulus is a degenerate conic of two lines. The plane section can never degenerate into two coincident lines because the lines of a regulus and its conjugate are distinct from each other. In like manner, the projection from a point on a line of the regulus is a degenerate cone of planes consisting of two pencils of planes whose axes are a ruler and a directrix of the regulus.

DEFINITION. The class of all points on the lines of a regulus is called a *surface of the second order* or a *quadric surface.* The planes on the lines of the regulus are called the *tangent planes* of the surface or of the regulus. The point of intersection of the two lines of the regulus and its conjugate in a tangent plane is called the *point of contact* of the plane. The lines through the point of contact in a tangent plane are called *tangent lines,* and the point of contact of the plane is also the *point of contact* of any tangent line.

The tangent lines at a point of a quadric surface include the lines of the two conjugate reguli through this point and all other lines through this point which meet the surface in no other point. Any other line, of course, meets the surface in two or no points, since a plane through the line meets the surface in a conic. The tangent lines are, by duality, also the lines through each of which passes only one tangent plane to the surface.

THEOREM 7. *The tangent planes at the points of a plane section of a quadric surface pass through a point and constitute a cone of planes. Dually, the points of contact of the cone of tangent planes through a point are coplanar and form a point conic.*

*Proof.* It will suffice to prove the latter of these two dual theorems. Let the vertex $P$ of the cone of tangent planes be not a point of the surface. Consider three tangent planes through $P$, and their points of contact. The three lines from these points of contact to $P$ are tangent lines of the surface and hence there is only one tangent plane through each of them. Hence they are lines of the cone of lines associated with the cone of tangent planes. Let $\pi$ be the plane through their points of contact. The section by $\pi$ of the cone of planes through $P$ is therefore the conic determined by the three points of contact and the two tangent lines in which two of the tangent planes meet $\pi$. The plane $\pi$, however, meets the regulus in a conic of which the three points of contact are points. The two lines of intersection with

$\pi$ of two of the tangent planes through $P$ are tangents to this conic, because they cannot meet it in more than one point each. The section of the surface and the section of the cone of planes then have three points and the tangents through two of them in common. Hence these sections are identical, which proves the theorem when $P$ is not on the surface.

If $P$ is on the surface, the cone of planes degenerates into two lines of the surface (or the pencils of planes on these lines), and the points of contact of these planes are all on the same two lines. Hence the theorem is true also in this case.

DEFINITION. If a point $P$ and a plane $\pi$ are so related to a regulus that all the tangent planes to the regulus at points of its section by $\pi$ pass through $P$ (and hence all the points of contact of tangent planes through $P$ are on $\pi$), then $P$ is called the *pole* of $\pi$ and $\pi$ the *polar* of $P$ with respect to the regulus.

COROLLARY. *A tangent plane to a regulus is the polar of its point of contact.*

THEOREM 8. *The polar of a point $P$ not on a regulus contains all points $P'$ such that the line $PP'$ meets the surface in two points which are harmonic conjugates with respect to $P$, $P'$.*

*Proof.* Consider a plane, $\alpha$, through $PP'$ and containing two lines $a$, $b$ of the cone of tangent lines through $P$. This plane meets the surface in a conic $C^2$, to which the lines $a$, $b$ are tangent. As the polar plane of $P$ contains the points of contact of $a$ and $b$, its section by $\alpha$ is the polar of $P$ with respect to $C^2$. Hence the theorem follows as a consequence of Theorem 13, Chap. V.

THEOREM 9. *The polar of a point of a plane $\pi$ with respect to a regulus meets $\pi$ in the polar line of this point with regard to the conic which is the section of the regulus by $\pi$.*

*Proof.* By Theorem 8 the line in which the polar plane meets $\pi$ has the characteristic property of the polar line with respect to a conic (Theorem 13, Chap. V). This argument applies equally well if the conic is degenerate. In this case the theorem reduces to the following

COROLLARY. *The tangent lines of a regulus at a point on it are paired in an involution the double lines of which are the ruler and directrix through that point. Each line of a pair contains the polar points of all the planes on the other line.*

THEOREM 10. *The polars with regard to a regulus of the points of a line $l$ are an axial pencil of planes projective with the pencil of points on $l$.*

*Proof.* In case the given line is a line of the regulus this reduces to Theorem 6. In any other case consider two planes through $l$. In each plane the polars of the points of $l$ determine a pencil of lines projective with the range on $l$. Hence the polars must all meet the line joining the centers of these two pencils of lines, and, being perspective with either of these pencils of lines, are projective with the range on $l$.

DEFINITION. A line $l'$ is *polar* to a line $l$ if the polar planes of the points of $l$ meet on $l'$. A line is *conjugate* to $l$ if it meets $l'$. A point $P'$ is *conjugate* to a point $P$ if it is on the polar of $P$. A line $p$ is *conjugate* to $P$ if it is on the polar of $P$. A plane $\pi'$ is *conjugate* to a plane $\pi$ if $\pi'$ is on the pole of $\pi$. A line $p$ is *conjugate* to $\pi$ if it is on the pole of $\pi$.

## EXERCISES

*Polar points and planes with respect to a regulus are denoted by corresponding capital Roman and small Greek letters. Conjugate elements of the same kind are denoted by the same letters with primes.*

**1.** If $\pi$ is on $R$, then $P$ is on $\rho$.

**2.** If $l$ is polar to $\bar{l}$, then $\bar{l}$ is polar to $l$.

**3.** If one element (point, line, or plane) is conjugate to a second element, then the second element is conjugate to the first.

**4.** If two lines intersect, their two polar lines intersect.

**5.** A ruler or a directrix of a regulus is polar to itself. A tangent line is polar to its harmonic conjugate with regard to the ruler and directrix through its point of contact. Any other line is skew to its polar.

**6.** The points of two polar lines are conjugate.

**7.** The pairs of conjugate points (or planes) on any line form an involution the double points (planes) of which (if existent) are on the regulus.

**8.** The conjugate lines in a flat pencil of which neither the center nor the plane is on the regulus form an involution.

**9.** The line of intersection of two tangent planes is polar to the line joining the two points of contact.

**10.** A line of the regulus which meets one of two polar lines meets the other.

**11.** Two one- or two-dimensional forms whose bases are not conjugate or polar are projective if conjugate elements correspond.

**12.** A line $l$ is conjugate to $l'$ if and only if some plane on $l$ is polar to some point on $l'$.

**13.** Show that there are two (proper or improper) lines $r$, $s$ meeting two given lines and conjugate to them both. Show also that $r$ is the polar of $s$.

**14.** If $a, b, c$ are three generators of a regulus and $a'$, $b'$, $c'$ three of the conjugate regulus, then the three diagonal lines joining the points

$$(bc') \text{ and } (b'c),$$
$$(c'a) \text{ and } (ca'),$$
$$(ab') \text{ and } (a'b)$$

meet in a point $S$ which is the pole of a plane containing the lines of intersection of the pairs of tangent planes at the same vertices.

**15.** The six lines $a$, $b$, $c$, $a'$, $b'$, $c'$ of Ex. 14 determine the following trios of simple hexagons

$$(bc'ab'ca'), \quad (ba'ac'cb'), \quad (bb'aa'cc'),$$
$$(bc'aa'cb'), \quad (bb'ac'ca'), \quad (ba'ab'cc').$$

The points $S$ determined by each trio of hexagons are collinear, and the two lines on which they lie are polar with regard to the quadric surface.*

**16.** The section of the figure of Ex. 14 by a plane leads to the Pascal and Brianchon theorems; and, in like manner, Ex. 15 leads to the theorem that the 60 Pascal lines corresponding to the 60 simple hexagons formed from 6 points of a conic meet by threes in 20 points which constitute 10 pairs of points conjugate with regard to the conic (cf. Ex. 19, p. 138).

**105. Projective conics.** Consider two sections of a regulus by planes which are not tangent to it. These two conics are both perspective with any axial pencil of a pair of axial pencils which generate the regulus (cf. § 76, Chap. VIII). The correspondence established between the conics by letting correspond pairs of points which lie on the same ruler is therefore projective. On the line of intersection, $l$, of the two planes, if it is not a tangent line, the two conics determine the same involution I of conjugate points. Hence, if one of them intersects this line in two points, they have these two points in common. If one is tangent, they have one common point and one common tangent. The projectivity between the two conics fully determines a projectivity between their planes in which the line $l$ is transformed into itself. The involution I belongs to the projectivity thus determined on $l$. The converse of these statements leads to a theorem which is exemplified in the familiar string models:

THEOREM 11. *The lines joining corresponding points of two projective conics in different planes form a regulus, provided the two conics determine the same involution,* I, *of conjugate points on the*

* Cf. Sannia, Lezioni di Geometria Projettiva (Naples, 1895), pp. 262–263.

*line of intersection, l, of the two planes; and provided the collineation between the two planes determined by the correspondence of the conics transforms l into itself by a projectivity to which* I *belongs (in particular, if the conics meet in two points which are self-corresponding in the projectivity).*

*Proof.* Let $L$ be the pole with regard to one conic of the line of intersection, $l$, of the two planes (fig. 109). Let $A$ and $B$ be two

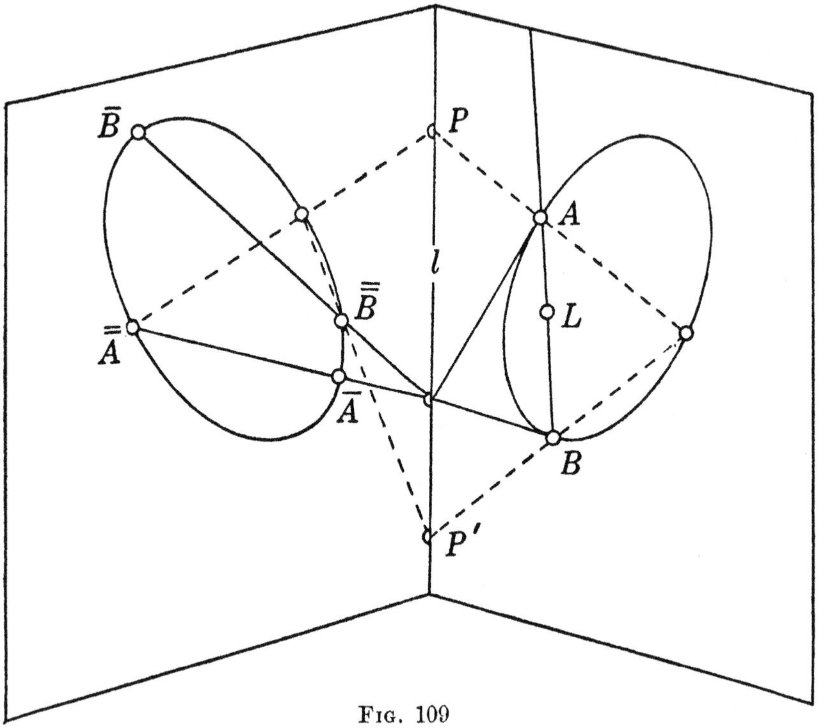


Fig. 109

points of this conic collinear with $L$ and not on $l$. The conic is generated by the two pencils $A[P]$ and $B[P']$ where $P$ and $P'$ are conjugates in the involution I on $l$ (cf. Ex. 1, p. 137). Let $\bar{A}$ and $\bar{B}$ be the points homologous to $A$ and $B$ on the second conic, and let $\bar{\bar{A}}$ be the point in which the second conic is met by the plane containing $A$, $\bar{A}$, and the tangent at $A$; and let $\bar{\bar{B}}$ be the point in which the second conic is met by the plane of $B$, $\bar{B}$, and the tangent at $B$.

The line $\bar{A}\bar{B}$ contains the pole of $l$ with regard to the second conic because this line is projective with $AB$. Since the tangents to the first conic at $A$ and $B$ meet on $l$, the complete quadrangle $\bar{A}\bar{\bar{A}}\bar{B}\bar{\bar{B}}$ has one diagonal point, the intersection of $\bar{A}\bar{\bar{A}}$ and $\bar{B}\bar{\bar{B}}$, on $l$; hence the opposite side of the diagonal triangle passes through the pole of $l$. Hence it intersects $\bar{A}\bar{B}$ in the pole of $l$. But the intersection of $\bar{\bar{A}}\bar{\bar{B}}$ with $\bar{A}\bar{B}$ is on this diagonal line. Hence $\bar{\bar{A}}\bar{\bar{B}}$ meets $\bar{A}\bar{B}$ in the pole of $l$. Hence the pencils $\bar{\bar{A}}[P]$ and $\bar{\bar{B}}[P']$ generate the second conic. Hence, denoting by $a$ and $b$ the lines $A\bar{\bar{A}}$ and $B\bar{\bar{B}}$, the pencils of planes $a[P]$ and $b[P']$ are projective and generate a regulus of which the two conics are sections.

The projectivity between the planes of the two conics established by this regulus transforms the line $l$ into itself by a projectivity to which the involution I belongs and makes the point $A$ correspond to $\bar{A}$. The projectivity between two conics is fully determined by these conditions (cf. Theorem 12, Cor. 1, Chap. VIII). Hence the lines of the regulus constructed above join homologous points in the given projectivity. Q.E.D.

It should be observed that if the two conics are tangent to $l$, the projectivity on $l$ fully determines the projectivity between the two conics. For if a point $P$ of $l$ corresponds to a point $Q$ of $l$, the unique tangent other than $l$ through $P$ to the first conic must correspond to the tangent to the second conic from $Q$. If the projectivity between the two conics is to generate a regulus, the projectivity on $l$ must be parabolic with the double point at the point of contact of the conics with $l$. For if another point $D$ is a double point of the projectivity on $l$, the plane of the tangents other than $l$, through $D$ to the two conics meets each conic in one and only one point, and, as these points are homologous, contains a straight line of the locus generated. As this plane contains only one point on either conic, it meets the locus in only one line, whereas a plane meeting a regulus in one line meets it also in another distinct line.

Since the parabolic projectivity on $l$ is fully determined by the double point and one pair of homologous points, the projectivity between the two conics is fully determined by the correspondent of one point, not on $l$, of the first conic.

To show that a projectivity between the two conics which is parabolic on $l$ does generate a regulus, let $A$ be any point of the first conic and $A'$ its correspondent on the second (fig. 110). Let the plane of $A'$ and the tangent at $A$ meet the second conic in $A''$. Denote the common point of the two conics by $B$, and consider the two conics as generated by the flat pencils at $A$ and $B$ and at $A''$ and $B$. The correspondence established between the two flat pencils at $B$ by letting correspond lines joining $B$ to homologous points of the two conics is perspective because the line $l$ corresponds to itself. Hence there is a pencil of planes whose axis, $b$, passes through $B$ and whose planes contain homologous pairs of lines of the flat pencils at $B$. The correspondence established in like manner between the flat pencil at $A$ and the flat pencil at $A''$ may be regarded as the product of the projectivity between the two planes, which carries the pencil at $A$ to the pencil at $A'$, followed by the projectivity between the pencils at $A'$ and $A''$ generated by the second conic. Both of these projectivities determine parabolic projectivities on $l$ with $B$ as invariant point. Hence their product determines on $l$ either a parabolic projectivity with $B$ as invariant point or the identity. This product transforms the tangent at $A$ into the line $A''A'$. As these lines meet $l$ in the same point, the projectivity determined on $l$ is the identity. Hence corresponding lines of the projective pencils at $A$ and $A''$ meet on $l$, and hence they determine a pencil of planes whose axis is $a = AA''$.

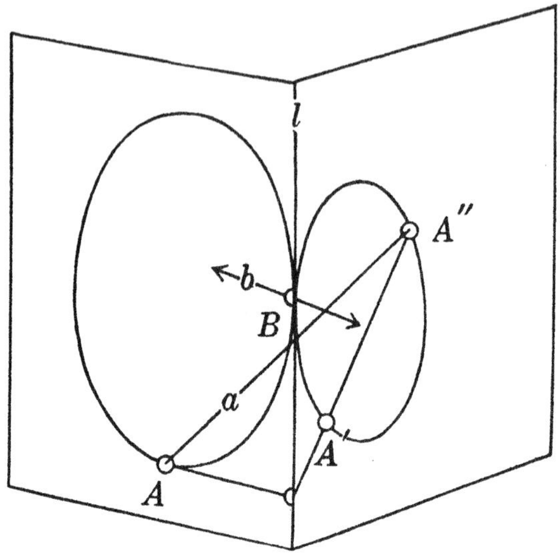

FIG. 110

The axial pencils on $a$ and $b$ are projective and hence generate a regulus the lines of which, by construction, pass through homologous points of the two conics. We are therefore able to supplement Theorem 11 by the following

COROLLARY 1. *The lines joining corresponding points of two projective conics in different planes form a regulus, if the two conics have a common tangent and point of contact and the projectivity determined between the two planes by the projectivity of the conics transforms their common tangent into itself and has the common point of the two conics as its only fixed point.*

The generation of a regulus by projective ranges of points on skew lines may be regarded as a degenerate case of this theorem and corollary. A further degenerate case is stated in the first exercise.

The proof of Theorem 11 given above is more complicated than it would have been if, under Proposition $K_2$, we had made use of the points of intersection of the line $l$ with the two conics. But since the discussion of linear families of lines in the following section employs only proper elements and depends in part on this theorem, it seems more satisfactory to prove this theorem as we have done. It is of course evident that any theorem relating entirely to proper elements of space which is proved with the aid of Proposition $K_n$ can also be proved by an argument employing only proper elements. The latter form of proof is often much more difficult than the former, but it often yields more information as to the constructions related to the theorem.

These results may be applied to the problem of passing a quadric surface through a given set of points in space. Proposition $K_2$ will be used in this discussion so as to allow the possibility that the two conjugate reguli may be improper though intersecting in proper points.

COROLLARY 2. *If three planes $\alpha$, $\beta$, $\gamma$ meet in three lines $a = \beta\gamma$, $b = \gamma\alpha$, $c = \alpha\beta$ and contain three conics $A^2$, $B^2$, $C^2$, of which $B^2$ and $C^2$ meet in two points $P$, $P'$ of $a$, $C^2$ and $A^2$ meet in two points $Q$, $Q'$ of $b$, and $A^2$ and $B^2$ meet in two points $R$, $R'$ of $c$, then there is one and but one quadric surface* * *containing the points of the three conics.*

*Proof.* Let $M$ be any point of $C^2$. The conic $B^2$ is projected from $M$ by a cone which meets the plane $\alpha$ in a conic which intersects $A^2$ in two points, proper or improper or coincident, other than $R$ and $R'$. Hence there are two lines $m$, $m'$, proper or improper or coincident, through $M$ which meet both $A^2$ and $B^2$. The projectivity determined between $A^2$ and $B^2$ by either of these lines generates a regulus, or, in a special case, a cone of lines, the lines of which must pass through all points of $C^2$ because they pass through $P$, $P'$, $Q'$, $Q$, and $M$, all of which are points of $C^2$.

The conjugate of such a regulus also contains a line through $M$ which meets both $A^2$ and $B^2$. Hence the lines $m$ and $m'$ determine conjugate reguli if they are distinct. If coincident they evidently determine a cone. The three conics being proper, the quadric must contain proper points even though the lines $m$, $m'$ are improper.

* In this corollary and in Theorem 12 the term *quadric surface* must be taken to include the points on a cone of lines as a special case.

If six points 1, 2, 3, 4, 5, 6 are given, no four of which are coplanar,* there evidently exist two planes, $\alpha$ and $\beta$, each containing three of the points and having none on their line of intersection.

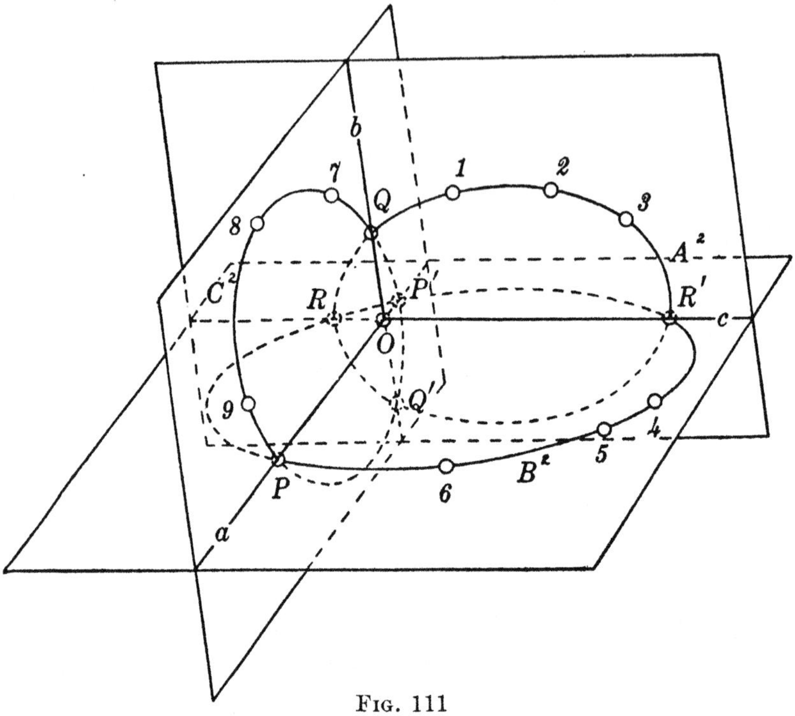


FIG. 111

Assign the notation so that 1, 2, 3 are in $\alpha$. A quadric surface which contains the six points must meet the two planes in two conics $A^2$, $B^2$ which meet the line $\alpha\beta = c$ in a common point-pair or point of contact; and every point-pair, proper or improper or coincident, of $c$ determines such a pair of conics.

Let us consider the problem of determining the polar plane $\omega$ of an arbitrary point $O$ on the line $c$. The polar lines of $O$ with regard to a pair of conics $A^2$ and $B^2$ meet $c$ in the same point and hence determine $\omega$. If no two of the points 1, 2, 3, 4, 5, 6 are collinear with $O$, any line $l$ in the plane $\alpha$ determines a unique conic $A^2$ with regard to which it is polar to $O$, and which passes through 1, 2, 3. $A^2$ determines a unique conic $B^2$ which passes through 4, 5, 6 and meets $c$ in the same points as $A^2$; and with regard to this conic $O$

* The construction of a quadric surface through nine points by the method used in the text is given in Rohn and Papperitz, Darstellende Geometrie, Vol. II (Leipzig, 1896), §§ 676, 677.

has a polar line $m$. Thus there is established a one-to-one correspondence $\Pi$ between the lines of $\alpha$ and the lines of $\beta$. This correspondence is a collineation. For consider a pencil of lines $[l]$ in $\alpha$. The conics $A^2$ determined by it form a pencil. Hence the point-pairs in which they meet $c$ are an involution. Hence the conics $B^2$ determined by the point-pairs form a pencil, and hence the lines $[m]$ form a pencil. Since every line $l$ meets its corresponding line $m$ on $c$, the correspondence $\Pi$ is not only a collineation but is a perspectivity, of which let the center be $C$. Any two corresponding lines $l$ and $m$ are coplanar with $C$. *Hence the polar planes of $O$ with regard to quadrics through* 1, 2, 3, 4, 5, 6 *are the planes on $C$.*

This was on the assumption that no two of the points 1, 2, 3, 4, 5, 6 are collinear with $O$. If two are collinear with $O$, every polar plane of $O$ must pass through the harmonic conjugate of $O$ with regard to them. This harmonic conjugate may be taken as the point $C$.

Now if nine points are given, no four being in the same plane, the notation may be assigned so that the planes $\alpha = 123$, $\beta = 456$, $\gamma = 789$ are such that none of their lines of intersection $a = \beta\gamma$, $b = \gamma\alpha$, $c = \alpha\beta$ contains one of the nine points. Let $O$ be the point $\alpha\beta\gamma$ (or a point on the line $\alpha\beta$ if $\alpha$, $\beta$, and $\gamma$ are in the same pencil). By the argument above the polars of $O$ with regard to all quadrics through the six points in $\alpha$ and $\beta$ must meet in a point $C$. The polars of $O$ with regard to all quadrics through the six points in $\beta$ and $\gamma$ must similarly pass through a point $A$, and the polars with regard to all quadrics through the six points in $\gamma$ and $\alpha$ must pass through a point $B$.

If $A$, $B$, and $C$ are not collinear, the plane $\omega = ABC$ must be the polar of $O$ with regard to any quadric through the nine points. The plane $\omega$ meets $\alpha$, $\beta$, and $\gamma$ each in a line which must be polar to $O$ with regard to the section of any such quadric. But this determines three conics $A^2$ in $\alpha$, $B^2$ in $\beta$, and $C^2$ in $\gamma$, which meet by pairs in three point-pairs on the lines $a$, $b$, $c$. Hence if $\alpha$, $\beta$, $\gamma$ are not in the same pencil, it follows, by Corollary 2, that there is a unique quadric through the nine points. If $\alpha$, $\beta$, $\gamma$ have a line in common, the three conics $A^2$, $B^2$, $C^2$ meet this line in the same point-pair. Consider a plane $\sigma$ through $O$ which meets the conics $A^2$, $B^2$, $C^2$ in three point-pairs. These point-pairs are harmonically conjugate to $O$ and the trace, $s$, on $\sigma$ of the plane $\omega$. Hence they lie on a conic $D^2$, which, with $A^2$ and $B^2$, determines a unique quadric. The section of this quadric by the plane $\gamma$ has in common with $C^2$ two point-pairs and the polar pair $O$, $s$. Hence the quadric has $C^2$ as its section by $\gamma$.

In case $A$, $B$, and $C$ are collinear, there is a pencil of planes $\omega$ which meet them. There is thus determined a family of quadrics which is called a *pencil* and is analogous to a pencil of conics. In case $A$, $B$, and $C$ coincide, there is a bundle of possible planes $\omega$ and a quadric is determined for each one. This family of quadrics is called a *bundle*. Without inquiring at present under what conditions on the points $1, 2, \cdots, 9$ these cases can arise, we may state the following theorem:

THEOREM 12. *Through nine points no four of which are coplanar there passes one quadric surface or a pencil of quadrics or a bundle of quadrics.*

## EXERCISES

**1.** The lines joining homologous points of a projective conic and straight line form a regulus, provided the line meets the conic and is not coplanar with it, and their point of intersection is self-corresponding.

**2.** State the duals of Theorems 11 and 12.

**3.** Show that two (proper or improper) conjugate reguli pass through two conics in different planes having two points (proper or improper or coincident) in common and through a point not in the plane of either conic. Two such conics and a point not in either plane thus determine one quadric surface.

**4.** Show how to construct a regulus passing through six given points and a given line.

**106. Linear dependence of lines.** DEFINITION. If two lines are coplanar, the lines of the flat pencil containing them both are said to be *linearly dependent* on them. If two lines are skew, the only lines linearly dependent on them are the two lines themselves. On three skew lines are *linearly dependent* the lines of the regulus, of which they are rulers. If $l_1, l_2, \cdots, l_n$ are any number of lines and $m_1, m_2, \cdots, m_k$ are lines such that $m_1$ is linearly dependent on two or three of $l_1, l_2, \cdots, l_n$, and $m_2$ is linearly dependent on two or three of $l_1, l_2, \cdots, l_n, m_1$, and so on, $m_k$ being linearly dependent on two or three of $l_1, l_2, \cdots, l_n, m_1, m_2, \cdots, m_{k-1}$, then $m_k$ is said to be *linearly dependent* on $l_1, l_2, \cdots, l_n$. A set of $n$ lines no one of which is linearly dependent on the $n-1$ others is said to be *linearly independent.*

As examples of these definitions there arise the following cases of linear dependence of lines on three linearly independent lines which may be regarded as degenerate cases of the regulus. (1) If lines $a$

and $b$ intersect in a point $P$, and a line $c$ skew to both of them meets their plane in a point $Q$, then in the first place all lines of the pencil $ab$ are linearly dependent on $a$, $b$, and $c$; since the line $QP$ is in this pencil, all lines of the pencil determined by $QP$ and $c$ are in the set. As these pencils have in common only the line $QP$ and do not contain three mutually skew lines, the set contains no other lines. Hence in this case the lines linearly dependent on $a$, $b$, $c$ are the flat pencil $ab$ and the flat pencil $(c, QP)$. (2) If one of the lines, as $a$, meets both of the others, which, however, are skew to each other, the set of linearly dependent lines consists of the flat pencils $ab$ and $ac$. This is the same as case (1). (3) If every two intersect but not all in the same point, the three lines are coplanar and all lines of their plane are linearly dependent on them. (4) If all three intersect in the same point and are not coplanar, the bundle of lines through their common point is linearly dependent on them. The case where all three are concurrent and coplanar does not arise because three such lines are not independent.

This enumeration of cases may be summarized as follows:

THEOREM 13. DEFINITION. *The set of all lines linearly dependent on three linearly independent lines is either a regulus, or a bundle of lines, or a plane of lines, or two flat pencils having different centers and planes but a common line. The last three sets of lines are called degenerate reguli.*

DEFINITION. The set of all lines linearly dependent on four linearly independent lines is called a *linear congruence.* The set of all lines linearly dependent on five linearly independent lines is called a *linear complex.**

**107. The linear congruence.** Of the four lines $a$, $b$, $c$, $d$ upon which the lines of the congruence are linearly dependent, $b$, $c$, $d$ determine, as we have just seen, either a regulus, or two flat pencils with different centers and planes but with one common line, or a bundle of lines, or a plane of lines. The lines $b$, $c$, $d$ can of course be replaced by any three which determine the same regulus or degenerate regulus as $b$, $c$, $d$.

* The terms *congruence* and *complex* are general terms to denote *two- and three-parameter families* of lines respectively. For example, all lines meeting a curve or all tangents to a surface form a complex, while all lines meeting two curves or all common tangents of two surfaces are a congruence.

So in case $b$, $c$, $d$ determine a nondegenerate regulus of which $a$ is not a directrix, the congruence can be regarded as determined by four mutually skew lines. In case $a$ is a directrix, the lines linearly dependent on $a$, $b$, $c$, $d$ clearly include all tangent lines to the regulus $bcd$, whose points of contact are on $a$. But as $a$ is in a flat pencil with any tangent whose point of contact is on $a$ and one of the rulers, the family of lines dependent on $a$, $b$, $c$, $d$ is the family dependent on $b$, $c$, $d$ and a tangent line which does not meet $b$, $c$, $d$. Hence in either case the congruence is determined by four skew lines.

If one of the four skew lines meets the regulus determined by the other three in two distinct points, $P$, $Q$, the two directrices $p$, $q$ through these points meet all four lines. The line not in the regulus determines with the rulers through $P$ and $Q$, two flat pencils of lines which join $P$ to all the points of $q$, and $Q$ to all the points of $p$. From this it is evident that *all lines meeting both $p$ and $q$ are linearly dependent on the given four.* For if $P_1$ is any point on $p$, the line $P_1Q$ and the ruler through $P_1$ determine a flat pencil joining $P_1$ to all the points of $q$; similarly, for any point of $q$. No other lines can be dependent on them, because if three lines of any regulus meet $p$ and $q$, so do all the lines.

If one of the four skew lines is tangent to the regulus determined by the other three in a point $P$, the family of dependent lines includes the regulus and all lines of the flat pencil of tangents at $P$. Hence it includes the directrix $p$ through $P$ and hence all the tangent lines whose points of contact are on $p$. By Theorem 6 this family of lines can be described as the set of all lines on homologous pairs in a certain projectivity $\Pi$ between the points and planes of $p$. Any two lines in this set, if they intersect, determine a flat pencil of lines in the set. Any regulus determined by three skew lines $l$, $m$, $n$ of the set determines a projectivity between the points and planes on $p$, but this projectivity sets up the same correspondence as $\Pi$ for the three points and planes determined by $l$, $m$, and $n$. Hence by the fundamental theorem (Theorem 17, Chap. IV) the projectivity determined by the regulus $lmn$ is the same as $\Pi$, and all lines of the regulus are in the set. Hence, *when one of four skew lines is tangent to the regulus of the other three, the family of dependent lines consists of a regulus and all lines tangent to it at points of a directrix.* The directrix is itself in the family.

If no one of the four skew lines meets the regulus of the other three in a proper point, we have a case studied more fully below.

In case $b$, $c$, $d$ determine two flat pencils with a common line, $a$ may meet the center $A$ of one of the pencils. The linearly dependent lines, therefore, include the bundle whose center is $A$. The plane of the other flat pencil passes through $A$ and contains three nonconcurrent lines dependent on $a$, $b$, $c$, $d$. Hence the family of lines also includes all lines of this plane. The family of all lines through a point and all lines in a plane containing this point has evidently no further lines dependent on it. This is a degenerate case of a congruence. If $a$ is in the plane of one of the flat pencils, we have, by duality, the case just considered. If $a$ meets the common line of the two flat pencils in a point distinct from the centers, the two flat pencils may be regarded as determined by their common line $d'$ and by lines $b'$ and $c'$, one from each pencil, not meeting $a$. Hence the family of lines includes those dependent on the regulus $ab'c'$ and its directrix $d'$. This case has already been seen to yield the family of all lines of the regulus $ab'c'$ and all lines tangent to it at points of $d'$.

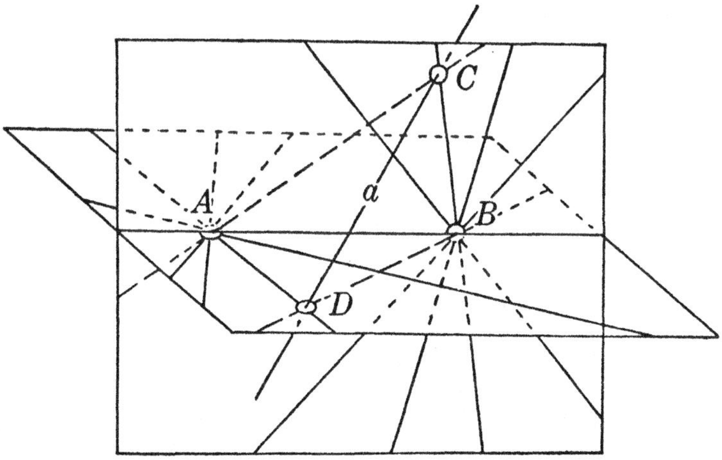


FIG. 112

If $a$ does not meet the common line, it meets the planes of the two pencils in points $C$ and $D$. Call the centers of the pencils $A$ and $B$ (fig. 112). The first pencil consists of the lines dependent on $AD$ and $AB$, the second of those dependent on $AB$ and $BC$. As $CD$ is the line $a$, the family of lines is seen to consist of the lines which are linearly dependent on $AB$, $BC$, $CD$, $DA$. Since any point of $BD$ is joined by lines of the family to $A$ and $C$, it is joined by lines of the family to every point of $AC$. Hence this case gives the family of all lines meeting both $AC$ and $BD$.

In case $b$, $c$, $d$ determine a bundle of lines, $a$, being independent of them, does not pass through the center of the bundle. Hence the family of dependent lines includes all lines of the plane of $a$ and the center of the bundle as well as the bundle itself.

Lastly, if $b$, $c$, $d$ are coplanar, we have, by duality, the same case as if $b$, $c$, $d$ were concurrent. We have thus proved

THEOREM 14. *A linear congruence is either* (1) *a set of lines linearly dependent on four linearly independent skew lines, such that no one of them meets the regulus containing the other three in a proper point; or* (2) *it is the set of all lines meeting two skew lines; or* (3) *it is the set of all rulers and tangent lines of a given regulus which meet a fixed directrix of the regulus; or* (4) *it consists of a bundle of lines and a plane of lines, the center of the bundle being on the plane.*

DEFINITION. A congruence of the first kind is called *elliptic;* of the second kind, *hyperbolic;* of the third kind, *parabolic;* of the fourth kind, *degenerate.* A line which has points in common with all lines of a congruence is called a *directrix* of the congruence.

COROLLARY. *A parabolic congruence consists of all lines on corresponding points and planes in a projectivity between the points and planes on a line. The directrix is a line of the congruence.*

To study the general nondegenerate case, let us denote four linearly independent and mutually skew lines on which the other lines of the congruence depend by $a$, $b$, $c$, $d$, and let $\pi_1$ and $\pi_2$ be two planes intersecting in $a$. Let the points of intersection with $\pi_1$ and $\pi_2$ of $b$, $c$, and $d$ be $B_1$, $C_1$, and $D_1$ and $B_2$, $C_2$, and $D_2$ respectively. By letting the complete quadrilateral $a$, $B_1C_1$, $C_1D_1$, $D_1B_1$ correspond to the complete quadrilateral $a$, $B_2C_2$, $C_2D_2$, $D_2B_2$, there is established a projective collineation $\Pi$ between the planes $\pi_1$ and $\pi_2$ in which the lines $b$, $c$, $d$ join homologous points (fig. 113).

Among the lines dependent on $a$, $b$, $c$, $d$ are the lines of the reguli $abc$, $acd$, $adb$, and all reguli containing $a$ and two lines from any of these three reguli. But all such reguli meet $\pi_1$ and $\pi_2$ in lines (e.g. $B_1D_1$, $B_2D_2$) because they have $a$ in common with $\pi_1$ and $\pi_2$. Furthermore, the lines of the fundamental reguli join points which correspond in Π (Theorem 5 of this chapter and Theorem 18, Chap. IV). Hence the reguli which contain $a$ and lines shown by means of such reguli to be dependent on $a$, $b$, $c$, $d$ are those generated by the projectivities determined by Π between lines of $\pi_1$ and $\pi_2$.

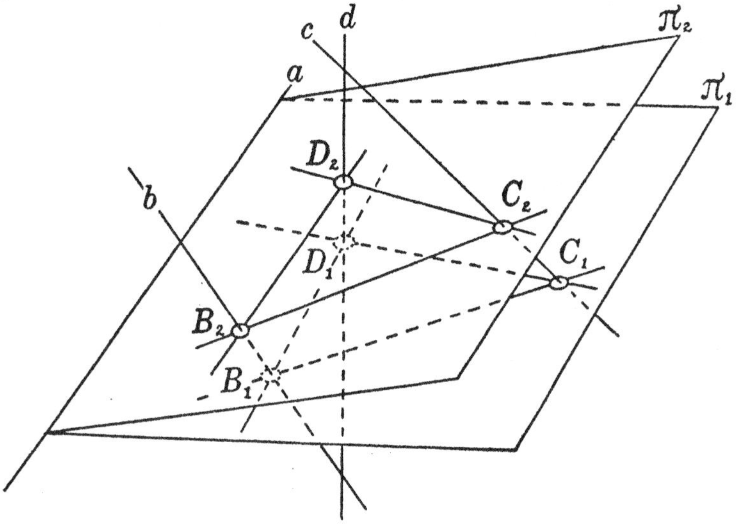


FIG. 113

Now consider reguli containing triples of the lines already shown to be in the congruence, but not containing $a$. Three such lines, $l$, $m$, $n$, join three noncollinear points $L_1$, $M_1$, $N_1$ of $\pi_1$ to the points $L_2$, $M_2$, $N_2$ of $\pi_2$ which correspond to them in the collineation Π. The regulus containing $l$, $m$, and $n$ meets $\pi_1$ and $\pi_2$ in two conics which are projective in such a way that $L_1$, $M_1$, $N_1$ correspond to $L_2$, $M_2$, $N_2$. The projectivity between the conics determines a projectivity between the planes, and as this projectivity has the same effect as Π on the quadrilateral composed of the sides of the triangle $L_1M_1N_1$ and the line $a$, it is identical with Π. Hence the lines of the regulus $lmn$ join points of $\pi_1$ and $\pi_2$ which are homologous under Π and are therefore among the lines already constructed.

Among the lines linearly dependent on the family thus far constructed are also such as appear in flat pencils containing two intersecting lines of the family. If one of the two lines is $a$, the other must meet $a$ in a double point of the projectivity determined on $a$ by Π. If neither of the two lines is $a$, they must meet $\pi_1$ and $\pi_2$, the first in points $P_1$, $P_2$ and the second in points $Q_1$, $Q_2$, and these four points are clearly distinct from one another. But as the given lines of the congruence, $P_1P_2$ and $Q_1Q_2$, intersect, so must also the lines $P_1Q_1$ and $P_2Q_2$ of $\pi_1$ and $\pi_2$ intersect, and the projectivity determined between $P_1Q_1$ and $P_2Q_2$ by $\Pi$ is a perspectivity. Hence the common point of $P_1Q_1$ and $P_2Q_2$ is a point of $a$ and is transformed into itself by $\Pi$. Hence, if lines of the family intersect, $\Pi$ has at least one double point on $a$, which means, by § 105,* that the line $a$ meets the regulus $bcd$ and the congruence has one or two directrices. Thus two lines of a nondegenerate congruence intersect only in the parabolic and hyperbolic cases; and from our previous study of these cases we know that the lines of a congruence through a point of intersection of two lines form a flat pencil.

We have thus shown that all the lines linearly dependent on $a$, $b$, $c$, $d$, with the exception of a flat pencil at each double point of the projectivity on $a$, are obtained by joining the points of $\pi_1$ and $\pi_2$ which are homologous under $\Pi$. From this it is evident that any four linearly independent lines of the congruence could have been taken as the fundamental lines instead of $a$, $b$, $c$, $d$. These two results are summarized as follows:

THEOREM 15. *All the lines of a linear congruence are linearly dependent on any linearly independent four of its lines. No lines not in the congruence are linearly dependent on four such lines.*

THEOREM 16. *If two planes meet in a line of a linear congruence and neither contains a directrix, the other lines of the congruence meet the planes in homologous points of a projectivity. Conversely, if two planes are projective in such a way that their line of intersection corresponds to itself, the lines joining homologous points are in the same linear congruence.*

* If there are two double points, $E$, $F$, on $a$, the conic $B_1C_1D_1EF$ must be transformed by $\Pi$ into the conic $B_2C_2D_2EF$, and the lines joining corresponding points of these conics must form a regulus contained in the congruence. As $E$ and $F$ are on lines of the regulus $bcd$, there are two directrices $p$, $q$ of this regulus which meet $E$ and $F$ respectively. The lines $p$ and $q$ meet all four of the lines $a$, $b$, $c$, $d$. Hence they meet all lines linearly dependent on $a$, $b$, $c$, $d$.

In the parabolic case the regulus $bcd$ must be met by $a$ in the single invariant point $H$ of the parabolic projectivity on $a$, because the conic tangent to $a$ at $H$ and passing through $B_1C_1D_1$ must be transformed by $\Pi$ into the conic tangent to $a$ at $H$ and passing through $B_2C_2D_2$; and the lines joining homologous points of these conics must form a regulus contained in the congruence. As $H$, a point of $a$, is on a line of the regulus $bcd$, there is one and only one directrix $p$ of this regulus which meets all four of $a$, $b$, $c$, $d$ and hence meets all lines of the congruence.

The dual of Theorem 16 may be stated in the following form:

THEOREM 17. *From two points on the same line of a linear congruence the latter is projected by two projective bundles of planes. Conversely, two bundles of planes projective in such a way that the line joining their centers is self-corresponding, generate a linear congruence.*

DEFINITION. A regulus all of whose rulers are in a congruence is called a *regulus of the congruence* and is said *to be in* or *to be contained in* the congruence.

COROLLARY. *If three lines of a regulus are in a congruence, the regulus is in the congruence.*

In the hyperbolic (or parabolic) case the regulus $bcd$ (in the notation already used) is met by $a$ in two points (or one point), its points of intersection with the directrices (or directrix). In the elliptic case the regulus $bcd$ cannot be met by $a$ in proper points, because if it were, the projectivity $\Pi$, between $\pi_1$ and $\pi_2$, would have these points as double points. Hence no line of the congruence meets a regulus of the congruence without being itself a generator. Hence through each point of space, without exception, there is one and only one line of the congruence. The involution of conjugate points of the regulus $bcd$ on the line $a$ is transformed into itself by $\Pi$, and the same must be true of any other regulus of the congruence, if it does not contain $a$. Since there is but one involution transformed into itself by a noninvolutoric projectivity on a line (Theorem 20, Chap. VIII), we have that the same involution of conjugate points is determined on any line of the congruence by all reguli of the congruence which do not contain the given line. This is entirely analogous to the hyperbolic case, and can be used to gain a representation in terms of proper elements of the improper directrices of an elliptic congruence.

The three kinds of congruences may be characterized as follows:

THEOREM 18. *In a parabolic linear congruence each line is tangent at a fixed one of its points to all reguli of the congruence of which it is not a ruler. On each line of a hyperbolic or elliptic congruence all reguli of the congruence not containing the given line determine the same involution of conjugate points. Through each point of space there is one and only one line of an elliptic congruence. For hyperbolic and parabolic congruences this statement is true except for points on a directrix.*

### EXERCISES

**1.** All lines of a congruence can be constructed from four lines by means of reguli all of which have two given lines in common.

**2.** Given two involutions (both having or both not having double points) on two skew lines. Through each point of space there are two and only two lines which are axes of perspectivity projecting one involution into the other, i.e. such that two planes through conjugate pairs of the first involution pass through a conjugate pair of the second involution. These lines constitute two congruences.

**3.** All lines of a congruence meeting a line not in the congruence form a regulus.

**4.** A linear congruence is self-polar with regard to any regulus of the congruence.

**5.** A degenerate linear congruence consists of all lines meeting two intersecting lines.

**108. The linear complex.** THEOREM 19. *A linear complex consists of all lines linearly dependent on the edges of a simple skew pentagon.**

*Proof.* By definition (§ 106) the complex consists of all lines linearly dependent on five independent lines. Let $a$ be one of these which does not meet the other four, $b', c', d', e'$. The complex consists of all lines dependent on $a$ and the congruence $b'c'd'e'$. If this congruence is degenerate, it consists of all lines dependent on three sides of a triangle $cde$ and a line $b$ not in the plane of the triangle (Theorems 14, 15). As $b$ may be any line of a bundle, it may be chosen so as to meet $a$; $c$ may be chosen so as to meet $b$, and $e$ may be so chosen as to meet $a$. Thus in this case the complex depends on five lines $a, b, c, d, e$ not all coplanar, forming the edges of a simple pentagon.

If the congruence is not degenerate, the four lines $b'', c'', d'', e''$ upon which it depends may (Theorem 15) be chosen so that no two of them intersect, but so that two and only two of them, $b''$ and $e''$, meet $a$. Thus the complex consists of all lines linearly dependent on the two flat pencils $ab''$ and $ae''$ and the two lines $c''$ and $d''$. Let $b$ and $e$ be the lines of these pencils (necessarily distinct from each other and from $a$) which meet $c''$ and $d''$ respectively. The complex then consists of all lines dependent on the flat pencils $ab, bc'', ae, ed''$.

* The edges of a simple skew pentagon are five lines in a given order, not all coplanar, each line intersecting its predecessor and the last meeting the first.

Finally, let $c$ and $d$ be two intersecting lines distinct from $b$ and $e$, which are in the pencils $bc''$ and $ed''$. The complex consists of all lines linearly dependent on the flat pencils $ab$, $bc$, $cd$, $de$, $ea$. Not all the vertices of the pentagon $abcde$ can be coplanar, because then all the lines would be in the same degenerate congruence.

THEOREM 20. DEFINITION. *There are two classes of complexes such that all complexes of either class are projectively equivalent. A complex of one class consists of a line and all lines of space which meet it. These are called special complexes. A complex of the other class is called general. No four vertices of a pentagon which determines it are coplanar.*

*Proof.* Given any complex, by the last theorem there is at least one skew pentagon $abcde$ which determines it. If there is a line $l$ meeting the five edges of this pentagon, this line must meet all lines of the complex, because any line meeting three linearly independent lines of a regulus (degenerate or not) meets all lines of it. Moreover, if the line $l$ meets $a$ and $b$ as well as $c$ and $d$, it must either join their two points of intersection or be the line of intersection of their common planes. If $l$ meets $e$ also, it follows in either case that four of the vertices of the pentagon are coplanar, two of them being on $e$. (That all five cannot be coplanar was explained at the end of the last proof.) Conversely, if four of the five vertices of the skew pentagon are coplanar, two and only two of its edges are not in this plane, and the line of intersection of the plane of the two edges with the plane of the other three meets all five edges.

Hence, *if and only if four of the five vertices are coplanar, there exists a line meeting the five lines.* Since any two skew pentagons are projectively equivalent, if no four vertices are coplanar (Theorem 12, Chap. III), any two complexes determined by such pentagons are projectively equivalent. Two simple pentagons are also equivalent if four vertices, but not five, of each are coplanar, because any simple planar four-point can be transformed by a collineation of space into any other, and then there exists a collineation holding the plane of the second four-point pointwise invariant and transforming any point not on the plane into any other point not on the plane. Therefore all complexes determined by pentagons of this kind are projectively equivalent. But these are the only two kinds of skew pentagons. Hence there are two and only two kinds of complexes.

In case four vertices of the pentagon are coplanar, we have seen that there is a line $l$ meeting all its edges. Since this line was determined as the intersection of the plane of two adjacent edges with the plane of the other three, it contains at least two vertices. It cannot contain three vertices because then all five would be coplanar. As one of the two planes meeting on $l$ contains three independent lines, all lines of that plane are lines of the complex. The line $l$ itself is therefore in the complex as well as the two lines of the other plane. Hence all lines of both planes are in the complex. Hence all lines meeting $l$ are in the complex. But as any regulus three of whose lines meet $l$ has all its lines meeting $l$, the complex satisfies the requirements stated in the theorem for a special complex.

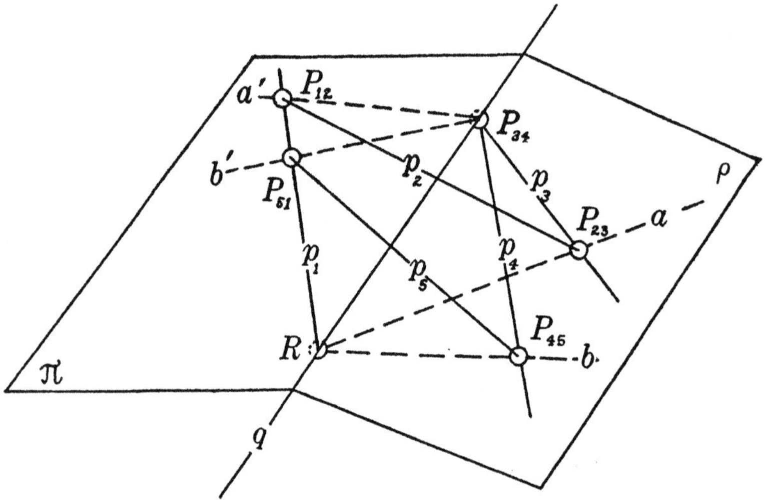


FIG. 114

A more definite idea of the general complex may be formed as follows. Let $p_1p_2p_3p_4p_5$ (fig. 114) be a simple pentagon upon whose edges all lines of the complex are linearly dependent. Let $q$ be the line of the flat pencil $p_3p_4$ which meets $p_1$, and let $R$ be the point of intersection of $q$ and $p_1$. Denote the vertices of the pentagon by $P_{12}$, $P_{23}$, $P_{34}$, $P_{45}$, $P_{51}$, the subscripts indicating the edges which meet in a given vertex.

The four independent lines $p_1p_2p_3q$ determine a congruence of lines all of which are in the complex and whose directrices are $a = RP_{23}$ and $a' = P_{12}P_{34}$. In like manner, $qp_4p_5p_1$ determine a congruence whose directrices are $b = RP_{45}$ and $b' = P_{34}P_{51}$. The complex consists of all lines linearly dependent on the lines of these two congruences. The directrices of the two congruences intersect at $R$ and $P_{34}$ respectively and determine two planes, $ab = \rho$ and $a'b' = \pi$, which meet on $q$.

Through any point $P$ of space not on $\rho$ or $\pi$ there are two lines $l$, $m$, the first meeting $a$ and $a'$, and the second meeting $b$ and $b'$ (fig. 115). All lines in the flat pencil $lm$ are in the complex by definition. This flat pencil meets $\rho$ and $\pi$ in two perspective ranges of

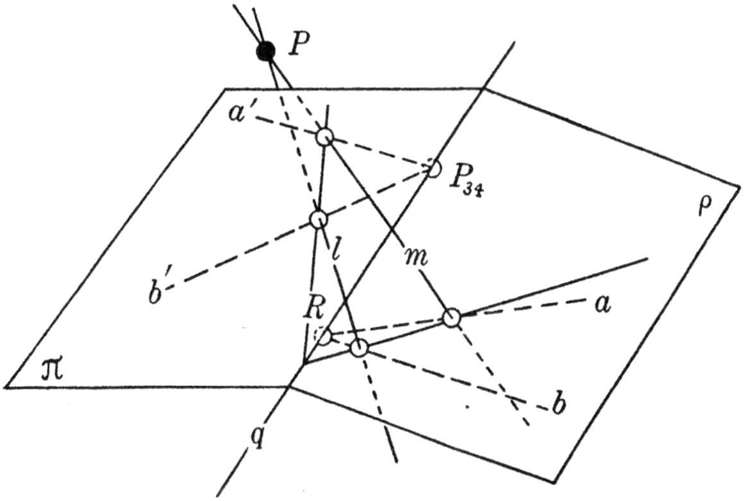


FIG. 115

points and thus determines a projectivity between the flat pencil $ab$ and the flat pencil $a'b'$, in which $a$ and $a'$, $b$ and $b'$ correspond and $q$ corresponds to itself. The projectivity thus determined between the pencils $ab$ and $a'b'$ is the same for all points $P$, because $a, b, q$ always correspond to $a', b', q'$. *Hence the complex contains all lines in the flat pencils of lines which meet homologous lines in the projectivity determined by*

$$abq \barwedge a'b'q.$$

*Denote this set of lines by* S. We have seen that it has the property that all its lines through a point not on $\rho$ or $\pi$ are coplanar. If a point $P$ is on $\rho$ but not on $q$, the line $PR$ has a corresponding line $p'$ in the pencil $a'b'$ and hence S contains all lines joining $P$ to points of $p'$. Similarly, for points on $\pi$ but not on $q$. By duality every plane not on $q$ contains a flat pencil of lines of S.

Each of the flat pencils not on $q$ has one line meeting $q$. Hence each plane of space not on $q$ contains one and only one line of S meeting $q$. Applying this to the planes through $P_{34}$ not containing $q$, we have that any line through $P_{34}$ and not on $\rho$ is not in the set S. Let $l$ be any such line. All lines of S in each plane through $l$ form a flat pencil $P$, and the centers of all these pencils lie on a line $l'$, because all lines through two points of $l$ form two flat pencils each of which contains a line from each pencil $P$. Hence the lines of S meeting $l$ form a congruence whose other directrix $l'$ evidently lies on $p$. The point of intersection of $l'$ with $q$ is the center of a flat pencil of lines of S all meeting $l$. Hence all lines of the plane $lq$ form a flat pencil. Since $l$ is any line on $P_{34}$ and not on $\pi$, this establishes that each plane and, by duality, each point on $q$, as well as not on $q$, contains a flat pencil of lines of S.

We can now prove that *the complex contains no lines not in* S To do so we have to show that all lines linearly dependent on lines of S are in S. If two lines of S intersect, the flat pencil they determine is by definition in S. If three lines $m_1$, $m_2$, $m_3$ of S are skew to one another, not more than two of the directrices of the regulus containing them are in S. For if three directrices were in S, all the tangent lines at points of these three lines would be in S, and hence any plane would contain three nonconcurrent lines of S. Let $l$ be a directrix of the regulus $m_1m_2m_3$, which is not in S. By the argument made in the last paragraph all lines of S meeting $l$ form a congruence. But this congruence contains all lines of the regulus $m_1m_2m_3$, and hence all lines of this regulus are in S. Hence the set of lines S is identical with the complex.

THEOREM 21 (SYLVESTER'S THEOREM *). *If two projective flat pencils with different centers and planes have a line $q$ in common which is self-corresponding, all lines meeting homologous pairs of lines in these two pencils are in the same linear complex. This complex consists of these lines together with a parabolic congruence whose directrix is $q$.*

*Proof.* This has all been proved in the paragraphs above, with the exception of the statement that $q$ and the lines meeting $q$ form a linear congruence. Take three skew lines of the complex meeting $q$; they determine with $q$ a congruence C all of whose lines are in the complex. There cannot be any other lines of the complex meeting $q$, because there would be dependent on such lines and on the congruence C all lines meeting $q$, and hence all lines meeting $q$ would be in the given complex, contrary to what has been proved above.

* Cf. Comptes Rendus, Vol. LII (1861), p. 741.

Another theorem proved in the discussion above is:

THEOREM 22. DEFINITION OF NULL SYSTEM. *All the lines of a linear complex which pass through a point P lie in a plane* $\pi$, *and all the lines which lie in a plane* $\pi$ *pass through a point P. In case of a special complex, exception must be made of the points and planes on the directrix. The point P is called the null point of the plane* $\pi$ *and* $\pi$ *is called the null plane of P with regard to the complex. The correspondence between the points and planes of space thus established is called a null system or null polarity.*

Another direct consequence, remembering that there are only two kinds of complexes, is the following:

THEOREM 23. *Any five linearly independent lines are in one and only one complex. If the edges of a simple pentagon are in a given complex, the pentagon is skew and its edges linearly independent. If the complex is general, no four vertices of a simple pentagon of its lines are coplanar.*

THEOREM 24. *Any set of lines,* K, *in space such that the lines of the set on each point of space constitute a flat pencil is a linear complex.*

*Proof.* (*a*) If two lines of the set K intersect, the set contains all lines linearly dependent on them, by definition.

(*b*) Consider any line $a$ not in the given set K. Two points $A, B$ on $a$ have flat pencils of lines of K on different planes; for if the planes coincided, every line of the plane would, by (*a*), be a line of K. Hence the lines of K through $A$ and $B$ all meet a line $a'$ skew to $a$. From this it follows that all the lines of the congruence whose directrices are $a, a'$ are in K. Similarly, if $b$ is any other line not in K but meeting $a$, all lines of K which meet $b$ also meet another line $b'$. Moreover, since any line meeting $a$, $b$, and $b'$ is in K and hence also meets $a'$, the four lines $a, a', b, b'$ lie on a degenerate regulus consisting of the flat pencils $ab$ and $a'b'$ (Theorem 13). Let $q$ (fig. 115) be the common line of the pencils $ab$ and $a'b'$. Through any point of space not on one of the planes $ab$ and $a'b'$ there are three coplanar lines of K which meet $q$ and the pairs $aa'$ and $bb'$. Hence K consists of lines meeting homologous lines in the projectivity

$$qab \barwedge qa'b',$$

and therefore is a complex by Theorem 21.

Corollary. *Any (1, 1) correspondence between the points and the planes of space such that each point lies on its corresponding plane is a null system.*

Theorem 25. *Two linear complexes have in common a linear congruence.*

*Proof.* At any point of space the two flat pencils belonging to the two complexes have a line in common. Obviously, then, there are three linearly independent lines $l_1$, $l_2$, $l_3$ common to the complexes. All lines in the regulus $l_1 l_2 l_3$ are, by definition, in each complex. But as there are points or planes of space not on the regulus, there is a line $l_4$ common to the two complexes and not belonging to this regulus. All lines linearly dependent on $l_1$, $l_2$, $l_3$, $l_4$ are, by definition, common to the complexes and form a congruence. No further line could be common or, by Theorem 23, the two complexes would be identical.

Corollary 1. *The lines of a complex meeting a line $l$ not in the complex form a hyperbolic congruence.*

*Proof.* The line is the directrix of a special complex which, by the theorem, has a congruence in common with the given complex. The common congruence cannot be parabolic because the lines of the first complex in a plane on $l$ form a flat pencil whose center is not on $l$, since $l$ is not in the complex.

Corollary 2. *The lines of a complex meeting a line $l$ of the complex form a parabolic congruence.*

*Proof.* The centers of all pencils of lines in this congruence must be on $l$ because $l$ is itself a line of each pencil.

Definition. A line $l$ is a *polar* to a line $l'$ with regard to a complex or null system, if and only if $l$ and $l'$ are directrices of a congruence of lines of the complex.

Corollary 3. *If $l$ is polar to $l'$, $l'$ is polar to $l$. A line is polar to itself, if and only if it is a line of the complex.*

Theorem 26. *A null system is a projective correspondence between the points and planes of space.*

*Proof.* The points on a line $l$ correspond to the planes on a line $l'$ by Corollaries 1 and 2 of the last theorem. If $l$ and $l'$ are distinct, the correspondence between the points of $l$ and planes of $l'$ is a perspectivity. If $l = l'$, the correspondence is projective by the corollary of Theorem 14.

## EXERCISES

**1.** If a point $P$ is on a plane $\rho$, the null plane $\pi$ of $P$ is on the null point $R$ of $\rho$.

**2.** Two pairs of lines polar with regard to the same null system are always in the same regulus (degenerate, if a line of one pair meets a line of the other pair).

**3.** If a line $l$ meets a line $m$, the polar of $l$ meets the polar of $m$.

**4.** Pairs of lines of the regulus in Ex. 2 which are polar with regard to the complex are met by any directrix of the regulus in pairs of points of an involution. Thus the complex determines an involution among the lines of the regulus.

**5.** Conversely (Theorem of Chasles), the lines meeting conjugate pairs of lines in an involution on a regulus are in the same complex. Show that Theorem 21 is a special case of this.

**6.** Find the lines common to a linear complex and a regulus not in the complex.

**7.** Three skew lines $k$, $l$, $m$ determine one and only one complex containing $k$ and having $l$ and $m$ as polars of each other.

**8.** If the number of points on a line is $n+1$, how many reguli, how many congruences, how many complexes are there in space? How many lines are there in each kind of regulus, congruence, complex?

**9.** Given any general complex and any tetrahedron whose faces are not null planes to its vertices. The null planes of the vertices constitute a second tetrahedron whose vertices lie on the planes of the first tetrahedron. The two tetrahedra are mutually inscribed and circumscribed each to the other* (cf. Ex. 6, p. 105).

**10.** A null system is fully determined by associating with the three vertices of a triangle three planes through these vertices and having their one common point in the plane of the triangle but not on one of its sides.

**11.** A tetrahedron is self-polar with regard to a null system if two opposite edges are polar.

**12.** Every line of the complex determined by a pair of Möbius tetrahedra meets their faces and projects their vertices in projective throws of points and planes.

**13.** If a tetrahedron $T$ is inscribed and circumscribed to $T_1$ and also to $T_2$, the lines joining corresponding vertices of $T_1$ and $T_2$ and the lines of intersection of their corresponding planes are all in the same complex.

**14.** A null system is determined by the condition that two pairs of lines of a regulus shall be polar.

**15.** A linear complex is self-polar with regard to a regulus all of whose lines are in the complex.

**16.** The lines from which two projective pencils of points on skew lines are projected by involutions of planes are all in the same complex. Dualize.

* This configuration was discovered by Möbius, Journal für Mathematik, Vol. III (1828), p. 273. Two tetrahedra in this relation are known as *Möbius tetrahedra.*

**109. The Plücker line coördinates.** Two points whose coördinates are

$$(x_1, x_2, x_3, x_4)$$
$$(y_1, y_2, y_3, y_4)$$

determine a line $l$. The coördinates of the two points determine six numbers

$$p_{12}=\begin{vmatrix} x_1 & x_2 \\ y_1 & y_2 \end{vmatrix}, \quad p_{13}=\begin{vmatrix} x_1 & x_3 \\ y_1 & y_3 \end{vmatrix}, \quad p_{14}=\begin{vmatrix} x_1 & x_4 \\ y_1 & y_4 \end{vmatrix},$$
$$p_{34}=\begin{vmatrix} x_3 & x_4 \\ y_3 & y_4 \end{vmatrix}, \quad p_{42}=\begin{vmatrix} x_4 & x_2 \\ y_4 & y_2 \end{vmatrix}, \quad p_{23}=\begin{vmatrix} x_2 & x_3 \\ y_2 & y_3 \end{vmatrix},$$

which are known as the *Plücker coördinates of the line.* Since the coördinates of the two points are homogeneous, the ratios only of the numbers $p_{ij}$ are determined. Any other two points of the line determine the same set of line coördinates, since the ratios of the $p_{ij}$'s are evidently unchanged if $(x_1, x_2, x_3, x_4)$ is replaced by $(x_1+\lambda y_1, x_2+\lambda y_2, x_3+\lambda y_3, x_4+\lambda y_4)$. The six numbers satisfy the equation*

$$p_{12}p_{34}+p_{13}p_{42}+p_{14}p_{23}=0. \tag{1}$$

This is evident on expanding in terms of two-rowed minors the identity

$$\begin{vmatrix} x_1 & x_2 & x_3 & x_4 \\ y_1 & y_2 & y_3 & y_4 \\ x_1 & x_2 & x_3 & x_4 \\ y_1 & y_2 & y_3 & y_4 \end{vmatrix} \equiv 0.$$

Conversely, if any six numbers, $p_{ij}$, are given which satisfy Equation (1), then two points $P=(x_1, x_2, x_3, 0)$, $Q=(y_1, 0, y_3, y_4)$ can be determined such that the numbers $p_{ij}$ are the coördinates of the line $PQ$. To do this it is simply necessary to solve the equations

$$\begin{aligned} -x_2y_1&=p_{12}, & x_3y_4&=p_{34}, \\ x_1y_3-x_3y_1&=p_{13}, & -x_2y_4&=p_{42}, \\ x_1y_4&=p_{14}, & x_2y_3&=p_{23}, \end{aligned}$$

which are easily seen to be consistent if and only if

$$p_{12}p_{34}+p_{13}p_{42}+p_{14}p_{23}=0.$$

Hence we have

THEOREM 27. *Every line of space determines and is determined by the ratios of six numbers* $p_{12}, p_{13}, p_{14}, p_{34}, p_{42}, p_{23}$ *subject to the*

* Notice that in Equation (1) the number of inversions in the four subscripts of any term is always even.

*condition* $p_{12}p_{34}+p_{13}p_{42}+p_{14}p_{23}=0$, *such that if* $(x_1, x_2, x_3, x_4)$ *and* $(y_1, y_2, y_3, y_4)$ *are any two points on the line,*

$$p_{12}=\begin{vmatrix} x_1 & x_2 \\ y_1 & y_2 \end{vmatrix}, \quad p_{13}=\begin{vmatrix} x_1 & x_3 \\ y_1 & y_3 \end{vmatrix}, \quad p_{14}=\begin{vmatrix} x_1 & x_4 \\ y_1 & y_4 \end{vmatrix},$$

$$p_{34}=\begin{vmatrix} x_3 & x_4 \\ y_3 & y_4 \end{vmatrix}, \quad p_{42}=-\begin{vmatrix} x_2 & x_4 \\ y_2 & y_4 \end{vmatrix}, \quad p_{23}=\begin{vmatrix} x_2 & x_3 \\ y_2 & y_3 \end{vmatrix}.$$

COROLLARY. *Four independent coördinates determine a line.*

In precisely similar manner two planes $(u_1, u_2, u_3, u_4)$ and $(v_1, v_2, v_3, v_4)$ determine six numbers such that

$$q_{12}=\begin{vmatrix} u_1 & u_2 \\ v_1 & v_2 \end{vmatrix}, \quad q_{13}=\begin{vmatrix} u_1 & u_3 \\ v_1 & v_3 \end{vmatrix}, \quad q_{14}=\begin{vmatrix} u_1 & u_4 \\ v_1 & v_4 \end{vmatrix},$$

$$q_{34}=\begin{vmatrix} u_3 & u_4 \\ v_3 & v_4 \end{vmatrix}, \quad q_{42}=\begin{vmatrix} u_4 & u_2 \\ v_4 & v_2 \end{vmatrix}, \quad q_{23}=\begin{vmatrix} u_2 & u_3 \\ v_2 & v_3 \end{vmatrix}.$$

The quantities $q_{ij}$ satisfy a theorem dual to the one just proved for the $p_{ij}$'s.

THEOREM 28. *The* $p$ *and* $q$ *coördinates of a line are connected by the equations* $p_{12}:p_{13}:p_{14}:p_{34}:p_{42}:p_{23}=q_{34}:q_{42}:q_{23}:q_{12}:q_{13}:q_{14}$.

*Proof.* Let the $p$ coördinates be determined by the two points $(x_1, x_2, x_3, x_4)$, $(y_1, y_2, y_3, y_4)$, and the $q$ coördinates by the two planes $(u_1, u_2, u_3, u_4)$, $(v_1, v_2, v_3, v_4)$. These coördinates satisfy the four equations

$$u_1x_1+u_2x_2+u_3x_3+u_4x_4=0,$$
$$v_1x_1+v_2x_2+v_3x_3+v_4x_4=0,$$
$$u_1y_1+u_2y_2+u_3y_3+u_4y_4=0,$$
$$v_1y_1+v_2y_2+v_3y_3+v_4y_4=0.$$

Multiplying the first equation by $-v_1$ and the second by $u_1$ and adding, we obtain

$$q_{12}x_2+q_{13}x_3+q_{14}x_4=0.$$

In like manner, from the third and fourth equations we obtain

$$q_{12}y_2+q_{13}y_3+q_{14}y_4=0.$$

Combining the last two equations similarly, we obtain

$$q_{13}p_{23}-q_{14}p_{42}=0,$$

or,

$$\frac{q_{13}}{q_{14}}=\frac{p_{42}}{p_{23}}.$$

By similar combinations of the first four equations we find

$$p_{12}:p_{13}:p_{14}:p_{34}:p_{42}:p_{23}=q_{34}:q_{42}:q_{23}:q_{12}:q_{13}:q_{14}.$$

### EXERCISE

Given the tetrahedron of reference, the point (1, 1, 1, 1), and a line $l$, determine six sets of four points each, whose cross ratios are the coördinates of $l$.

**110. Linear families of lines.** THEOREM 29. *The necessary and sufficient condition that two lines $p$ and $p'$ intersect, and hence are coplanar, is*

$$p_{12}p'_{34}+p_{13}p'_{42}+p_{14}p'_{23}+p_{34}p'_{12}+p_{42}p'_{13}+p_{23}p'_{14}=0,$$

*where $p_{ij}$ are the coördinates of $p$ and $p'_{ij}$ of $p'$.*

*Proof.* If the first line contains two points $x$ and $y$, and the second two points $x'$ and $y'$, the lines will intersect if and only if these four points are coplanar; that is to say, if and only if

$$0=\begin{vmatrix} x_1 & x_2 & x_3 & x_4 \\ y_1 & y_2 & y_3 & y_4 \\ x'_1 & x'_2 & x'_3 & x'_4 \\ y'_1 & y'_2 & y'_3 & y'_4 \end{vmatrix} = p_{12}p'_{34}+p_{13}p'_{42}+p_{14}p'_{23}+p_{34}p'_{12}+p_{42}p'_{13}+p_{23}p'_{14}.$$

THEOREM 30. *A flat pencil of lines consists of the lines whose coördinates are $\lambda p_{ij}+\mu p'_{ij}$, if $p$ and $p'$ are two lines of the pencil.*

*Proof.* The lines $p$ and $p'$ intersect in a point $A$ and are perspective with a range of points $\lambda C+\mu D$. Hence their coördinates may be written

$$\begin{vmatrix} a_1 & a_2 \\ \lambda c_1+\mu d_1 & \lambda c_2+\mu d_2 \end{vmatrix}, \text{ etc.,}$$

which may be expanded in the form

$$\lambda\begin{vmatrix} a_1 & a_2 \\ c_1 & c_2 \end{vmatrix}+\mu\begin{vmatrix} a_1 & a_2 \\ d_1 & d_2 \end{vmatrix}=\lambda p_{12}+\mu p'_{12}, \text{ etc.}$$

THEOREM 31. *The lines whose coördinates satisfy one linear equation*

$$(1) \qquad a_{12}p_{12}+a_{13}p_{13}+a_{14}p_{14}+a_{34}p_{34}+a_{42}p_{42}+a_{23}p_{23}=0$$

*form a linear complex. Those whose coördinates satisfy two independent linear equations form a linear congruence, and those satisfying three independent linear equations form a regulus. Four independent linear equations are satisfied by two (distinct or coincident) lines, which may be improper.*

*Proof.* If $(b_1, b_2, b_3, b_4)$ is any point of space, the points $(x_1, x_2, x_3, x_4)$ which lie on lines through $b_1, b_2, b_3, b_4$ satisfying (1) must satisfy

$$a_{12}\begin{vmatrix} b_1 & b_2 \\ x_1 & x_2 \end{vmatrix} + a_{13}\begin{vmatrix} b_1 & b_3 \\ x_1 & x_3 \end{vmatrix} + a_{14}\begin{vmatrix} b_1 & b_4 \\ x_1 & x_4 \end{vmatrix} + a_{34}\begin{vmatrix} b_3 & b_4 \\ x_3 & x_4 \end{vmatrix} + a_{42}\begin{vmatrix} b_4 & b_2 \\ x_4 & x_2 \end{vmatrix} + a_{23}\begin{vmatrix} b_2 & b_3 \\ x_2 & x_3 \end{vmatrix} = 0,$$

or

$$(2) \quad (a_{12}b_2 + a_{13}b_3 + a_{14}b_4)x_1 + (-a_{12}b_1 + a_{23}b_3 - a_{42}b_4)x_2 \\ + (-a_{13}b_1 - a_{23}b_2 + a_{34}b_4)x_3 + (-a_{14}b_1 + a_{42}b_2 - a_{34}b_3)x_4 = 0,$$

which is the equation of a plane. Hence the family of lines represented by (1) has a flat pencil of lines at every point of space, and so, by Theorem 24, is a linear complex.

Since two complexes have a congruence of common lines, two linear equations determine a congruence. Since a congruence and a complex have a regulus in common, three linear equations determine a regulus.

If the four equations

$$a'_{12}p_{12} + a'_{13}p_{13} + a'_{14}p_{14} + a'_{34}p_{34} + a'_{42}p_{42} + a'_{23}p_{23} = 0,$$
$$a''_{12}p_{12} + a''_{13}p_{13} + a''_{14}p_{14} + a''_{34}p_{34} + a''_{42}p_{42} + a''_{23}p_{23} = 0,$$
$$a'''_{12}p_{12} + a'''_{13}p_{13} + a'''_{14}p_{14} + a'''_{34}p_{34} + a'''_{42}p_{42} + a'''_{23}p_{23} = 0,$$
$$a^{\text{IV}}_{12}p_{12} + a^{\text{IV}}_{13}p_{13} + a^{\text{IV}}_{14}p_{14} + a^{\text{IV}}_{34}p_{34} + a^{\text{IV}}_{42}p_{42} + a^{\text{IV}}_{23}p_{23} = 0,$$

are independent, one of the four-rowed determinants of their coefficients is different from zero, and the equations have solutions of the form *

$$p_{12} = \lambda p'_{12} + \mu p''_{12}, \quad p_{13} = \lambda p'_{13} + \mu p''_{13}, \cdots.$$

If one of these solutions is to represent the coördinates of a line, it must satisfy the condition

$$p_{12}p_{34} + p_{13}p_{42} + p_{14}p_{23} = 0,$$

which gives a quadratic equation to determine $\lambda/\mu$. Hence, by Proposition $K_2$, there are two (proper, improper, or coincident) lines whose coördinates satisfy four linear equations.

COROLLARY 1. *The lines of a regulus are of the form*

$$p_i = \lambda_1 p'_i + \lambda_2 p''_i + \lambda_3 p'''_i$$

*where $p'$, $p''$, $p'''$ are lines of the regulus. In like manner, the lines of a congruence are of the form*

$$p_i = \lambda_1 p'_i + \lambda_2 p''_i + \lambda_3 p'''_i + \lambda_4 p^{\text{IV}}_i,$$

* Cf. Bocher, Introduction to Higher Algebra, Chap. IV.

*and of a complex of the form*

$$p_i = \lambda_1 p_i' + \lambda_2 p_i'' + \lambda_3 p_i''' + \lambda_4 p_i^{\text{IV}} + \lambda_5 p_i^{\text{V}}.$$

*All of these formulas must be taken in connection with*

$$p_{12}p_{34} + p_{13}p_{42} + p_{14}p_{23} = 0.$$

COROLLARY 2. *As a transformation from points to planes the null system determined by the complex whose equation is*

$$a_{12}p_{12} + a_{13}p_{13} + a_{14}p_{14} + a_{34}p_{34} - a_{24}p_{42} + a_{23}p_{23} = 0$$

*is*

$$\begin{aligned} u_1 &= 0 + a_{12}x_2 + a_{13}x_3 + a_{14}x_4, \\ u_2 &= -a_{12}x_1 + 0 + a_{23}x_3 + a_{24}x_4, \\ u_3 &= -a_{13}x_1 - a_{23}x_2 + 0 + a_{34}x_4, \\ u_4 &= -a_{14}x_1 - a_{24}x_2 - a_{34}x_3 + 0. \end{aligned}$$

The first of these corollaries simply states the form of the solutions of systems of homogeneous linear equations in six variables. The second corollary is obtained by inspection of Equation (2) the coefficients of which are the coördinates of the null plane of the point $(b_1, b_2, b_3, b_4)$.

Corollary 1 shows that the geometric definition of linear dependence of lines given in this chapter corresponds to the conventional analytic conception of linear dependence.

**111. Interpretation of line coördinates as point coördinates in $S_5$.** It may be shown without difficulty that the method of introducing homogeneous coördinates in Chap. VII is extensible to space of any number of dimensions (cf. Chap. I, § 12). Therefore the set of all sets of six numbers

$$(p_{12}, p_{13}, p_{14}, p_{34}, p_{42}, p_{23})$$

can be regarded as homogeneous point coördinates in a space of five dimensions, $S_5$. Since the coördinates of a line in $S_3$ satisfy the quadratic condition

$$(1) \qquad p_{12}p_{34} + p_{13}p_{42} + p_{14}p_{23} = 0,$$

they may be regarded as forming the points of a quadratic locus or spread,* $L_4^2$, in $S_5$. The lines of a linear complex correspond to the points of intersection with this spread of an $S_4$ that is determined by one linear equation. The lines of a congruence correspond, therefore, to the intersection with $L_4^2$ of an $S_3$, the lines of a regulus to the

* This is a generalization of a conic section.

intersection with $L_4^2$ of an $S_2$, and any pair of lines to the intersection with $L_4^2$ of an $S_1$.

Any point $(p'_{12}, p'_{13}, p'_{14}, p'_{34}, p'_{42}, p'_{23})$ of $S_5$ has as its polar* $S_4$, with regard to $L_4^2$,

$$(2) \qquad p'_{34}p_{12} + p'_{42}p_{13} + p'_{23}p_{14} + p'_{12}p_{34} + p'_{13}p_{42} + p'_{14}p_{23} = 0,$$

which is the equation of a linear complex in the original $S_3$. *Hence any point in* $S_5$ *can be thought of as representing the complex of lines represented by the points of* $S_5$ *in which its polar* $S_4$ *meets* $L_4^2$.

Since a line is represented by a point on $L_4^2$, a special complex is represented by a point on $L_4^2$, and all the lines of the special complex by the points in which a tangent $S_4$ meets $L_4^2$.

The points of a line, $a + \lambda b$, in $S_5$ represent a set of complexes whose equations are

$$(3) \qquad (a_{34} + \lambda b_{34})p_{12} + (a_{42} + \lambda b_{42})p_{13} + \cdots = 0,$$

and all these complexes have in common the congruence common to the complexes $a$ and $b$. Their congruence, of course, consists of the lines of the original $S_3$ represented by the points in which $L_4^2$ is met by the polar $S_3$ of the line $a + \lambda b$.

A system of complexes, $a + \lambda b$, is called a *pencil of complexes*, and their common congruence is called its *base* or *basal congruence*. It evidently has the property that the null planes of any point with regard to the complexes of the pencil form an axial pencil whose axis is a line of the basal congruence. Dually, the null points of any plane with regard to the complexes of the pencil form a range of points on a line of the basal congruence.

The cross ratio of four complexes of a pencil may be defined as the cross ratio of their representative points in $S_5$. From the form of Equation (3) this is evidently the cross ratio of the four null planes of any point with regard to the four complexes.

A pencil of complexes evidently contains the special complexes whose directrices are the directrices of the basal congruence. Hence

* Equation (2) may be taken as the definition of a *polar* $S_4$ of a point with regard to $L_4^2$. Two points are *conjugate with regard to* $L_4^2$ if the polar $S_4$ of one contains the other. The polar $S_4$'s of the points of an $S_i$ $(i = 1, 2, 3, 4)$ all have an $S_{4-i}$ in common which is called the *polar* $S_{4-i}$ of the $S_i$. These and other obvious generalizations of the polar theory of a conic or a regulus we take for granted without further proof.

there are two improper, two proper, one, or a flat pencil of lines which are the directrices of special complexes of the pencil. These cases arise as the representative line $a + \lambda b$ meets $L_4^2$ in two improper points, two proper points, or one point, or lies wholly on $L_4^2$. Two points in which a representative line meets $L_4^2$ are the double points of an involution the pairs of which are conjugate with regard to $L_4^2$.

Two complexes $p$, $p'$ whose representative points are conjugate with regard to $L_4^2$ are said to be *conjugate* or *in involution*. They evidently satisfy Equation (2) and have the property that the null points of any plane with regard to them are harmonically conjugate with regard to the directrices of their common congruence. Any complex $a$ is in involution with all the special complexes whose directrices are lines of $a$.

Let $a_1$ be an arbitrary complex and $a_2$ any complex conjugate to (in involution with) it. Then any representative point in the polar $S_3$ with regard to $L_4^2$ of the representative line $a_1a_2$ represents a complex conjugate to $a_1$ and $a_2$. Let $a_3$ be any such complex. The representative points of $a_1$, $a_2$, $a_3$ form a self-conjugate triangle of $L_4^2$. Any point of the representative plane polar to the plane $a_1a_2a_3$ with regard to $L_4^2$ is conjugate to $a_1a_2a_3$. Let such a point be $a_4$. In like manner, $a_5$ and $a_6$ can be determined, forming a self-polar 6-point of $L_4^2$, the generalization of a self-polar triangle of a conic section. The six points are the representatives of six complexes, each pair of which is in involution.

It can be proved that by a proper choice of the six points of reference in the representative $S_5$, the equation of $L_4^2$ may be taken as any quadratic relation among six variables. Hence the lines of a three-space may be represented analytically by six homogeneous coördinates subject to any quadratic relation. In particular they may be represented by $(x_1, x_2, \cdots, x_6)$, where

$$x_1^2 + x_2^2 + x_3^2 + x_4^2 + x_5^2 + x_6^2 = 0.*$$

In this case, the six-point of reference being self-polar with regard to $L_4^2$, its vertices represent complexes which are two by two in involution.

* These are known as Klein's coördinates. Most of the ideas in the present section are to be found in F. Klein, Zur Theorie der Liniencomplexe des ersten und zweiten Grades, Mathematische Annalen, Vol. II (1870), p. 198.

## EXERCISES

**1.** If a pencil of complexes contains two special complexes, the basal congruence of the pencil is hyperbolic or elliptic, according as the special complexes are proper or improper.

**2.** If a pencil of linear complexes contains only a single special complex, the basal congruence is parabolic.

**3.** If all the complexes of a pencil of linear complexes are special, the basal congruence is degenerate.

**4.** Define a pencil of complexes as the system of all complexes having a common congruence of lines and derive its properties synthetically.

**5.** The polars of a line with regard to the complexes of a pencil form a regulus.

**6.** The null points of two planes with regard to the complexes of a pencil generate two projective pencils of points.

**7.** If $C = 0$, $C' = 0$, $C'' = 0$ are the equations of three linear complexes which do not have a congruence in common, the equation $C + \lambda C' + \mu C'' = 0$ is said to represent a *bundle* of complexes. The lines common to the three fundamental complexes $C$, $C'$, $C''$ of the bundle form a regulus, the conjugate regulus of which consists of all the directrices of the special complexes of the bundle.

**8.** Two linear complexes $\Sigma a_{ij}p_{ij} = 0$ and $\Sigma b_{ij}p_{ij} = 0$ are in involution if and only if we have

$$a_{12}b_{34} + a_{13}b_{42} + a_{14}b_{23} + a_{34}b_{12} + a_{42}b_{13} + a_{23}b_{14} = 0.$$

**9.** Using Klein's coördinates, any two complexes are given by $\Sigma a_i x_i = 0$ and $\Sigma b_i x_i = 0$. These two are in involution if $\Sigma a_i b_i = 0$.

**10.** The six fundamental complexes of a system of Klein's coördinates intersect in pairs in fifteen linear congruences all of whose directrices are distinct. The directrices of one of these congruences are lines of the remaining four fundamental complexes, and meet, therefore, the twelve directrices of the six congruences determined by these four complexes.

# INDEX

The numbers refer to pages

# NOTES AND CORRECTIONS

Page 34. In the definition of *projection*, after "$P$," in the last line on the page, insert ", together with the lines and planes of $F$ through $P$,".

Page 34. In the definition of *section*, after "$\pi$," in the last line on the page, insert "together with the lines and points of $F$ on $\pi$,".

Page 35. In the definition of section of a plane figure $F$ by a line $l$, the section should include also all the points of $F$ that are on $l$.

Page 44, line 5 from bottom of page. The triple system referred to does not, of course, satisfy $E_3$. It is not difficult, however, to build up a system of triples which does satisfy all the assumptions $A$ and $E$. Such a finite $S_3$ would contain 15 "points" and 15 "planes" (of which the given triple system is one) and 35 "lines" (triples). See Ex. 3, p. 25, and Ex. 15, p. 203.

Page 47, Theorem 3. Add the restriction that the line $l$ must not contain a vertex of either quadrangle.

Page 49. In the definition of *quadrangular set*, after "a line $l$" insert ", not containing a vertex of the quadrangle,".

Page 52, Ex. 1. The latter part should read: ". . . of an edge joining two vertices of the five-point with the face containing the other three vertices?"

Page 53, Exs. 14, 15, 16. The term *circumscribed* may be explicitly defined as follows: A simple $n$-point is said to be circumscribed to another simple $n$-point if there is a one-to-one reciprocal correspondence between the lines of the first $n$-point and the points of the second, such that each line passes through its corresponding point. The second $n$-point is then said to be *inscribed* in the first.

Page 53, Ex. 16. The theorem as stated is inaccurate. If $m$ is the smallest exponent for which $2^m \equiv \pm 1$, mod. $n$, the vertices of the plane section may be divided into $\frac{n-1}{2}$ simple $n$-points, which fall into $\frac{n-1}{2m}$ cycles of $m$ $n$-points each, such that the $n$-points of each cycle circumscribe each other cyclically. Thus, when $n = 17$, there are two cycles of 4 $n$-points, the $n$-points of each cycle circumscribing each other cyclically.

Page 85, Theorem 9. If the quadrangular set contains one or two diagonal points of the determining quadrangle, these diagonal points must be among the five or four given points.

Page 88, Theorem 12. To complete the proof of this theorem the perspectivity mentioned must be used in both directions — i.e. it also makes the points of $R_1$ or $R_2$ perspective with the points of $R^2$ on $l$.

Page 99, Theorem 22. See note to p. 53, Exs. 14, 15, 16.

Page 119, Ex. 6. The latter part of this exercise requires a quadratic construction. See Chap. IX.

Page 137, Ex. 7 (Miscellaneous Exercises). The two points must not be collinear with a vertex; or, if collinear with a vertex, they must be harmonic with respect to the vertex and the opposite side.

Page 165, last paragraph. The point $(-1, 1)$ forms an exception in the definition of homogeneous coördinates subject to the condition $x_1 + x_2 = 1$. An exceptional point (or points) will always exist if homogeneous coördinates are subjected to a nonhomogeneous condition.

Page 168, Ex. 10. The points $A$, $B$, $C$, $D$ must be distinct.

Page 182, bottom of page. We assume that the center of the pencil of lines is not on the axis of the pencil of points (cf. the footnote on p. 183).

Page 186. While the second sentence of Theorem 7 is literally correct, it may easily be misunderstood. If the left-hand member of the equation of one of the lines $m = 0$, $n = 0$, or $p = 0$ be multiplied by a constant $\rho$, the value of $k$ may be changed without changing the conic. In fact, by choosing $\rho$ properly, $k$ may be given an arbitrary value ($\neq 0$) for any conic.

As pointed out in the review of this book by H. Beck, Archiv der Mathematik, Vol. XVIII (1911), p. 85, the equation of the conic may be written as follows: Let $(a_1, a_2, a_3)$ be an arbitrary point in the plane of the conic, and let

$$m_x = m_1x_1 + m_2x_2 + m_3x_3,$$
$$n_x = n_1x_1 + n_2x_2 + n_3x_3,$$
$$p_x = p_1x_1 + p_2x_2 + p_3x_3;$$

then the equation of the conic may be written

$$k_2 m_a n_a p_x^2 - k_1 p_a^2 m_x n_x = 0.$$

When the equation is written in this form, there is one and only one conic for every value of the ratio $\frac{k_1}{k_2}$.

Page 301. The first sentence is not correct under our original definition of section by a plane. We have accordingly changed this definition (cf. note to p. 34).

Page 301. In the sentence before Theorem 7 the tangent lines referred to are not lines of the quadric surface.

Page 303, Ex. 5. The tangent line must not be a line of the surface.

Page 303, Ex. 7. The line must not be a tangent line.

Page 304. Theorem 11 should read: "... form a regulus or a cone of lines, provided ...". In case the collineation between the planes of the conics leaves every point of $l$ invariant, the lines joining corresponding points of the two conics form a cone of lines. In this case $\overline{A} = \overline{\overline{A}}$ and $\overline{B} = \overline{\overline{B}}$, and the lines $a$ and $b$ intersect.

Page 306, line 7. After "sections," insert ", unless $a$ and $b$ intersect, in which case they generate a cone of lines" (cf. note to p. 304).

Page 308, proof of Corollary 2. Let $A_1^2$ be the projection on $\alpha$ of $B^2$ from the point $M$. $A_1^2$ might have double contact with $A^2$ at $R$ and $R'$, or might have contact of the second order at $R$ or $R'$. However, if $C^2$ is not degenerate, it is possible to choose $M$ for which neither of these happens. For if all conics obtained from $[M]$ had either of the above properties, they would form a pencil of conics of which $A^2$ is one. There would then exist a point $M$ for which $A_1^2$ and $A^2$ would coincide. $C^2$ would in this case have to contain three collinear points and would then be degenerate.

Page 310, paragraph beginning "Now if nine points ...". It is obvious that no line of intersection of two of the planes $\alpha$, $\beta$, $\gamma$ will contain one of the nine points, no matter how the notation is assigned.

Page 315, line 12 from bottom of page. Neither $\pi_1$ nor $\pi_2$ must contain a directrix.

Page 319, Ex. 2. If the two involutions have double points, the points on the lines joining the double points are to be excepted in the second sentence.

Printed in Dunstable, United Kingdom

91220100R00199